Petroleum and Marine Technology
Information Guide

Petroleum and Marine Technology Information Guide

A bibliographic sourcebook and directory of services

Edited by

A. Myers

Information Scientist, Heriot-Watt University Library

J. Whittick

C-CORE, Centre for Cold Ocean Resources Engineering, Memorial University of St John's, Newfoundland

D. Edmonds

Consultant, Instant Library, Loughborough

and

H.A. Richardson

ASR Books, Lancaster

CRC Press
Taylor & Francis Group
Boca Raton London New York

CRC Press is an imprint of the
Taylor & Francis Group, an **informa** business

A CHAPMAN & HALL BOOK

CRC Press
Taylor & Francis Group
6000 Broken Sound Parkway NW, Suite 300
Boca Raton, FL 33487-2742

First issued in paperback 2019

ISBN-13: 978-0-419-18210-8 (hbk)
ISBN-13: 978-0-367-45005-2 (pbk)

A catalogue record for this book is available from the British Library

Library of Congress Cataloging-in-Publication data available

Publisher's Note
The publisher has gone to great lengths to ensure the quality of this reprint but points out that some imperfections in the original may be apparent.

Visit the Taylor & Francis Web site at
http://www.taylorandfrancis.com

and the CRC Press Web site at
http://www.crcpress.com

*Dedicated to the memory of
Tony (H A) Richardson*

*who originally conceived and published this work
and whose presence and advice were sorely
missed in the production of this edition*

Index to advertisers

Advertisement Management:

Edwin Bray
Group Advertisement Manager
E & FN Spon Ltd
7 St Peter's Place
Brighton, East Sussex BN1 5TB
United Kingdom

Contents

Editors and contributors

A. Myers	Information Scientist, Heriot-Watt University Library
J. Hutcheon	Librarian, Department of Petroleum Engineering, Heriot-Watt University
J. Whittick	C-CORE, Memorial University of St John's Newfoundland
B. Rodden	C-CORE, Memorial University of St John's Newfoundland
S. Oue	C-CORE, Memorial University of St John's Newfoundland
M. Reynolds	Information Consultant, Lincolnshire
D. Edmonds	Consultant, Instant Library, Loughborough
G. Lee	Consultant, Instant Library, Loughborough
S. and H.A. Richardson	ASR Books, Lancaster

Preface

This is the fourth edition of a work which has undergone several metamorphoses in its history. Originally, in 1981, Tony Richardson of ASR Books and Arnold Myers of the Institute of Offshore Engineering (Heriot-Watt University) collaborated to produce the *Current Bibliography of Offshore Technology and Offshore Literature Classification*. After two editions the work was expanded with the incorporation of the *Guide to Information Services in Marine Technology* and the addition of Diana Edmonds' section on online databases in the oil and gas industry. It became the *Offshore Information Guide*.

We were all very proud when Arnold Myers won the 1991 Jason Farradane Award of the Institute of Information Scientists "*specifically for your publication The Offshore Information Guide but also . . . your widely recognized contribution to information services within the international oil and gas industry*". Our aim has always been to make the Guide as useful a working tool as possible.

Development has continued with this new edition, as reflected in its new title: *Petroleum and Marine Technology Information Guide*. As well as comprehensive revision, the Guide has expanded both its subject and geographical coverage. Petroleum exploration, reservoirs and production are now included in the Bibliography (thanks to Jane Hutcheon) The section on information services has been completely revised following an extensive questionnaire survey, planned with the help of Margaret Reynolds, and carried out in North America by the staff of C-Core (Memorial University of St John's, Newfoundland) under the leadership of Judith Whittick, and by ASR in Europe. Diana Edmonds and her team at Instant Library have expanded the already exhaustive online section to include databases on CD-ROMS. The Directory has been streamlined to control its continued growth. An unsung but useful feature of the Guide is the variety of listings and indexes, which, like the Directory, have been increased with each edition.

The most momentous change, however, was Tony Richardson's unexpected illness and death just as the contributors were completing their work and production should have been getting into full swing. Happily an old colleague of Tony's, David Inglis, successfully proposed that E & FN Spon should take on responsibility for the title. So the Guide is in good hands and can look forward to a new lease of life.

There are many people to whom, as a user of this Guide, you should be grateful, but without Chapman & Hall, E & FN Spon and David Inglis this book may never have reached your desk.

Section A

BIBLIOGRAPHY

with
Offshore and Petroleum
Literature Classification

Arnold Myers and Jane Hutcheon

INTRODUCTION

This bibliography is an expansion and update of that in the 1988 *Offshore Information Guide* to include Petroleum Engineering. The subject area of marine technology, for the purposes of this Guide, is defined as:

The technology directly associated with the rational exploitation of the non-living resources of the sea and seabed

This definition covers man's activities in the sea, including oil and gas exploration and production, mineral extraction and wave energy conversion, but excludes fishing and the use of the sea as a means of transportation.

In the *Bibliography* we have endeavoured to list the material that should be purchased by a library setting up an offshore technology section in order to provide a thorough basic reference collection. In such a rapidly developing subject, a detailed bibliography citing individual technical reports, conference papers and journal articles would be inappropriate; however, there is a core of publications which provides the bulk of the material identified in searches for information on specific topics. The contents reflect the state of the literature actually published, not what users might like to have available.

Most of the items in the *Bibliography* were published in the last five years: only items likely to be of continuing value have been brought forward from previous editions. A large proportion of the publications have arisen from North Sea oil and gas activities and most are in English. For the specialist library, the bibliography acts as a check-list of essential material: details of publishers are given in *Section D: Directory* to expedite acquisition. The simple classification system used in previous editions to arrange the entries in the bibliography has been increased to accommodate the expanded petroleum engineering content of this edition; this could readily be adapted to the classification of the books and pamphlets in an offshore technology library.

A characteristic of offshore technology is that developments are reported more promptly and in more detail at conferences than through other publishing media. An offshore technology library can be judged on the completeness of its holding of conference papers.

In this edition we have appended a list of conferences and seminars expected to give rise to published papers. At the time of going to press this list includes meetings held but with the papers not yet available, and forthcoming meetings.

Certain reports series and journals have been listed where regular acquisition should be considered. In addition, some important reports have been individually listed. Diana Edmonds and Gail Lee have also lised UK Stockbroker Research Reports in *Section C: Databases & CD-Roms*, which users worldwide will find invaluable as reference sources.

Detailed searches for information on specific topics can be carried out using abstracting and indexing journals (entries A0033-A0049). Such in-depth searches are carried out by information scientists attached to various organisations involved with offshore technology, many of which are listed in *Section B: Organisations Providing Information Services*.

The policy adopted in selection is assessment for relevance, currency and quality. We have included all material we believe would be of value to enquirers in our fields. Other things being equal, recent publications are of more value than older material, but whether or not a book is in print is not a criterion in itself: the actual *usage* of books in a library is chronologically scattered. For obvious statistical reasons large compilations of material such as Offshore Technology Conferences and Offshore Mechanics and Arctic Engineering Proceedings will be used over a longer time-span than the more specific titles.

Each title is entered in the bibliography with the publisher's imprint as recorded on the title page of the most recent edition, in accordance with international bibliographic standards, but often imprints and foreign rights are subsequently transferred to other publishers. The comprehensive listing of publishers addresses given in *Section D: Directory* will help to locate items.

Some books and sets of papers are available for sale only for short periods so that, at the time of use of the bibliography, some items may only be obtainable by inter-library loan. Users of the bibliography wishing to keep up-to-date can do so by referring to the *Offshore Engineering Information Bulletin* published by Heriot-Watt University Library.

Arnold Myers
Information Scientist, Heriot-Watt University Library
Jane Hutcheon
Petroleum Engineering Department, Heriot-Watt University

Edinburgh, July 1992

SECTION A CONTENTS

GENERAL

BIBLIOGRAPHY

A0001
Arctic offshore engineering, ships and marine technology: a bibliography
Penney P W, et al
Newcastle: University of Newcastle upon Tyne, Dept of Naval Architecture & Ship-building
October 1985, 71 pages

A0002
Bibliography of material on oil and gas, 6th ed
Dickson M, compiler
Aberdeen: Aberdeen City Libraries, Commerical & Technical Dept
1987, 398 pages

A0003
Bibliography of nautical books, 7th edition
Obin A
Southampton: Warsash Nautical Bookshop
1992

A0004
Bibliography of North Sea regulations, 5th edition; volume 1: UK
Mirzoeff J
Beckenham: T A Hollobone & Co Ltd
1992

A0005
Bibliography of North Sea Regulations, 5th edition; volume 2: Norway, Denmark, Netherlands and Ireland
Mirzoeff J
Beckenham: T A Hollobone & Co Ltd
1992

A0006
Bibliography of the oil and gas industry 1980–1989
Deakin E B and Cappel J J
Denton, TX: Institute of Petroleum Accounting, University of North Texas
1990, vi + 254 pages

A0007
Chinese petroleum: an annotated bibliography
Chang R
Boston, MA: G K Hall & Co
1982, 204 pages

A0008
Department of Energy publications 1988
Pope R
London: Department of Energy Library
1989, 43 pages

A0009
Department of Energy publications 1989
Martin S
London: Department of Energy
1990, 62 pages

A0010
Department of Energy publications 1990
Campling J and Gibson R
London: Department of Energy
1991, 83 pages

A0011
Department of Energy publications 1991
Gibson R
London: Department of Energy
1992, 63 pages

A0012
Documentation control in the energy industries, conference; London, 18 October 1988
Etherton J J (editor) for Information for Energy Group
London: Institute of Petroleum
1988, 60 pages

A0013
European seas: a bibliography of atlases and charts
Simpson P
Luxembourg: Office for Official Publications of the European Communities
1989, 153 pages [Report EUR 12070 EN]

A0014
Glossary for the gas industry
Arlington VA: American Gas Association
1990, 129 pages

A0015
Guide to statistical sources on the petroleum industry
Karayannis-Bacon H, et al
Oxford: Oxford University Press
1988, 200 pages

A0016
Guide to the petroleum reference literature
Pearson B C and Ellwood K B
Littleton: Libraries Unlimited
1987, xii + 193 pages

A0017
Guide to petroleum statistical information
API Central Abstracting & Information Services
New York: American Petroleum Institute
1992

A0018
Information management in the oil and gas industry: a new vision for productivity and safety
Chatterton P
London: Financtial Times Business Information
1991 [Financial Times Management Reports]

A0019
Information offshore, revised edition
Glasgow: Offshore Supplies Office
1990, 45 pages

A0020
The Law of the Sea: a select bibliography 1989
New York: United Nation Office for Ocean Affairs and the Law of the Sea
1991 [LOS/LIB/5]

A0021
The Law of the Sea: a select bibliography 1990
New York: United Nations Office for Ocean Affairs and the Law of the Sea
1991 [LOS/LIB/6]

A0022
Marine affairs bibliography: a comprehensive index to marine law and policy literature cumulation: 1980–85
Wiktor C L and Foster L A
Dordrecht, Lancaster: Martinus Nijhof (Kluwer)
1987, 685 pages

A0023
Marine civil engineering bibliography
Bognor Regis: Dillingham & Assoiciates
1990, 16 pages

A0024
Offshore Information Conference papers, 1984
Schanche G and Myers A, editors
Edinburgh: Institute of Offshore Engineering
1984, 166 pages

A0025
Offshore information conference papers 1986
Myers A (editor)
Edinburgh: Institute of Offshore Engineering
1987, 166 pages

A0026
Offshore information conference papers; Bergen, 21–23 September 1988
Myers A (editor)
Edinburgh: Institute of Offshore Engineering
1989, 291 pages

A0027
Offshore information conference papers; Glasgow, 26–28 September 1990
Myers A (editor)
Edinburgh: Institute of Offshore Engineering
1990, 212 pages

A0028
Offshore information guide: publications, databases, services, directory
Myers A et al
Berkhamsted: ASR Books
1988, viii + 117 + 28 + 39 + 78 pages

A0029
Offshore oil and gas: a bibliography
Dempster A
Glasgow: Offshore Supplies Office
1988, 32 pages

A0030
The offshore technology bibliography: a bibliography of offshore oil and gas production engineering technology and techniques
Gale L
Bedford: BHRA
1989, 648 pages

A0031
Sources of information for the offshore industry conference; London, 12 June 1986
Information for Energy Group and Association of British Offshore Industries,
London: Institute of Petroleum
1986, 68 pages

A0032
The underwater industry bibliography
Bevan J and Hunter T
London: Submex
1988, 67 pages

ABSTRACTING & INDEXING JOURNALS

A0033
Aquatic Pollution & Environmental Quality:
Aquatic Sciences & Fisheries Abstracts, Part 3
Ca,mbridge MA: Cambridge Scientific Abstracts
1992 = Volume 22 [monthly]
[Part 3 commenced with 1990 = Volume 20.
Also available on CD-ROM]

A0034
Current Energy Information
London: Department of Energy
[weekly bulletin of periodical contents pages]

A0035
Engineering Index
New York: Engineering Index Inc
1992 = Volume 91 [monthly]

A0036
Fluid Abstracts: Civil Engineering
Barking: Elsevier Science
1992 = Volume 2 [monthly]
[incorporates Offshore Engineering Abstracts]

A0037
Geological Abstracts
Norwich: Elsevier Geo-Abstrafcts
[bi-monthly]

A0038
Offshore Engineering Information Bulletin
Edinburgh: Heriot-Watt University
January 1992 = No 196 [monthly]

A0039
International Petroleum Abstracts
Chichester: John Wiley (for Inst of Petroleum)
1988 = Volume 20 [quarterly]
[incorporating Offshore Abstracts]

A0040
Marine Pollution Research Titles
Plymouth: Plymouth Marine Laboratory
1992 = Volume 19 [monthly]

A0041
Marine Science Contents Tables
Rome: Food and Agriculture Organisation of the United Nations (HMSO)
1992 = Volume 27 [monthly]

A0042
OEIC Bulletin
St John's, Newfoundland: Memorial University of Newfoundland Ocean Engineering Information Centre
1992 = Volume 12 [monthly]

A0043
Ccean Technology and Engineering
Springfield, Virginia: National Technical Information Service
1992 = Volume 92 [weekly]

A0044
Ocean Technology, Policy and Non-living Resources: Aquatic Sciences and Fisheries Abstracts, part 2
Cambridge MA: Cambridge Scientific Abstracts
1992 = Volume 22 [monthly]
[Part 2 commenced with 1978 = Volume 8.
Also available on CD-ROM]

A0045
Oceanic Abstracts
Cambridge MA: Cambridge Scientific Abstracts
1992 = Volume 26 [bi-monthly]

A0046
Oil-index
Stavanger: Norwegian Petroleum Directorate
1992 = Volume 13 [quarterly]
[before 1980 published in Norwegian only]

A0047
Petroleum Abstracts
Tulsa: University of Tulsa, Information Services Dept
1992 = Volume 32 [weekly]

A0048
Petroleum: Energy Business News Index
New York: American Petroleum Institute
[monthly]

A0049
Water Resources Abstracts
Reston VA: US Geological Survey
[monthly] [Also available on CD-ROM]

TERMINOLOGY

A0050
An A–Z of offshore oil and gas, 2nd ed
Whitehead H
Houston: Gulf Publishing Company
1983, 438 pages

A0051
Aquatic sciences and fisheries thesaurus,
revision 1
Fagetti E, Privett D W and Sears J R L
Bethesda, MD: Cambridge Scientific Abstracts
(for Aquatic Science & Fisheries Information
System)
1987, 418 pages
[ASFIS Reference Series No 6]

A0052
Collins dictionary of seas and oceans
Charton B
Glasgow: Collins
1989, 458 pages

A0053
Concise marine almanack
Mangone G J
New York: Crane Russack
1991

A0054
A dictionary for the petroleum industry
Austin, TX: University of Texas, PETEX
1991

A0055
Dictionary of drilling and boreholes:
English/French
Moureau M and Brace G (editors)
Paris: Éditions Technip
1990, 436 pages

A0056
Dictionary of marine technology
Taylor D A
London: Butterworth
1989, 256 pages

A0057
Dictionary of oil and gas abbreviations
Ledbury: Oilfield Publications
1987, 88 pages

A0058
Dictionary of petroleum exploration, drilling and
production
Hyne N J
Tulsa, OK: PennWell
1991, 635 pages

A0059
Dictionary of seismic prospecting:
English/French
French Oil & Gas Industry Association
Paris: Éditions Technip
1987, 328 pages

A0060
The diver's reference dictionary
San Pedro, CA: Best Publishing
1986, 131 pages

A0061
Elsevier's dictionary of water and hydraulic
engineering (English, French, Spanish, Dutch
and German)
van der Tuin J D, compiler (Delft Hydraulics)
Amsterdam: Elsevier Science Publishers
1987, xvi + 450 pages [5117 terms]

A0062
Elsevier's maritime dictionary:
English-French-Arabic
Bakr M
Amsterdam: Elsevier Science Publishers
1987, 664 pages [12,024 entries]

A0063
Elsevier's oil and gas field dictionary
(in English/American, French, Spanish, Italian,
Dutch and German, with Arabic supplement)
Chaballe L Y, et al, compilers
Netherlands: Elsevier Science Publishers
1980, xii + 672 pages [4843 terms]

A0064
English-French petroleum dictionary
Arnould M and Zubini F
Paris: Éditions Technip
(Graham & Trotman Ltd)
1982, 288 pages

A0065
English/Spanish glossary of the petroleum
industry, 2nd ed
Tulsa: PennWell
1982, 378 pages

A0066
Exploration and production thesaurus:
exploration, development and production of
crude oil and natural gas, 10th edition
Tulsa: University of Tulsa
1991, looseleaf

A0067
Glossary of geology, 3rd edition
Bates R L and Jackson J A
Alexandria, VA: American Geological Institute
1987, x + 788 pages

A0068
Glossary of onshore and offshore pipelines;
English/French
Paris: Éditions Technip
1979, 320 pages

A0069
Handbook of oil industry terms and phrases,
4th ed
Langenkamp R D
Tulsa: PennWell
1984, 360 pages

A0070
Handbook of terminology for the use of divers
and inspectors on offshore structures
Department of Energy
London: HMSO
1987 vii + 190 pages [Offshore Technology
Report OTH 87277]

A0071
Illustrated glossary of petroleum geochemistry
Miles J A
Oxford: Clarendon Press
1989, ix + 137 pages

A0072
Illustrated petroleum reference dictionary, 3rd
ed
Langenkamp R D, editor
Tulsa: PennWell
1985, 696 pages

A0073
Infoil II Thesaurus
A Myers, editor of English version
*London: HMSO (for UK Department of Energy
and Norwegian Petroleum Directorate)*
1988 [OTI 88 505]

A0074
The international system of units (SI) in
oceanography: report of IAPSO Working Group
on symbols, units and nomenclature in physical
oceanography (SUN)
Paris: UNESCO
1985, xiii + 124 pages

A0075
Landman's encyclopedia, 3rd edition
Hankinson R L and Hankinson R L Jr (editors)
Houston, TX: Gulf Publishing
1988, 496 pages

A0076
The marine encyclopaedic dictionary
Sullivan E
Colchester: Lloyd's of London Press
1988, 468 pages

A0077
Ocean and marine dictionary
Tver D F
Centreville, MD: Cornell Maritime
1979, ix + 358 page

A0078
Oil and gas dictionary: an encyclopaedic
dictionary of economic and financial concepts
Stevens P (editor)
London: Macmillan
1988

A0079
An oil and gas handbook, 2nd edition
*Edinburgh: Bank of Scotland International
Division*
1989, 91 pages

A0080
Oil & Gas Journal databook 1991
Tulsa, OK: PennWell
1991, 350 pages

A0081
Oil & Gas Journal energy statistics sourcebook
Tulsa, OK: PennWell
1991, 450 pages

A0082
Oil and gas terms: annotated manual of legal,
engineering, tax words and phrases, 6th ed
Williams H R and Meyers C J
New York: Bender
1987

A0083
Onshore/offshore oil and gas multilingual
glossary
*Commisssion of the European Communities,
Terminology Bureau*
London: Graham & Trotman Ltd
1979, 490 pages

A0084
The petroleum dictionary
Tver D F and Berry R W
Wokingham: Van Nostrand Reinhold
1980, 374 pages

A0085
Petroleum fundamentals glossary
Austin, TX: University of Texas, PETEX
1990, 120 pages

A0086
Thesaurus of the National Maritime Research
Center, version 1.0
Feldman R
*Kings Point, NY: US Dept of Transportation
National Maritime Research Center*
1986, 264 pages [NMRC 101]

DIRECTORIES

Organisations and Personnel

A0087
ANEP '92: European petroleum yearbook:
handbook and reference work for the oil and
natural gas industries, Volume 25
Hamburg: Urban Verlag
1992, c 500 pages

A0088
Arab oil and gas directory 1990
Paris: Arab Petroleum Research Center
1990, 620 pages [annual]

A0089
Energy world yearbook 1992
Institute of Energy
Royston: Energy Publications
1992, 272 pages

A0090
1990 European oil and gas management
directory
Paris: Éditions Technip
1989, 688 pages

A0091
Financial Times oil and gas international
yearbook 1992: 92nd edition
London: Longman
1992, xxv + 544 pages pages

A0092
Guide to North Sea operators, 4th edition
Leitch M
London: Financial Times Management Reports
1991

A0093
1991–1992 Handbook of UK government
departments and other agencies concerned
with the offshore industry
Richardson B G
London: MTD Ltd
1991, 185 pages

A0094
International directory of marine science
libraries and information centres
Winn C P
*Woods Hole MA: Woods Hole Oceanographic
Institution*
1987

A0095
ICC financial survey and directory: oil and
petroleum producers
London: ICC Inforfmation Group
[annual]

A0096
The Institution of Water & Environmental
Management Yearbook 1991, incorporating the
National Water Industry Handbook 1991
Norwich: Business Information Ltd
1991, 252 pages

A0097
OCS Directory; Federal and State agencies
involved in the Outer Continental Shelf Oil and
Gas Program
*Vienna, VA: US Dept of the Interior, Minerals
Management Service*
1988, 148 pages

A0098
Oil and gas databook for developing countries,
2nd edition
Fee D
London: Graham & Trotman
1989, 220 pages

A0099
Oil and gas USSR directory 1991
Pickering R H
Hastings: Arguments & Facts International
1991, 120 pages

A0100
Oil companies and operators worldwide
directory 1990–91
Petroconsultants
Ledbury: OPL
1990, 100 pages

A0101
The PetroCompanies Asia/Pacific/Americas
data supplement 1991/92
London: PetroCompanies Ltd
1991

A0102
The PetroCompanies European data
supplement 1991/92
London: PetroCompanies Ltd
1991

A0103
The PetroCompanies foundation book 1991/92:
corporate profiles of 76 International oil and gas
companies
London: PetroCompanies Ltd
1991

A0104
Petrodirectory
London: PetroCompanies Ltd
1991

A0105
Petroguide 1991–92
London: Petroguide Ltd
1991

A0106
Petroleum Economist energy finance handbook
London: Petroleum Economist
1991

A0107
SPE membership directory
Richardson TX: Society of Petroleum Engineers
[Annual, incorporated in May issue of Journal
of Petroleum Technology]

A0108
A review of postgraduate education and
training in marine technology, 2nd edition
Caldwell J B
London: MTD Ltd
1988, 99 pages

A0109
Who's who in USSR oil and gas industry
*Moscow Board of Scientific & Technical Society
of Oil & Gas Industry*
*New York: Victoria International Petroleum of
USA*
1991

A0110
Who's Who in world oil and gas, 10th edition
London: Longman
1991, xvi + 506 pages

A0111
Worldwide petroleum phone/fax/telex directory
Tulsa OK: PennWell
1991, 600 pages [available on diskette]

A0112
Yearbook 1990–91 and directory of members:
25th anniversary issue
*International Association on Water Pollution
Research & Control*
London: Kogan Page
1991, xvi + 431 pages

A0113
Yearbook for the Norwegian Petroleum Society
Oslo: Norwegian Petroleum Society
annual [in Norwegian]

Research

A0114
Department of Energy Offshore Supplies Office
research and development projects
Marine Technology Support Unit
Glasgow: Offshore Supplies Office
1992, xvii + 126 pages

A0115
Directory of individual projects completed
between July 1985 and June 1989
Clements R J
London: Marine Technology Directorate
1990, 181 pages

A0116
Directory of completed projects to 1.7.92
London: MTD Ltd
due 1992

A0117
Directory of current projects to 1.7.89
Clements R J (compiler)
London: Marine Technology Directorate
1989, 209 pages

A0118
Directory of current projects
London: MTD Ltd
1991, 220 pages

A0119
European Communities oil and gas
technological development projects: fourth
status report
Joulia J P et al (for EEC Commission)
London, Lancaster: Graham & Trotman
1990, 468 pages

A0120
International marine science funding guide
Fenwick J et al
Woods Hole, MA: WHOI Sea Grant Program
1990, vii + 161 pages

A0121
Marine science and technology in the United
Kingdom: report to government
Coordinating Committee on Marine Science
and Technology
London: HMSO
1990, xi + 162 pages

A0122
Marine Science and Technology (MAST) R&D
Programme 1989–92: research contracts, lst
edition
Weydert M and Boissonnas J
*London: HMSO for European Communities
Commission*
1991, xii + 53 pages [Marine Science &
Technology Series EUR 13437]

A0123
Offshore safety research and development
programme
London: HMSO
1990 [OTI 90 548]

A0124
Offshore safety research and development
programme: project handbook 1991
Marine Technology Support Unit
London: HMSO
1991 [OTI 91 553]

A0125
Oil and gas R&D: research and development
facilities in Scotland for the international oil and
gas industry
Aberdeen: Scottish Development Agency
1988, 32 pages

A0126
Renewable energy R&D programmes: progress
report for the Department of Energy 1989–1990
Energy Technology Support Unit
London: HMSO
1990, 135 pages [ETSU R 56]

A0127
Research and development in the oil industry
London: Shell International Petroleum Co
1991, 9 pages [Shell Briefing Service 4 1991]

A0128
Research and development projects
MATSU AEA Technology (compiler)
Glasgow: Offshore Supplies Office
1990, xiii + 104 pages

A0129
Strategic plan for offshore technology research
and development for 1990–1994
*Oslo: Royal Norwegian Council for Scientific &
Industrial Research*
1990, 27 pages [text in Norwegian]

A0130
A summary of marine structures research and
development in Canada
German & Milne Inc
*St John's, Newfoundland: National Research
Council Canada Institute for Marine Dynamics*
1988 [Laboratory memorandum LMHYD 65]

Equipment

A0131
Anchor handling tug and supply vessels of the
world
Ledbury: OPL Ltd
1992, xix + 349 pages

A0132
British offshore oil and gas fields and installations 1988
Maclachlan M (editor)
Biggar: North Sea Books
1988, various pagings

A0133
Construction vessels of the world
Ledbury: OPL
1991, 700 pages

A0134
Diving support vessels of the world
Ledbury: OPL
1990, 347 pages

A0135
European continental shelf guide, 1991–92 edition
Ledbury: OPL
1991,763 pages

A0136
Guide to offshore support vessels, 2nd edition
Ledbury: OPL
1990, 750 pages

A0137
International petroleum encyclopedia 1990
Tulsa OK: PennWell
1990, 400 pages

A0138
List of existing government-funded facilities
Interdepartmental Committee on Large Experimental Facilities
London:
1988, 214 entries

A0139
Mobile drilling units of the world, 2nd edition (formerly: Dayton's guide to mobile drillling units)
Ledbury: OPL
1992 800 pages

A0140
The Noble Denton towing vessel register, 3rd edition
Ledbury: OPL
1992, 350 pages

A0141
North Sea facts, 1990–91 edition
Ledbury: OPL
1990, 500 pages

A0142
North Sea field development guide, 4th edition
Ledbury: OPL
1992, lxiii + 1173 pages

A0143
Offshore mobile drilling rig data: electronic rig stats
Tulsa OK: PennWell
5¼" diskettes
[one time, weekly subscription or monthly subscription - routinely updated]

A0144
The offshore service vessel register, 14th edition
London: Clarkson Research Studies
1992

A0145
Oil patch directory Europe: the on-disk directory of vendors to the oil and gas industry
Canterbury: Douglas-Westwood Ltd
1991, 1 diskette

A0146
Petroleum offshore licences and licencees
London: Department of Energy
1988

A0147
The register book of Det norske Veritas Classification A/S, classification of ships and mobile offshore units
Høvik: Det norske Veritas
1990, 900 pages

A0148
Register of offshore units, submersibles and diving systems 1992–93
London: Lloyd's Register of Shipping
due 1992

A0149
ROV Review, 3rd ed
Spring Valley CA: Windate Enterprises
1988

A0150
Scottish petroleum annual 1990–91
Strachan T (editor)
Aberdeen: Aberdeen Petroleum Publishing
1990

A0151
Standby vessels of the world
Ledbury: OPL Ltd
1991, 300 pages

A0152
Survey vessels of the world
Ledbury: OPL Ltd
due 1992

A0153
Undersea vehicles directory 1990–91, 4th edition
Busby Associates Ltd
Dunkirk MD: DS Oceans Ltd
1990, 448 pages

A0154
World field development guide, volume 1: Africa, Mediterranean, Middle East and Africa
Ledbury: OPL Ltd
1991, 403 pages

A0155
World field development guide, volume 2: Indian sub-continent, SE Asia, Australasia and Far East
Ledbury: OPL Ltd
1991, 366 pages

Firms and products

A0156
A guide to suppliers of marine equipment in Britain
London: British Marine Equipment Council
1991, 27 pages

A0157
Asia-Pacific/Africa Middle East petroleum directory
Tulsa OK: PennWell
1990, 470 pages

A0158
Asia Pacific oil and gas directory 1991–92, 11th edition
Petromin
Ledbury: OPL Ltd
1991, 600 pages

A0159
1991 Canadian oil industry directory
Tulsa OK: PennWell
1991, 260 pages

A0160
Composite catalog of oil field equipment and services, 39th revision 1990–91
Houston TX: Gulf Publishing
1990, 2 volumes

A0161
Danish offshore guide and yearbook 1987
Esbjerg: Bjørndal & Gundestrup
1987, 216 pages

A0162
Danish offshore suppliers l986–87
Copenhagen: Federation of Danish Industries
1986, 135 pages

A0163
Das europäische Wirschafthandbuch, Jahrbuch 1991: Öl und Gas
Essen: Glückauf
1990, 1350 pages

A0164
Esbjerg offshore contacts 1987 edition
Birkerød: Ole Camae
1987, 113 pages

A0165
European continental shelf guide 1991–92
Ledbury: OPL
1991, 500 pages

A0166
European petroleum directory
Tulsa OK: PennWell
1990, 460 pages

A0167
Guide du Pétrole
(offshore-gaz-pétrochemie)
Paris: Éditions Technip
Annual [bilingual French-English]

A0168
German offshore and ocean industries 1989/90
Dusseldorf: Mecon for VSM and VDMA
1989, 133 pages

A0169
Kelly's oil and gas industry directory 1991, 7th edition
East Grinstead: Reed Information Services
1990, 158 pages

A0170
Landward Oil and gas yearbook
London: IBC Technical Services
Annual

A0171
National Ocean Industries Association directory of membership
Washington DC: NOIA
updated, but circa 1984 xii + 106 pages

A0172
Grampian business directory 1989
Aberdeen: NESDA
1989, 264 pages

A0173
Netherland offshore catalogue 1991
Rotterdam: Van Kouteren's (for Netherlands Industrial Council for Oceanology, IRO)
1991, 94 pages

A0174
Noroil contacts, spring 1992 edition
Kingston upon Thames: Offshore Directories
1992

A0175
North of England offshore directory
Newcastle upon Tyne: Northern Development Company
1987, 40 pages

A0176
North Sea oil and gas directory, 20th edition
Kingstom upon Thames: Judith Patten Public Relations
1992

A0177
Ocean resources: a new directory of companies in the Strathclyde region with products or services to offer the oil, marine, fishing and ocean resources industries
Industrial Development Unit
Glasgow: Strathclyde Regional Council

A0178
1988–89 Offshore buyers guide
Offshore
1988, Volume 48(9), 74 pages

A0179
Offshore buyers guide, Norway
Stavanger: Oil Information Services Norway AS
1991 = Volume 22 [6 per year]

A0180
Offshore index '92, 17th edition
Oslo: Selvig Publishing
1992, 128 pages

A0181
Offshore industrial directory 1991–92
Offshore & Industrial Benefits Division
St John's, Newfoundland: Newfoundland & Labrador Dept of Development
1991, 247 pages + update

A0182
The offshore industry: Norway
British Embassy, Oslo
London: Department of Trade & industry
1989, 21 leaves

A0183
Offshore Northern Seas, 9th international conference exhibition catalogue
Stavanger: Offshore Northern Seas Foundation
1990, 352 pages

A0184
Offshore services and equipment directory, 10th edition
San Diego CA: Greene Dot
1987, 160 pages

A0185
Oil and gas companies directory
Paris: Éditions Technip
annual

A0186
1992 Petroleum software directory
Tulsa OK: PennWell
1991, 400 pages

A01867
Process engineering index: issue 90-ol, June 1990–January 1991
Bracknell: Technical Indexes Ltd
1990, 363 pages

A0188
Scandinavian on and offshore handbook 1987
Copenhagen: Thomson Communications (Scandinavia)
1987, 288 pages

A0189
Scottish oil directory
Aberdeen: Scottish Enterprise
1991

A0190
Sea Technology buyers' guide/directory 1992–93
Arlington VA: Compass
due 1992

A0191
Singapore marine and ocean engineering directory 1987/88
Singapore: Singapore Association of Shipbuilders & Repairers
1988

A0192
Software directory for the offshore industry, 2nd edition
London: Marine Technology Directorate
due 1992

A0193
Soviet Union oil and gas infrastructure
Ledbury: OPL
1992, 450 pages

A0194
Strathclyde oil register 1987–1988
Glasgow: Strathclyde Regional Council Industrial Development Unit
1987, 82 pages

A0195
Swedish ocean technology 1987–88
Stockholm: Sveriges Mekanfrbund (for Swedocean)
1986, 136 pages

A0196
UK oil and gas technology in Indonesia
Glasgow: Offshore Supplies Office
1990, 48 pages

A0197
1991 USA oil industry directory
Tulsa TX: PennWell
1991

A0198
1991 USA oilfield service supply and manufacturers directory
Tulsa OK: PennWell
1991

A0199
Wireline logging tool catalog
Houston TX: Gulf Publishing
1986, 460 pages

A0200
1991 Worldwide natural gas industry directory
Tulsa OK: PennWell
1991, 300 pages

A0201
1991 Worldwide offshore contractors and equipment directory, 23rd edition
Tulsa OK: PennWell
1991, 496 pages

PERIODICALS

A0202
AAPG Bulletin
Tulsa OK: AAPG
1992 = Volume 76 [monthly]

A0203
Aberdeen Petroleum Report
Aberdeen: Aberdeen Petroleum Publishing
1992 = Volume 12 [weekly]

A0204
Applied Ocean Research
Southampton: CML Publications
1992 = Volume 14 [quarterly]

A0205
Arab Oil & Gas
Paris: Arab Petroleum Research Center
1969– [bi-weekly]

A0206
Asian Oil & Gas
Tokyo: Asian Oil & Gas Publications
[seven per year]

A0207
Basic Petroleum Data Book
Washington DC: API
1992 = Volume 12 [four monthly]

A0208
BMT Abstracts
Wallsend: BMT Cortec Ltd
1992 = Volume 47 [monthly]

A0209
British offshore business opportunities services
London: Norconsultants [monthly]

A0210
C-Core News
St John's, Newfoundland: Centre for Cold Ocean Resources Engineering, Memorial University of Newfoundland
1992 = Volume 17 [three per year]

A0211
CNWM Far East Oil Service (SE Asia & Australasia)
Edinburgh: County NatWest WoodMac
1984–
[monthly reports and regular updates; mapping service available to subscibers only]

A0212
CNWM North Sea Service
Ediburgh: County NatWest Woodmac
1974–
[monthly reports and regular updates;
mapping service available to subscribers only]

A0213
CNWM North West Europe Oil Service
Edinburgh: County NatWest WoodMac
1986–
[monthly reports and regular updates;
mapping service available to subscribers only]

A0214
CNWM West Africa Oil Service
Edinburgh: County NatWest WoodMac
1990–
[monthly reports and regular updates;
mapping service available to subscribers only]

A0215
Daily Oil News Digest
Warminster, UK: McCarthy Information Ltd
[daily]

A0216
Deep-Sea Research: Part A, Oceanographic
Research Papers; Part B, Oceanographic
Literature Review
Oxford: Pergamon Press
1992 = Volume 39 [monthly]

A0217
Department of Trade & Industry News Release
London: Department of Trade & Industry
[irregular]

A0218
Drilling Contractor
Houston: Drillling Contractor Publ Inc
1992 = Volume 53 [bi-monthly]

A0219
Drilling News: an international newsletter
Bedford: BHRA
1989– [monthly]

A0220
Eastern Offshore News
*St John's Newfoundland: Canadian Petroleum
Association*
1988 = Volume 10 [two per year]

A0221
East European Energy Report
London: Financial Times
1991– [monthly]

A0222
Erdoel und Kohle Erdgas
Hamburg: Urban-Verlag
1992 = Volume 108 [monthly]

A0223
Energy Economics
Letchworth, UK: Butterworth/Heinemann
1992 = [quarterly]

A0224
Energy Economist
London: Financial Times
[monthly]

A0225
Energy Trends
London: Department of Trade & Industry
[monthly]

A0226
Engineering Structures: the journal of
earthquake, wind and ocean engineering
Guildford: Butterworth
1992 = Volume 14 [quarterly]

A0227
Enviro-Marine Newsletter
Manchester: Marintech North West
Autumn 1991 = Issue 1

A0228
Euroil
Stavanger: Noroil Publishing
1992 = Volume 20 [monthly]

A0229
European Energy Report
London: FTBI
[monthly]

A0230
European Environment Review
London: Graham & Trotman
1992 = Volume 6 [quarterly]

A0231
European Offshore Petroleum Newsletter
Stavanger: Noroil Publishing
1976– [weekly]

A0232
EuroPetroleum Magazine
Aberdeen: Aberdeen Petroleum Publishing
1992 = Volume 4 [quarterly]

A0233
Extraction of Mineral Oil and Natural Gas Quarterly
Business Statistics Office
London: HMSO [quarterly]

A0234
First Break
Oxford: Blackwell
1992 = Volume 10 [monthly]

A0235
Geobyte
Tulsa OK: AAPG
1992 = Volume 7 [bi-monthly]

A0236
Geologie en Mijnbouw
Amsterdam: Elsevier
1992 = Volume 71 [quarterly]

A0237
Geo-marine letters: an international journal of marine geology
New York: Springer
1988 = Volume 8 [quarterly]

A0238
Golob's Oil Pollution Bulletin
Cambridge MA: World Information Systems
1990– [bi-weekly]

A0239
IEA Oil Market Report
London: IEA/OECD [monthly]

A0240
IMO News
London: International Maritime Organization [quarterly]

A0241
International Gas Report
London: FTBI [bi-weekly]

A0242
JAMSTEC Technical Reports
Yokosuka-shi: Japan Marine Science & Technology Centre
[two per year]

A0243
Japan Petroleum & Energy Trends
Tokyo: Japan Petroleum Energy Consultants
1966– [bi-weekly]

A0244
Journal of Canadian Petroleum Technology
Montreal: Canadian Institute of Mining & Metallurgy
1992 = Volume 31 [bi-monthly]

A0245
Journal of Hyperbaric Medicine
Bethesda MA: Undersea & Hyperbaric Medical Society [quarterly]

A0246
Journal of Oceanic Engineering
New York: Institute of Electrical & Electronic Engineers
1992 = Volume 17 [quarterly]

A0247
Journal of Petroleum Science & Engineering
Amsterdam: Elsevier
1992 = Volumes 7 & 8 [8 issues / year]

A0248
Journal of Petroleum Technology
Dallas: Society of Petroleum Engineers of AIME
1992 = Volume 44 [monthly]

A0249
Journal of the Waterway, Port, Coastal and Ocean Division
New York : American Society of Civil Engineers
1992 = Volume 122 [quarterly]

A0250
Lloyd's List
Colchester: Lloyd's of London Press
[six issues per week]

A0251
Log Analyst
Houston Tx: SPWLA
1992 = Volume 33 [bi-monthly]

A0252
Marine & Petroleum Geology
Guildford: Butterworths
1992 = Volume 9 [quarterly]

A0253
Marine Engineers Review
London: Institute of Marine Engineers
1992 = Volume 23 [monthly]

A0254
Marine Geophysical Researches
Dordrecht: Kluwer
1992 = Volume 14 [quarterly]

A0255
Marine Geotechnology
New York/London: Taylor & Francis
1992 = Volume 12 [irregular]

A0256
Marine Mining
New York: Crane Russack
1992 = Volume 11 [irregular]

A0257
Marine Policy
Guildford: Butterworths
1992 = Volume 16 [quarterly]

A0258
Marine Pollution Bulletin
New York: Pergamon
197– [24 per year]

A0259
Marine Technology Research News
London: Hollobone Hibbert
[irregular]

A0260
Marine Technololgy Society Journal
Washington DC: Marine Technology Society
1992 = Volume 26 [quarterly]

A0261
Naval Architect
London: Royal Institution of Naval Architects
[bi-monthly]

A0262
North Sea Letter
Financial Times [weekly]

A0263
North Sea Monitor
Amsterdam: North Sea Monitor
1983– [quarterly]

A0264
North Sea Observer
Oslo: North Sea Observer
1992 = Volume 15 [monthly]

A0265
North Sea Report
now part of CNWM North Sea Service

A0266
Norwegian offshore business opportunities service
London: Norconsultants
1988– [monthly]

A0267
Norwegian Oil Review
Oslo: Norsk Olje Revy
1992 = Volume 18 [monthly]

A0268
Ocean Construction Locator
Houston: Offshore Data Services
[monthly]

A0269
Ocean Construction Report
Houston: Offshore Data Services
[weekly]

A0270
Ocean Engineering
Oxford: Pergamon Press
1988 = Volume 15 [bi-monthly]

A0271
Ocean and coastal management
formerly Ocean and shoreline management
Barking: Elsevier Applied Science
1992 = Volume 17-18 [bi-monthly]

A0272
Ocean Industry
Houston: Gulf Publishing
1992 = Volume 27 [monthly]

A0273
Ocean Oil Weekly report
Tulsa: PennWell
1992 = Volume 26 [weekly]

A0274
Ocean Oil Weekly Roundup
Tulsa: PennWell
1992 = Volume 26 [weekly]

A0275
Offshore (incorporating Oilman)
Houston: PennWell
1992 = Volume 52 [monthly]

A0276
Offshore Business
Canterbury: Smith Rea Energy Associates
1983– [4 issues per year]

A0277
Offshore Engineer
London: Thomas Telford
1975– [monthly]

A0278
Offshore Fleet Economics
Houston TX: Offshore Data Service
1984– [bi-weekly]

A0279
Offshore Mechanics and Arctic Engineering Journal
Fairfield, NJ: ASME
1992 = Volume 5 [quarterly]

A0280
Offshore Oil International
Aberdeen: Sharnel
1992 = Volume 9 [fortnightly]

A0281
Offshore Research Focus
London: MTD
[bi-monthly]

A0282
Offshore Rig Location Report
Houston: Offshore Data Services
[monthly]

A0283
Offshore Rig Newsletter
Houston: Offshore Data Services
[monthly]

A0284
Offshore Rig Report
Tulsa: PennWell
1992 = Volume 8 [monthly]

A0285
Oil & Chemical Pollution
Barking: Elsevier Applied Science
1992 = Volume 8 [six per volume]

A0286
Oil & Gas Finance and Accountancy
London: Langham Publications
1992 = Volume 7 [quarterly]

A0287
Oil & Gas Journal
Tulsa: PennWell
1992 = Volume 90 [weekly]

A0288
Oil & Gas Law & Taxation Review
Oxford: ESC Publishing
1992 = Volume 10 [monthly]

A0289
Oil and Gas News
Singapore: Oil & Gas News
[50 per year]

A0290
Oil Gas European Magazine (international edition)
Hamburg: Urban Verlag
1992 = Volume 18 [two per year]

A0291
Oil News Service
Dollar, Clackmannanshire: Oil News Service
1992 = Volume 20 [weekly]

A0292
Oil Spill Intelligence Report
Boston MA: Cutter Information
1992 = Volume 15 [weekly]

A0293
OPEC Bulletin
Vienna: OPEC
[bi-monthly]

A0294
OPEC Review
Oxford: Pergamon [monthly]

A0295
Petroleum Economist
London: Petroleum Economist Ltd
1992 = Volume 59 [monthly]

A0296
Petroleum Engineer (International)
Denver: Hart Publications
1992 = Volume 64 [12 issues per year]

A0297
Petroleum Industry Newsletter
London: KPMG Peat Marwick KcLintock
[quarterly]

A0298
Petroleum Management
now incorporated into Petroleum Economist

A0299
Petroleum Review
London: Institute of Petroleum
1992 = Volume 46 [monthly]

A0300
Petroleum Times
Maidstone: Whitehall Press
1992 = Volume 96 [monthly]

A0301
Pipeline & Gas Journal
Houston: Oildom Publishing Co of Texas
1992 = Volume 219 [monthly]

A0302
Pipes & Pipelines International
Beaconsfield: Scientific Surveys
1992 = Volume 37 [bi-monthly]

A0303
Pipe Line Industry
Houston: Gulf Publishing Company
Jan/June 1992 = Vol 76

A0304
Platt's Weekly
New York: McGraw-Hill
1992 = Volume 5 [weekly]

A0305
Press releases
London: DTI Press Office
1986– [approximately 250 per annum]

A0306
PSTI Technical Bulletin
*Edinburgh: Petroleum Science & Technology
Institute*
1991 = Number 1 [irregular]

A0307
Quarterly journal of technical papers
London: Institute of Petroleum
1986– [quarterly]

A0308
Rig Market Forecast
London: FTBI
1988– [monthly]

A0309
Royal Bank / Radio Scotland Oil Index
Edinburgh: Royal Bank of Scotland
[monthly]

A0310
Safety at Sea International
Redhill UK: International Trade Publ Ltd
1967– [monthly]

A0311
Schlumberger Oilfield Review
Amsterdam: Elsevier
1992 = Volume 4 [quarterly]

A0312
Sea Technology
Arlington: Compass Publications
1992 = Volume 33 [monthly]

A0313
SPE Drilling Engineering
Richardson TX: SPE
1992 = Volume 7 [quarterly]

A0314
SPE Formation Evaluation
Richardson TX: SPE
1992 = Volume 7 [quarterly]

A0315
SPE Production Engineering
Richardson TX: SPE
1992 = Volume 7 [quarterly]

A0316
SPE Reservoir Engineering
Richardson TX: SPE
1992 = Volume 7 [quarterly]

A0317
SPE Review
Richardson TX: SPE
[monthly]

A0318
Spill Technology Newsletter
Ottawa: Environment Canada
1992 = Volume 17 [quarterly]

A0319
Subsea Engineering News
Swindon: Knighton Enterprises Ltd
1992 = Volume 9 [fortnightly]

A0320
Transport in Porous Media
London: Graham & Trotman
1992 = Volume 7 [bi-monthly]

A0321
Undersea Biomedical Research
*Bethesda MA: Undersea & Hyperbaric Medical
Society*
1976– [bi-monthly]

A0322
Underwater Systems Design
London: Underwater Systems Design
1992 = Volume 14 [bi-monthly]

A0323
Underwater Technology
London: Society for Underwater Technology
1992 = Volume 18 [quarterly]

A0324
Veritas Forum
Høvik: Det norske Veritas
[six per year]

A0325
Weekly Exploration Service
London: Petroleum Information
[weekly]

A0326
World Oil
Houston: Gulf Publishing
Jan/Jne 1992 = Volume 214 [monthly]

A0327
Worldwide Offshore Rigfinder
Tulsa: Ocean Oil Weekly Report
[monthly - twice February and August]

TECHNICAL REPORT SERIES

A0328
BMT Reports
Feltham: British Maritime Technology

A0329
Delft Hydraulics Communications
Delft: Delft Hydraulics Laboratory

A0330
Environmental Studies Research Funds
(ERSF) reports
*Calgary: Pallister Resource Management for
Canada Oil & Gas Lands Administration
(COGLA)*

A0331
Joint Group of Experts on the Scientific
Approach to Marine Pollution (GESAMP),
reports and studies
*[variously published by IMO, FAO, UNESCO,
WMO, WHO, IAGA, UN, UNEP]*

A0332
IOS Reports
Wormley: Institute of Oceanographic Sciences

A0333
ICES papers
*Copenhagen: International Council for the
Exploration of the Seas*

A0334
Technical Information Papers
*London: International Tanker Owners Pollution
Federation*

A0335
MIT Sea Grant College programme reports
*Cambridge MA: Massachusetts Institute of
Technology*

A0336
Naval architecture & ocean engineering reports
*Glasgow: Glasgow University Marine
Technology Programme*

A0337
NCEL Reports
*Port Hueneme: Naval Civil Engineering
Laboratory*

A0338
NOSC Reports
San Diego: Naval Ocean Systems Centre

A0339
NUSC Reports
*New London: Naval Underwater Systems
Centre*

A0340
DnV Technical Reports
Høvik: Det norske Veritas

A0341
NHL Reports
*Trondheim: Norwegian Hydrodynamic
Laboratories*

A0342
Bulletins of the Norwegian Hydrotechnical
Laboratory
Trondheim: Norsk Hydrotechnisk Laboratorium
[published in annual compilations]

A0343
NUTEC Reports
*Bergen: Norwegian Underwater Technology
Centre*

A0344
Outer Continental Shelf (OCS) Reports
*Minerals Management Service, US Department
of the Interior*

A0345
OCS Studies
*Minerals Management Service, US Department
of the Interior*

A0346
Offshore Business
Canterbury: Smith Rea Energy Analysts

A0347
Offshore Technology Information / Reports
London: HMSO for Department of Energy

A0348
SINTEF Reports
*Trondheim: Foundation for Scientific &
Induistrial Research at the Norwegian Institute
of Technology*

> **Note:**
> details of some Stockbroker Research
> Reports appear at the end of Section C,

MAPS

A0349
Asian oil and gas maps
Houston: Gulf Publishing
[series of six]

A0350
Atlas for marine policy in Southeast Asian seas
Morgan J R and Valencia M J, editors
Berkeley: University of California Press
1983, large format, 145 pages

A0351
Atlas of the seas around the British Isles
*Lowestoft: Ministry of Agriculture, Fisheries &
Food, Directorate of Fisheries Research*
1981 [outsize]
[with later Supplement]

A0352
Atlas of the US exclusive economic zone,
Atlantic continental margin
EEZ-Scan 87 Scientific Staff
Reston VA: US Geological Survey
1991, 174 pages
[US Geological Survey Miscellaneous
Investigation Series 1-2054]

A0353
Energy map of Europe
*London: Petroleum Economist / ABN-AMRO
Bank*
1991, one sheet

A0354
Energy map of the Far East
London: Petroleum Economist
1992, one sheet

A0355
Energy map of the Middle East
London: Petroleum Economist / SAMAREC
1991, one Sheet

A0356
Energy map of the Soviet Union
London: Petroleum Economist
1991, one sheet

A0357
Energy map of Sub-Saharan Africa
London: Petroleum Economist / ESKOM
1991, one sheet

A0358
Energy map of the United Arab Emirates
London: Petroleum Economist
1991, one sheet

A0359
Energy map of the world
London: Petroleum Economist / Chase
1991, one sheet

A0360
Floor of the oceans, based on bathymetric
studies
Heezen B C and Tharp M
New York: American Geophysical Society
c 1979

A0361
France - base maps
Ledbury: OPL
c 1988

A0362
Geological maps
Trondheim: Continental Shelf Institute
various dates
[series of maps]

A0363
Geological map of Iraq and South Western Iran
Ledbury: OPL
1991, one sheet

A0364
Logistics and supply in Europe: crude oil & oil
product pipelines in East and West Europe
Paris: Enerfinance
1992, 1 sheet, map

A0365
Lloyd's maritime atlas, 15th edition
Colchester: Lloyd's of London Press
1986

A0366
Marine scientific research boundaries
Ross D A and Landry T A
Woods Hole: Woods Hole Oceanographic
Institution
1986, one sheet

A0367
The Mediterranean oil and gas activity and
concession map
Ledbury: Oilfield Publications
1990, two sheets

A0368
Mexico and Central America map
Tulsa, OK: PennWell
1991, one sheet

A0369
Middle East oil and gas map
Tulsa, OK: PennWell
1991, one sheet

A0370
Netherlands maps (including German offshore
area)
Ledbury: Oilfield Publications
1987

A0371
North Sea map 1991 (small scale 1:2,000,000)
Ledbury: OPL
1991, one sheet

A0372
The North Sea oil and gas activity and
concession map (1:1,000,000), 15th edition
Ledbury: Oilfield Publications
1991, one sheet

A0373
North Sea pipeline atlas, 1987 edition
Ledbury: Oilfield Publications
1987, 88 pages

A0374
North Sea - structural elements relief map
Naersnes, Norway: Nopec a s
c 1988, one sheet

A0375
North Sea subsea atlas 1990
Ledbury, OPL
1991

A0376
Northwest Europe offshore activity map
Mitcham, UK: Claygate Services Ltd
updated twice yearly, two sheets

A0377
North Western European continental shelf
Stavanger: Noroil, for Den norske Creditbank
1985, one sheet

A0378
Norway map
Ledbury: OPL
1989, one sheet

A0379
Norwegian Sea, N62° map
Ledbury: Oilfield Publications
1987

A0380
OCS Maps
Minerals Management Service,
US Department of the Interior

A0381
Offshore mid Norway - structural elements
relief map
Naersnes, Norway: Nopec a s
c 1988, one sheet

A0382
Petromin Southeast Asia petroleum map 1990
Riyadh, Saudi Arabia: Petromin
1991, one sheet

A0383
Polar oil and gas map
Tulsa, OK: PennWell
c 1988, one sheet

A0384
Reservoir rocks of the North Sea, map
Ledbury: OPL
1991, one sheet

A0385
Seabed sediments 1:250,000 series
Institute of Geological Sciences and
Continental Shelf Institute, Norway
Southampton: Ordnance Survey for IGS
1984

A0386
Sedimentary basins of the world and giant
hydrocarbon accumulations
Tulsa, OK: AAPG
1990, one sheet

A0387
South East Asia atlas (1:250,000 to 1:500,000)
Ledbury: OPL
1991, 45 pages [A2 size]

A0388
South East Asia [map]
Houston: Offshore
19893, one sheet

A0389
The South East Asia oil and gas activity and
concession map
Ledbury: Oilfield Publications
1989, two sheets

A0390
The Times Atlas of the Oceans
Couper A (editor)
London: Times Books
1983, 272 pages

A0391
UK oil and gas activity map, 1992
Alton: Department of Energy Publications
1992, 1 sheet

A0392
Wells and exploration agreements offshore
Newfoundland and Labrador, 31 May 1985
[map]
*St John's: Newfoundland Offshore Petroleum
Board*
1986, one sheet

A0393
West Africa atlas (1:500,000 and 1:250,000)
Ledbury: OPL in conjunction with
Petroconsultants
1990, 20 pages

A0394
West Africa oil and gas activity and concession
map set
Ledbury: Oilfield Publications
1990, 2 sheets

A0395
The world's coastline
Bird E C F and Schwartz M L (editors)
New York: Van Nostrand Reinhold
1985, 13 + 1071 pages

A0396
World production map
Tulsa, OK: PennWell
1992, one sheet

A0397
World sedimentary basins and related features
Tulsa, OK: PennWell
1984, one sheet

A0398
Zeepipe and NOGAT pipelines route map
Ledbury, OPL
1989, one sheet

OIL AND GAS

GENERAL

A0399
Advanced projects conference; Stavanger,
17–18 November 1987
Stavanger: ONS Foundation
1987

A0400
Advances in operations research in the oil and
gas industry, conference; Montréal, 13–14
June 1991
Breton M and Zaccour G (editors)
Paris: Éditions Technip

A0401
I Congresso AIOM: realizzazioni prospettive
nell' ingegneria offshore e marini = AIOM
Congress: status and prospectives of offshore
and marine engineering: Venice 4–6 June 1986
Milan: AIOM
1986, 600 pages

A0402
AMO '87, 2nd Atlantic meeting in offshore;
Nantes, 24–25 September 1987
*Nantes: Nantes Chamber of Commerce &
Industry*
1987, 161 pages

A0403
Asia-Pacific 6th conference; Sydney, 13–15
September 1989
*Richardson, TX: Society of Petroleum
Engineers*
1989, 290 pages

A0404
Asia-Pacific petroleum 7th conference;
Singapore, 23–25 September 1991
Richardson TX: Society of Petroleum Engineers
1991

A0405
Challenges and opportunity offshore:
WESCON '86; Glasgow 18–20 November 1986
Neil M, editor
Lathcron, Caithness: Whittles Publishing
1987, viii + 199 pages

A0406
Challenges of a new decade, proceedings;
Ventura, CA, 4–6 April 1990
Richardson TX: Society of Petroleum Engineers
1990, 604 pages

A0407
CORE: Canadian Offshore Resources
Exposition: sustainable petroleum development
on Canada's East Coast
Halifax, Nova Scotia:
1989

A0408
Deep Offshore Technology 1985, 3rd
international conference and exhibition:
Sorrento, 21–23 October 1985
Paris: ASTEO
1985, 2 volumes, 3 loose papers

A0409
DOT '87: 4th deep offshore technology
conference; Monte Carlo, 19–21 October 1987
Paris: GEP-ASTEO/DOT
1988, 2 volumes + 1 loose paper

A0410
DOT '89: 5th deep offshore technology
conference; Marbella, Spain, 16–18 October
1989
Amsterdam: Deep Ocean Technology
1989, 2 volumes + 2 loose papers

A0411
DOT '91: 6th deep offshore technology
conference; Monaco, 4–6 November 1991
Paris: Deep Ocean Technology
1991, 2 volumes

A0412
Deepwater and marginal oilfield development,
5th conference; Vilamoura, Portugal, 15–16
November 1990
Shillington: OCS
1990, 13 papers

A0413
Deepwater and marginal oilfield development, 7th conference; Santiago, Spain, 26–28 February 1992
Shillington: OCS
due 1992

A0414
Developments in deeper waters: offshore engineering group international symposium; Heathrow, 6–7 October 1986
London: Royal Institution of Naval Architects
1986–1987, Volume 1 (20 papers), Volume 2 (discussions)

A0415
The development of gas condensate fields, conference; London, 20–21 September 1988
London: IBC Technical Services
1988, 14 papers

A0416
Economics of floating production systems: cost–effectiveness of floating production systems, conference; London, 12–13 May 1987
London: Graham & Trotman for the Society for Underwater Technology
1987, 227 pages [Advances in Underwater Technology, Volume 13]

A0417
Energy for the future, conference; Charleston W VA, 1–4 November 1988
Richardson TX: Society of Petroleum Engineers
1988, 387 pages

A0418
Eurogas '90: European applied research conference on natural gas; Trondheim, May 1990
Trondheim: Tapir
1990, 736 pages

A0419
EUROPEC '86: European Petroleum Conference: facing the 90s, new challenges, new solutions; London 20–22 October 1986
Richardson TX: Society of Petroleum Engineers
1986, 465 pages

A0420
EUROPEC '88: building on expertise, the new momentum; London, 17–19 October 1988
Richardson TX: Society of Petroleum Engineers
1988, 562 pages

A0421
EUROPEC '90: increasing the margin; The Hague, 21–24 October 1990
Richardson TX: Society of Petroleum Engineers
1990

A0422
Floating production systems: blueprints for the 90s, 6th annual conference; Monte Carlo, 17–18 October 1991
Shillington: OCS
1991, 11 papers

A0423
4th Floating production system conference; London, 14–15 December 1988
London: IBC Technical Services
1988, 14 papers

A0424
6th Floating production systems conference; London, 10–11 December 1990
London: IBC Technical Services
1990

A0425
Gas supply: a critical balance of politics, the market place and technology, symposium; Dallas, TX, 13–15 June 1988
Richardson TX: Society of Petroleum Engineers
1988, 559 pages

A0426
Gastech '90: 14th international LNG/LPG conference; Amsterdam, 4–7 December 1990
Rickmansworth: Gastech Ltd
1990, 41 papers

A0427
1991 Gas technology symposium; Houston, 22–24 January 1991
Richardson TX: Society of Petroleum Engineers
1991, 352 pages

A0428
Hydrocarbons '86: offshore gas, the challenge of a changing market; Great Yarmouth, 1–3 October 1986
Chipping Norton: The Hawkedon Partnership
1986, 12 papers

A0429
Hydrocarbons '88: European gas, the way ahead; Great Yarmouth, 5–6 October 1988
Chipping Norton: Themedia Ltd
1988, 13 papers

A0430
Hydrocarbons '90; London, 9–10 October 1990
Chipping Norton: Themedia Ltd
1990, 12 papers

A0431
1991 International arctic conference;
Anchorage, 29–31 May 1991
Richardson TX: Society of Petroleum Engineers
1991, 892 pages

A0432
1988 International meeting on petroleum
engineering; Tianjin, China, 1–4 November
1988
Richardson TX: Society of Petroleum Engineers
1988, 952 pages

A0433
1990 International meeting on petroleum
engineering; China
Richardson TX: SPE
1990

A0434
International Petroleum conference, 7th; St
John's, Newfoundland, 17–19 June 1991
*Newfoundland: Newfoundland Ocean
Industries Association*
1991

A0435
1991 International symposium on oilfield
chemistry; Anaheim CA, 20–22 February 1991
*Richardson, TX: Society of Petroleum
Engineers*
1991, 504 pages

A0436
1991 International thermal operations
symposium; Bakersfield CA, 6–8 February 1991
Richardson TX: Society of Petroleum Engineers
1991, 240 pages

A0437
Low permeability symposium, SPE Rocky
Mountain regional meeting; Denver, 6–8 March
1989
Richardson TX: Society of Petroleum Engineers
1989, 752 pages

A0438
Low permeability symposium, SPE Rocky
Mountain regional meeting; April 1990
Richardson TX: SPE
1990

A0439
Low permeability reservoirs symposium, SPE
Rocky Mountain regional meeting; Denver,
15–17 April 1991
Richardson TX: SPE
1991, 757 pages

A0440
Marginal and deep water field development
conference and exhibition; London, 4–5 June
1987
Hitchin: Lorne & Maclean Marine
1987

A0441
Middle East oil show, 5th; Bahrain, 7–10 March
1987
Richardson TX: Society of Petroleum Engineers
1987, 742 pages

A0442
Middle East oil show, 6th; Bahrain, 11–14
March 1989
Richardson TX: SPE
I1989, 893 pages

A0443
Middle East Oil show; Bahrain, 16–19
November 1991
Richardson TX: SPE
1991, 998 pages

A0444
A minimum facilities approach to oil and gas
production; conference, 10–11 September 1990
*London: IBC Technical Services Ltd with
Management Associates International Ltd*
1990, 13 papers

A0445
Minimum facilities offshore, 2nd conference;
Aberdeen, 12–13 November 1991
London: IBC Technical Services
1991, 12 papers

A0446
Moex '92: 1st Mediterranean oil and gas
conference; Valetta, Malta, 28–31 January 1992
Kingston upon Thames: Spearhead Exhibtiions
1992, 1 volume

A0447
A multidisciplinary approach in exploration and
production R & D: European oil and gas
conference; Palermo, 9–12 October 1990
Imarisio G et al (editors)
London: Graham & Trotman
1991, xv + 602 pages

A0448
New technologies for the exploration and
exploitation of oil and gas resources, 2nd
symposium; Luxembourg, 5–7 December 1984
*London: Graham & Trotman (for the
Commission of the European Communities)*
1985, 2 volumes

A0449
New technologies for the exploration and
exploitation of oil and gas resources, 3rd
symposium; Luxembourg, 23–24 March 1988
*London: Graham & Trotman (for the
Commission of the European Communities)*
1988, 2 volumes

A0450
New technology, challenge of the nineties;
Columbus, Ohio, 31 October – 2 November
1990
Richardson TX: Society of Petroleum Engineers
1990, 312 pages

A0451
Newfoundland offshore: project issues; St
John's, 1989
*St John's, Newfoundland: Newfoundland
Ocean Industries Association*
1989, 407 pages

A0452
Nigeria oil and gas: hydrocarbons seminar;
London, 16 January 1990
London: Department of Trade & Industry
1990, iv (various pagings)

A0453
North Sea oil and gas beyond 2000?; London,
17 October 1988
*Lodge A E (editor) for Exploration & Production
Discussion Group
London: Institute of Petroleum*
1988, 81 pages

A0454
North Sea oil and gas conference; London 1–3
July 1990
London: Financial Times
1990, 16 papers

A0455
North Sea oil and gas conference; London2–3
July 1991
London: Financial Times
1991, 18 papers

A0456
Offshore and arctic operations symposium,
1988
*Ertas A (editor) et al
New York: American Society of Mechanical
Engineers*
1988, v + 137 pages [PD Volume 12]

A0457
Ofshore and arctic operations symposium, 1989
*Ertas A, Hui D and Urquhart R G (editors)
Fairfield NJ: American Society of Mechanical
Engineers*
1989 [PD Volume 26]

A0458
Offshore and arctic operations symposium,
1990
*Urquhart R G (editor)
New York: American Society of Mechanical
Engineers*
1990, 209 pages [PE Volume 29]

A0459
Offshore and arctic operations symposium,
1991
*Urquhart R G (editor)
Fairfield NJ: American Society of Mechanical
Engineers*
1991, 108 pages [PE volume 38]

A0460
Offshore and arctic operations symposium
1992; Houston, 26–30 January 1992
*Fairfield NJ: American Society of Mechanical
Engineers*
1992, 68 pages [PD volume 42]

A0461
Offshore and maritime technology transfer '86;
organised for The Greenwich Forum;
Heathrow, 30 September–1 October 1986
*Kingston upon Thames: Offshore Conference &
Exhibitions*
1986, 14 papers

A0462
Offshore contracting and contracts: 2nd
British-Scandinavian construction contract
seminar; Bergen
Bergen: Studia
1985

A0463
Offshore contracting in the North Sea, seminar;
Bergen, 26–27 May 1983
Page A C (editor)
Dundee: Centre for Petroleum & Mineral Law
Studies
1983, xiv + 171 pages

A0464
Offshore Europe '87; Aberdeen, 6–11
September 1987
Kingston upon Thames: Spearhead
1987, 2 volumes

A0465
Offshore Europe '89; Aberdeen, 5–8
September 1989
Kingston upon Thames: Spearhead Exhibitions
1989, 2 volumes

A0466
Offshore Europe '91: Aberdeen, 3–6
September 1991
Kingston upon Thames: Spearhead Exhibitions
(for Society of Petroleum Engineers)
1991, 2 volumes

A0467
Offshore Gøteborg international conference on
offshore and marine technology; Gøoteborg, 25
February–1 March 1985
1985, 5 volumes

A0468
OMAE 1990: 9th offshore mechanics and arctic
engineering conference, Volume 1, offshore
technology; Houston, 18–23 February 1990
New York: American Society of Mechanical
Engineers
1990, 2 volumes

A0469
OMAE 1991: 10th offshore mechanics and
arctic engineering conference, Volume 1,
offshore technology; Stavanger, 23–28 June
1991
New York: American Society of Mechanical
Engineers
1991, 2 volumes

A0470
Offshore natural gas technology: production
and processing, transport and reception;
Dubrovnik, Yugoslavia 8–12 October 1984
Geneva: UN Economic Commission for
Europe: HMSO
1984, 16 loose papers

A0471
Offshore Northern Seas (General conference:
Session F); Stavanger, 26–29 August 1986
Oslo: Norwegian Petroleum Society
1986, 7 papers

A0472
Offshore Northern Seas, 8th conference;
Stavanger, 23–24 August 1988
Stavanger: ONS
1988, 39 papers

A0473
Offshore research and development; London
23–24 March 1987
London: IBC Technical Services Ltd
1987, 15 papers

A0474
2nd Offshore research and development
conference; London 10–11 October 1991
London: IBC Technical Services
1991

A0475
Offshore satellite developments conference;
London, 13–14 May 1991
London: IBC Technical Services
1991, 15 papers

A0476
6th Offshore South-East Asia conference;
Singapore 28–31 January 1986
Singapore: Offshore South-East Asia
1986, 536 pages preprints, 2 loose papers,
102–page directory

A0477
7th Offshore South East Asia conference;
Singapore, 2–5 February 1988
Singapore: Offshore South East Asia
1988, v + 896 pages

A0478
8th Offshore South East Asia conference;
Singapore, 4–7 December 1990
Singapore: Singapore Exhibition Services
1990, 736 pages + 9 papers

A0479
Oil and gas prospects in Ireland; Cork, 21–22
May 1986
Rugby: Institution of Chemical Engineers
1986, 205 pages + 4 loose papers

A0480
1989 Petroleum computer conference; San
Antonio, TX, 25–28 June 1989
Richardson TX: Society of Petroleum Engineers
1989, 214 pages

A0481
1990 Petroleum computer conference; Denver,
25–28 June 1990
Richardson TX: SPE
1990, 340 pages

A0482
1991 Petroleum computer conference; Dallas,
17–20 June 1991
Richardson TX: SPE
1991, 294 pages

A0483
1988 Petroleum industry applications of
microcomputers; San Jose, CA, 27–29 June
1988
Richardson TX: SPE
1988, 216 pages

A0484
Petroleum technology in the 1990s; Bakersfield
CA, 5–7 April 1989
Richardson TX: Society of Petroleum Engineers
1989, 700 pages

A0485
Preparing for a new decade of energy;
Morgantown W VA, 24–27 October 1989
Richardson TX: Society of Petroleum Engineers
1989, 378 pages

A0486
Quality assurance for the offshore industry;
London 12–13 March 1987
London: IBC Technical Services (IBC)
1987

A0487
Quality assurance in the offshore oil and gas
industry
Rogerson J H (Editor)
London: Graham & Trotman
1988, 168 pages [Petroleum Engineering &
Development Studies, 3]

A0488
Quality improvement for the offshore industry;
Aberdeen, 14 November 1990
London: IIR Industrial
1990

A0489
Serving a competitive market: gas technology
symposium; Dallas, 7–9 June 1989
Richardson TX: Society of Petroleum Engineers
1989, 542 pages

A0490
SEAAOC '91: South East Asia Australia
offshore conference; Redfern, New South
Wales, 30 July–2 August 1991
*Redfern NSW: Ministry for Mines & Energy with
IBC Conferences*
1991, 19 papers

A0491
1987 SPE annual technical conference; Dallas,
27–30 September 1987
Richardson TX; Society of Petroleum Engineers
1987, 5 sections

A0492
1988 SPE annual technical conference;
Houston, 2–5 October 1988
Richardson TX; SPE
1988, 5 sections

A0493
1989 SPE annual technical conference; San
Antonio, TX, 8–11 October 1989
Richardson TX; SPE
1989, 5 sections

A0494
1990 SPE annual technical conference; New
Orleans, 23–26 September 1990
Richardson TX; SPE
1990, 5 sections

A0495
1991 SPE annual technical conference; Dallas,
6–9 October 1991
Richardson TX; SPE
1991, 5 sections

A0496
Technology in the Rockies, SPE Rocky
Mountain regional meeting; Casper, WY, 11–13
May 1988
Richardson TX: SPE
1988, 651 pages

A0497
Technology in the '90s: prospering in a
changing environment; Calgary 10–13 June
1990
*Calgary: SPE and Petroleum Society of the
Canadian Institute of Mining*
1990, 3 volume

A0498
Technology ... the balance for oil and
environment, 1991 SPE Western regional
meeting; Long Beach, CA, 20–22 March 1991
Richardson, TX: SPE
1991, 480 pages

A0499
Total quality management in the offshore
industry, conference; London, 7–8 September
1991
London: IBC Technical Services
1991

A0500
25 years petroleum exploration and exploitation
in Norway, conference; Stavanger, 9–11
December 1991
Oslo: Norwegian Petroleum Society
due 1992

A0501
Underwater technology '84: deepwater
technology; Bergen, 9–11 April 1984
Anderson J A (editor)
Bergen: NUTEC
1984, 474 + 21 pages

A0502
Underwater technology conference 1990:
subsea production systems, the search for
cost-effective technology; Bergen, 19–21
March 1990
Grandal B (editor)
Bergen: Underwater Technology Foundation
1990, 451 pages

A0503
UK Offshore – maintaining self-sufficiency;
Institution of Civil Engineers; London 9–10
October 1985
London: Thomas Telford
1986, 210 pages

A0504
The way forward for floating production
systems; London, 15–16 December 1987
London: IBC Technical Services
1987, 16 papers

A0505
Will UCKS gas condensate reserves be
developed? London, 17 November 1987
Lodge A E (editor)
London: Institute of Petroleum
1988, 132 pages

ECONOMICS

Finance & Commerce

A0506
Before the oil runs out: prospects for the North
Sea
London: Economist Intelligence Unit
1989

A0507
Bergen conference on oil and economics;
Bergen, 4–5 May 1988
Oslo: Norwegian Petroleum Society
1988, 8 papers

A0508
Bergen conference on oil and economics;
Bergen, 8–9 May 1989
Oslo: Norwegian Petroleum Society
1989, 2 papers

A0509
Economic analysis of horizontal drilling
investments
Crouse P C
Tulsa OK: Gulf
1991

A0510
Economics and finance
Richardson TX: SPE
1982 369 pages
[SPE Reprint Series 16]

A0511
The economics of incremental investments in
mature oil fields in the UK continental shelf
Kemp A G and Reading D
*Aberdeen: University of Aberdeen Department
of Economics*
1991, iv + 79 pages
[North Sea Study Occasional Paper 36]

A0512
Economic value of the oil and gas resources on
the Outer Continental Shelf
Rosenthal D H et al
Marine Resource Economics
1989, Volume 5(3) pages 171–189

A0513
Hydrocarbon economics and evaluation
symposium; Dallas, 2–3 March 1987
Richardson TX: Society of Petroleum Engineers
1987, 25 papers 240 pages

A0514
Hydrocarbon economics and evaluation
symposium; Dallas, 9–10 March 1989
Richardson TX: SPE
1989, 211 pages

A0515
Hydrcarbon economics and evaluation
symposium; Dallas, 11–12 April 1991
Richardson TX: SPE
1991, 23 papers 236 pages

A0516
Oil and the international economy
Heal G and Chichilniskly G
Oxford: Clarendon Press
1991, 158 pages

A0517
Oil economists' handbook, volume 1: statistics,
5th edition
Jenkins, G
London: Elsevier Applied Science
1989, xvi + 467 pages

A0518
Oil economists' handbook, volume 2:
dictionary, chronology and directory, 5th edition
Jenkins, G
London: Elsevier Applied Science
1989 vii + 396 pages

A0519
The 1986 oil price crisis: economic effects and
policy responses
Tulsa OK; PennWell
1986, 270 pages

A0520
Oil prices to 2000: the economics of the oil
market
Lynch M C
London: Economist Intelligence Unit
1989
[EIU Special Report 1160]

A0521
Petroleum economics and engineering (2nd
edition)
Abdel-Aal H K et al (editors)
Basel: Marcel Dekker
1992, 456 pages

A0522
The prize
Yergin D
London: Simon & Schuster
1991, xxxii + 877 pages

A0523
Recent modelling approaches in applied
energy economics
Bjerkholt O et al (editors)
New York: Chapman & Hall
1990, 268 pages

Costs

A0524
Cost reduction in offshore enginering,
conference; London, 10–11 February 1988
Offshore Engineering Society of the Institution
of Civil Engineers
London: Thomas Telford
1988, 130 pages

A0525
Cost reduction offshore: the way ahead;
conference; London, 16 November 1989
*Lodge A E (editor for Exploration & Production
Group)*
London: Institute of Petroleum
1990, 208 pages

A0526
Cost study: Norwegian continental sheld
Roistadas A
Trondheim: Norwegian Institute of Technology
1985, 21 pages [SINTEF Report ST17 A83034]

A0527
Lifetime cost model of fixed jacket platform and
economic delphi study design
Shi W B et al
*Glasgow: University of Glasgow Department of
Naval Architecture and Ocean Engineering*
1989, 42 pages

A0528
Transaction costs and trade between
multinational corporations: a study of offshore
oil production
Hallwood C P
Boston: London: Unwin Hyman
1990, xxviii + 202 pages

Commercial

A0529
Accounting for oil and gas exploration and
development activities: a survey of industry
practice in the UK
Oil Industry Accounting Committee
London: Ernst & Whinney
1983

Section A : Bibliography

A0530
Accounting methods for the oil and gas producer
Gage G H et al
New York: Arthur Young Extractive Industries Group
1983, 10 pages

A0531
Analysis of UK fabrication capacity and forecast demand
London: UK Module Constructors' Association Ltd
1989

A0532
Development and production prospects for oil and gas from the UK continental shelf after the Fulf crisis: a financial simulation
Kemp A G et al
Aberdeen: University of Aberdeen
1991, 47 pages [North Sea Study Occasional Paper 35]

A0533
Energy finance handbook 1991
London: Petroleum Economist
1991, 350 pages

A0534
Energy futures: trading oportunities for the 1980s
Treat J E et al
Tulsa OK: PennWell
1984, 168 pages

A0535
Financial reporting by the European oil industry
London: Arthur Andersen & Co
1988, 94 pages

A0536
Firm location and differential barriers to energy in the offshore oil supply industry
Cairns J A and Harris A H
1988, 8 pages

A0537
Forecasts of offshore activity and expenditure: UKCS and worldwide 1989–1993
Aberdeen; Scottish Development Agency, Oil & Gas Division
1989, 32 pages

A0538
Forecasts of offshore activity and expenditure UKCS and worldwide 1991–1996
Aberdeen; Scottish Enterprise, Oil & Gas Group
1991

A0539
FT guide to North Sea operators: a company-by-company review, 4th edition
Leitch M
London: Financial Times Business Information
1991

A0540
Fundamentals of oil and gas accounting
Gallun R A and Stevenson J W
Tulsa, OK: PennWell
1986, 408 pages

A0541
Hedging petroleum price exposure conference; London, 11 April 1991
Byfleet: IBC Financial Focus Ltd
1991, 9 papers

A0542
International purchasing contacts for oil and gas field exploration and development
London: Offshore Supplies Office
1986, 62 pages

A0543
An introduction to oil and gas agreements – the 1990 update, conference; London, 22–23 November 1990
Leicester: Langham Oil Conferences Ltd
1990, 7 papers

A0544
The Japanese offshore industry: technology and markets
Technical Communications
Oxford; Elsevier Advanced Technology Publications
1988, 148 leaves

A0545
Long range exploration planning: a guide to preparing goals, stratetgies & actions
Tulsa OK: PennWell
1985, 96 pages

A0546
Managing energy price risk conference; London, 25–26 June 1991
London: IIR Ltd
1991, 12 papers

A0547
The market for North Sea crude oil
Mabro R
Oxford: OUP/Oxford Institute for Energy Studies
1986, 372 pages

A0548
Netback pricing and oil price collapse of 1986
Mabro R
Tulsa OK: PennWell
1986, 45 pages

A0549
A new Europe – the coming changes for the oil
and gas industry and its contractors:
conference; Amsterdam, 17 October 1990
Leicester: Langham Oil Conferences Ltd
1990, 5 papers

A0550
Oil and gas accounting
Brock H R et al
Denton TX: Professional Development Institute
1990, 620 pages (4th edition)

A0551
Oil and gas: a survey of published accounts
London: KPMG Peat Marwick McLintock
1989, 96 pages

A0552
Oil and gas international purchasing contacts
London: Offshore Supplies Office
1989, 63 pages

A0553
Oil and gas project finance seminar; 5 October
1990
London: IBC Financial Focus
1990, 6 papers

A0554
Oil and gas reserve disclosures: annual survey
of over 200 public companies
Houston TX: Arthur Andersen & Co
Published annually

A0555
Oil and gas survey
New York: Price Waterhouse
Published annually

A0556
Oil business fundamentals
Langenkamp R D
Tulsa OK: PennWell
1982, 146 pages

A0557
Oil markets of the Pacific Rim: into the 1990s
McDonald P
London: FT Business Information
1990

A0558
Offshore contracting in the North Sea
conference proceedings
*Dundee: Dundee University Centre for
Petroleum and Mineral Law Studies*
1983, 171 pages

A0559
Options trading and oil futures markets
Chassard C and Halliwell M
Tulsa OK: PennWell
1986

A0560
Performance profiles of major energy
producers
*United States Department of Energy, Energy
Information Administration*
Washington DC: US Government Printing Office
Published annually

A0561
Petrocompanies 1991/92
London: Petrocompanies
1991 [loose leaf]

A0562
Petroleum accounting: price, procedures &
issues
Brock H R et al
Denton TX: Professional Development Institute
1985

A0563
Petroleum Argus oil prices worldwide 1989/90
Jenkins G (editor)
London: Wiley, with Petroleum Argus
1990, 240 pages

A0564
Petroleum exploitation strategy
*Fee D A (Commission of the European
Communities)*
London: Pinter
1988, 220 pages

A0565
The petroleum explorationists' guide to
contracts used in oil and gas operations
Mosburg L G, Jr (editor)
Tulsa OK: PennWell
1988,1200 pages

A0566
The preparation and operation of contracts in
the petroleum industry; health and safety
guidelines
Oil Industry Advisory Committee
London: HMSO
1987, 6 pages

A0567
Prospects for the world oil and gas industry
1990–1994
Aberdeen: Mackay Consultants
1990

A0568
Purchasing oil and gas
Robinson T
Newmarket: Energy Publications
1988, x + 100 pages

A0569
Standards pertaining to the estimating and
auditing of oil and gas reserve information
Richardson TX: SPE
1980, 12 pages

A0570
Striking out! New directions for offshore
workers and their unions: a discussion
document
Offshore Industry Liaison Committee
Aberdeen: Offshore Information Centre
1991, 103 pages

A0571
Survey of accounting practices in the oil and
gas industry
Deakin E B
Denton TX: Institute of Petroleum Accounting,
University of North Texas
1989, 112 pages

A0572
Trends in global offshore oil and gas business
1980–1990 and the implications for investors in
the marine equipment in the 1990s
Aberdeen: Petrodata Ltd
1990, 11 leaves

A0573
Worldwide offshore field prospects to 1995
Harrison P (editor)
London: Hollobone Hibbert & Associates
1990, 450 pages

Project Economics

A0574
Analysis of petroleum investments: risks, taxes
and time
Campbell J M
Tulsa OK: PennWell
1987, 400 pages

A0575
Decision analysis for petroleum exploration
Newendorp P D
Tulsa OK; PennWell
1976 668 pages

A0576
Engineering and project management
Robinson G (editor)
Canterbury: SREA
1991, iv + 44 pages [Offshore Business 34]

A0577
Evaluating and managing risk
Megill R E
Tulsa OK: PennWell
1985, 160 pages

A0578
An introduction to risk analysis, 2nd edition
Megill R E
Tulsa OK: PennWell
1984 288 pages

A0579
An introduction to exploration economics, 3rd
edition
Megill R E
Tulsa OK: PennWell
1988 250 pages

A0580
Oil property evaluation
Thompson R S and Wright J D
Tulsa OK: PennWell
1985, 250 pages

A0581
Petroleum evaluations and economic decisions
McCray A W
Englewood Cliffs, NJ; Prentice Hall
1975 xiv + 448 pages

GOVERNMENT

A0582
Acts, regulations and provisions for the petroleum activity ... 1 January 1992
Stavanger: Norwegian Petroleum Directorate
1991

A0583
Developments of the oil and gas resources of the United Kingdom
(The Brown Book)
Alton: Department of Energy Publications
1992, 152 pages

A0584
Energy law '90: changing energy markets – the legal consequences
International Bar Association
London: Graham & Trotman
1990, 736 pages

A0585
Energy law: a complete guide to international oil, gas and nuclear legislation
Mankabady S
London: Euromoney Books
1990, xxviii + 465 pages

A0586
Energy statistics yearbook 1989
UN Dept of International Economic & Social Affairs
New York: United Nations
1991, xviii + 440 pages

A0587
Gas regulation in Western Europe: a country-by-country guide
Cameron P
London: Financial Times Business Information
1990

A0588
Government and North Sea oil
Hann D
London: Macmillan
1986, ix + 175 pages

A0589
Natural gas in the European Community
Kalim Z
London: Financial Times Business Information
1991

A0590
Oil and gas law: the North Sea exploitation
Simmonds K R
Dobbs Ferry NY: Oceana
1988, [loose leaf]

A0591
OPEC: twenty-five years of prices and politics
Skeet I
Cambridge: Cambridge University Press
1991, 276 pages [Cambridge Energy Studies]

A0592
Petroleum investment policies in developing countries
Walde T and Beredjick N (editors)
London: Graham & Trotman
1988, 240 pages

A0593
Recent developments in UK oil and gas law, seminar; London, June 198
Dundee: Dundee University Centre for Petroleum and Mineral Law Studies
1986, 126 pages

A0594
United Kingdom oil and gas law
Daintith T and Willoughby G (editors)
London: Sweet & Maxwell
1984, [loose leaf updating]

Taxation

A0595
Advance petroleum revenue tax act 1986
London: HMSO
1986

A0596
A comparative analysis of the impact of the UK-Norwegian petroleum tax regimes on different oil fields
Kemp A G and Reading D
Aberdeen: University of Aberdeen Department of Economics
1990, 38 pages
[North Sea Study Occasional Paper 31]

A0597
The economics of North Sea oil taxation
Rowland C and Hann D
London: Macmillan
1987, 84 pages

A0598
Finance act 1990
London: HMSO
1990, vi + 192 pages

A0599
Finance act 1991
London: HMSO
1991, 80 pages

A0600
Guide to European oil taxation
London: Arthur Andersen & Co
1988, 36 pages

A0601
Guide to Irish offshore oil taxation
Dublin: Arthur Andersen & Co
1988, 8 pages

A0602
A guide to UK oil and gas taxation
London: KPMG Peat Marwick McLintock
1990, x + 135 pages

A0603
The impact of petroleum fiscal systems in
mature field life: a comparative study of the UK,
Norway, Indonesia, China, Egypt,
Nigeria and United States Federal Offshore
Kemp A G with Reading D
Aberdeen: University of Aberdeen Department
of Economics
1991, 46 pages
[North Sea Study Occasional Paper 32]

A0604
Income tax of natural resources
Burke F M Jr and Bowhay R W
Englewood Cliffs NJ: Prentice Hall
1985

A0605
Natural resources taxation: principles & policies
Dzienkowski J S and Peroni R J
Durham NC: Carolina Academic Press
1988, 651 pages

A0606
North Sea oil taxation: the development of the
North Sea tax system
Devereux M P and Morris C N
London: Institute for Fiscal Studies
1983 [IFS Report 6]

A0607
North Sea taxation for the 1990s
Bond S et al
London: Institute for Fiscal Studies
1987 [IFS Report Series 27]

A0608
Oil and gas taxation
Klingstedt J P
Tulsa OK: PennWell
1988, 588 pages

A0609
Oil and gas taxation 1989
Klingstedt J P et al
Denton TX: Professional Development Institute
1989, 520 pages

A0610
Oil taxation act 1975
London: HMSO
1975

A0611
Oil taxation act 1983
London: HMSO
1983

A0612
Oil taxation acts, 1991 edition
Inland Revenue
London: HMSO
19912, 584 pages

A0613
Petroleum revenue tax (nomination scheme for
disposal and appropriations) regulations 1987
London: HMSO
1987 [SI 1987 1338]

A0614
Regulation relating to the collection of fees
payable to treasury for supervision of the
petroleum activity
Stavanger: Norwegian Petroleum Directorate
1991

A0615
Taxation and the optimization of oil exploration
and production in the UKCS
Favero C A
Oxford: Oxford Institute for Energy Studies
1990 [Working Paper EE 12]

A0616
Taxes act 1988 (sections 492; 771; 772; 830)
London: HMSO
1988

A0617
United Kingdom oil and gas taxation and
accounting
Lilley P et al (editors)
London: Profesional Publishing
1983 [loose leaf updating]

A0618
UK oil taxation
London: Arthur Andersen & Co
1990, 100 pages

A0619
UK oil taxation
London; Longman
1991 in preparation
[Longman Law, Tax & Finance]

A0620
UK taxation of offshore oil and gas
Halltar R F et al
London: Butterworth
1985

A0621
United Kingdom taxation oil and gas, 6th edition
*Edinburgh: Bank of Scotland (International
Division)*
1989, 22 pages + update

A0622
UK taxation on the profits from the North Sea
oil, 5th edition
Price Waterhouse
London: Graham & Trotman
1991, 286 pages

Technical Legislation

A0623
Bibliography of North Sea regulations,
volume I: UK, 5th edition
London: Hollobone Hibbert & Assoc
1992 [Loose leaf]

A0624
Bibliography of North Sea regulations, volume
2: Norway, Denmark, Netherlands & Ireland,
5th edition
London: Hollobone Hibbert & Assoc
1989 [Loose leaf]

A0625
Continental Shelf operations notices
Department of Trade & Industry
London: HMSO
1972 onwards [regularly updated]

A0626
The depletion of oil resources: a comparison of
Saudi and British policies
Roberts S J
Oxford: OUP/Oxford Institute for Energy Studies
1986, 100 pages

A0627
The joint operating agreement: oil & gas law
Taylor M P G et al
London: Longman
1989, xi + 143 pages

A0628
Manual of legislative acts, rules and guidance
notes concerning North Sea offshore
developments – UK sector
Brentford: Weston Law Manual Services
1990, 4 volumes loose leaf [Regularly updated]

A0629
Maritime law
Hill C
Colchester: Lloyds of London Press
1989

A0630
OCS regulations related to minerals resource
activities on the outer continental shelf
Herndon VA: Minerals Management Services
1990 [OCS Report MMS 88-0025]

A0631
Offshore installations: guidance on design,
construction and certification, 4th edition
Department of Energy
London: HMSO
1990 [Loose leaf]

A0632
Oil and gas: Netherlands law and practice
Roggenkamp M
London: Chancery Law
1991, 250 pages

A0633
The oil supplies industry: a comparative study
of legislative restrictions and their impact
Cameron P
London: Financial Times Business Information
1986

A0634
The Petroleum (Production)(Landward Areas)
Regulations 1991
London: HMSO
1991

A0635
Regelverksamling for
petroleumviriksomheten...= Acts, regulations
and provision for the petroleum activity,
updated 1.1.88 edited &
Stavanger: Norwegian Petroleum Directorate
1988, 2 vols + updating service

A0636
UK marine oil pollution legislation
Annotations by Bates, J H
Colchester: Lloyd's of London Press
1987

A0637
Whitehead's UK offshore legislation guide
Osliff F (editor)
Croydon: Benn Technical Books
1983 [Loose leaf updating + supplements]

Licensing

A0638
Assessment of foreign oil and gas license
relinquishment policies
Ham C R (for Minerals Management Service)
Herndon VA: US Dept of the Interior
1991 [OCS Report MMS 91-0014]

A0639
Federal offshore statistics, 1990: leasing,
exploration, production and revenues
Barbagallo M B et al
Herndon VA: US Dept of the Interior
1991, xxi + 99 pages
[OCS Report MMS 91-0068]

A0640
Leasing energy resources on the outer
continental shelf
Minerals Management Service
Herndon VA: US Dept of the Interior
1987, vi + 49 pages

A0641
Oil and gas leasing/production program: annual
report, 1990
Minerals Management Service
Herndon VA: US Dept of the Interior
1991, xv + 57 pages
[OCS Report MMS 91-0041]

PETROLEUM ENGINEERING

Exploration

Regional geology

A0642
Basin analysis: principles & applications
Allen P A and Allen J R
Oxford: Blackwell Scientific Publications
1990, 464 pages

A0643
Basin analysis: quantitative methods, volume 1
Lerche I
London: Academic Press
1989, 562 pages

A0644
Basin analysis: quantitative methods, volume 2
Lerche I
London: Academic Press
1990, 611 pages

A0645
Chinese sedimentary basins
Zhu X (editor)
Amsterdam: Elsevier
1989, xviii + 328 pages
[Sedimentary Basins of the World 1]

A0646
The geology and tectonics of the Oman region
Robertson A H F et al (editors)
Bath: Geological Society
1990, 864 pages
[Geological Society Special Publication 49]

A0647
The geology of the Moray Firth
British Geological Survey
London: HMSO
1991, x + 96 pages
[UK Offshore Regional Report 3]

A0648
Geology of the north-west European
continental shelf, volume 1: the west British
shelf
Naylor D & Mouteney N
London: Graham & Trotman
1975, 162 pages

A0649
Geology of the north-west European
continental shelf, volume 2: the North Sea
Pegrum R M et al
London: Graham & Trotman
1975, 224 pages

A0650
The geology of offshore Ireland & West British
Naylor D and Shannon P M
London: Graham & Trotman
1982, xii + 161 pages

A0651
In search of oil, volume I: the search for oil and
its impediments
Grossling B F and Nielsen D T
London: FT Business Information
1985

A0652
In search of oil, volume II: country analyses
Grossling B F and Nielsen D T
London: FT Business Information
1985

A0653
Introduction to the geology of the North Sea,
3rd edition
Glennie K W (editor)
*Oxford: Blackwell Scientific (for Geological
Society)*
1990, xii + 402 pages

A0654
The Middle East: regional geology and
petroleum resources
Beydoun Z R
Beaconsfield: Scientific Press
1988, 292 pages

A0655
Venezuelan oil: development and chronology
Martinez A R
Amsterdam: Elsevier
1989, xviii + 248 pages

Petroleum geology

A0656
Basic petroleum geology
Link P K
Tulsa OK: OGCI
1987, 426 pages

A0657
Classic petroleum provinces
Brooks J (editor)
Bath: Geological Society
1990, 568 pages
[Geological Society Special Publication 50]

A0658
Deltas: sites and traps for fossil fuels
Whately M K G and Pickering K T (editors)
Oxford: Blackwell Scientific Publications
1989, 376 pages
[Geological Society Special Publications 41]

A0659
Elements of petroleum geology
Selley R E
Amsterdam: Elsevier
1983, 416 pages

A0660
Future petroleum provinces of the world:
Wallace E Pratt Memorial Conference;
Phoenix, December 1984
Halbouty M T (editor)
Tulsa OK: AAPG
1986, xi + 708 pages

A0661
Generation, accumulation and production of
Europe's hydrocarbons
Spencer A M (editor)
Oxford: Clarendon Press
1991, 600 pages
[Special Publication of the European
Association of Petroleum Geoscientists 1]

A0662
Geology of the Norwegian oil & gas fields: an
atlas of hydrocarbon discoveries
Spencer A M (for Norwegian Petroleum Society)
London: Graham & Trotman
1987, 464 pages

A0663
Geostatistics & petroleum geology
Hohn M E
Wokingham: Van Nostrand Rheinhold
1988 264 pages

A0664
Giant oil and gas fields of the decade:
1968–1978
Halbouty, M T (editor)
Tulsa OK: AAPG
1980, ix + 596 pages

A0665
Habitat of hydrocarbons on the Norwegian
continental shelf
*Spencer A M (editor, for Norwegian Petroleum
Society)*
London: Graham & Trotman
1987, 420 pages

A0666
Lacustrine petroleum source rocks
Fleet A J et al (editors)
Oxford: Blackwell Scientific Publications
1988, 408 pages
[Geological Society Special Publication 40]

A0667
Marine petroleum source rocks
Brooks J (editor)
Bath: Geological Society
1987, 444 pages
[Geological Society Special Publication 26]

A0668
North Sea oil and gas reservoirs
*Kleppe C (editor, for Norwegian Institute of
Technology)*
London: Graham & Trotman
1987, 448 pages

A0669
North Sea oil & gas reservoirs, conference;
Trondheim 8–11 May 1989
Buller A T et al (editors)
London: Graham & Trotman
1990, 464 pages

A0670
Oilfields of the world, 3rd edition
Tiratsoo E N
Beaconsfield: Scientific Press
1986, xvi + 392 pages + Supplement

A0671
Oil shales of the world: their origin, occurrence
and exploitation
Russell P L
Oxford: Pergamon Press
1990, 736 pages

A0672
Petroleum and tectonics in mobile belts
Letouzey J (editor)
Tulsa OK: Gulf Publishing
1991

A0673
Petroleum basin studies
Shannon P M and Naylor D
London: Graham & Trotman
1989

A0674
Petroleum formation and occurrence
Tissot B P and Welte D H
Berlin: Springer-Verlag
1988, xxi + 699 pages

A0675
Petroleum geology
Chapman R E
Amsterdam: Elsevier
1983, 416 pages

A0676
Petroleum geology
North F
London: Unwin & Hyman
1985, 620 pages

A0677
Petroleum geology of the continental shelf of
North-West Europe, conference; London, 4–6
March 1980
Illing L V and Hobson G B (editors)
*London: Heyden (John Wiley for Institute of
Petroleum)*
1981, xviii + 521 pages

A0678
Petroleum geology of North West Europe
Brooks J and Glennie K W
London: Graham & Trotman
1987, 1300 pages, 2 volumes

A0679
Petroleum migration
Fleet A J et al (editors)
Bath: Geological Society
1988, 391 pages
[Geological Society Special Publication 40]

A0680
Petroleum related rock mechanics
Fjaer E et al
Amsterdam: Elsevier
1992, 352 pages
[Developments in Petroleum Science 33]

A0681
Regional geology of the petroleum provinces,
volume 1: petroleum geology of the Middle East
Alsharhan A S and Nairn A E M
Amsterdam: Elsevier
in preparation

A0682
Rift systems; hydrocarbon habitat and potential
of rift-related basins
Stampfli G M and Grunau H R
London: Petroconsultants
1988

A0683
Rock mechanics
Charlez Ph A
Paris: Éditions Technip
1991, 360 pages

A0684
Sandstone petroleum reservoirs
Barwis J H et al
Berlin: Springer-Verlag
1990, 450 pages
[Casebooks in Earth Sciences]

A0685
Sedimentology and petroleum geology
Bjorlykke K (translator, Wahl B)
Berlin: Springer-Verlag
1989, xii + 363 pages

A0686
Tectonic events responsible for Britain's oil and
gas reserves
Hardman R F P and Brooks J (editors)
Bath: Geological Society
1991, 416 pages
[Geological Society Special Publication 55]

A0687
United Kingdom oil and gas fields
Abbotts I L (editor)
Bath: Geological Society
1991, 464 pages
[Geological Society Memoir 14]

Geophysical surveys

A0688
Acquiring better seismic data
Pritchett W C
London: Chapman & Hall
1989, 427 pages

A0689
Applied subsurface geological mapping
Tearpock D J and Bischke
Englewood Cliffs NJ: Prentice Hall
1991, 608 pages

A0690
Carbonate rock depositional models: a
microfacies approach
Carozzi A V
Englewood Cliffs NJ: Prentice Hall
1989, 576 pages

A0691
Concepts and techniques in oil and gas
exploration
Jain K C and De Figueiredo R J P
Tulsa OK: SEG
1982, 289 pages

A0692
Crosswell seismology and reverse VSP
Hardage B A
Castelnau-le-Lez: Geophysical Press
1992, xvi + 304 pages

A0693
Exploration stratigraphy
Visher G S
Tulsa OK: OGCI
1984, 344 pages

A0694
Geologic analysis of naturally fractured
reservoirs
Nelson R A
Houston TX: Gulf
1985, 320 pages [Contributions in Petroleum
Geology and Engineering 1]

A0695
Geology and exploration of oil- and gas-bearing
ancient deltas
Akramkhodzhaev A M et al
Rotterdam: A A Balkema
1989, 213 pages [Russian Translation Series
69]

A0696
Geophysical methods
Sherriff R E
Englewood Cliffs NJ: Prentice Hall
1989, 592 pages

A0697
Interpretating seismic data
Coffeen J A
Tulsa OK: PennWell
1984, 260 + 194 pages

A0698
An introduction to seismic interpretations:
reflection seismics in petroleum exploration,
2nd edition
McQuillin R et al
London: Graham & Trotman
1985, 256 pages

A0699
Mass spectrometric characterization of shale
oils
Aczel T (editor)
Philadelphia PA: ASTM
1986, 149 pages [ASTM STP 902]

A0700
North Sea seismicity
Department of Energy
London: HMSO
1986 [OTH 86 219]

A0701
Oil and gas prospecting
Robinson G (editor)
Canterbury: Smith Rea Energy Analysts
1989, iv + 28 pages
[Offshore Business 27]

A0702
Oil and gas traps: aspects of their
seismostratigraphy, morphology and
development
Jenyon M K
Chichester: Wiley
1990, 416 pages

A0703
The potential of deep seismic profiling for
hydrocarbon exploration, IFP Exploration and
Production Research Conference]
Pinet B and Bois C (editors)
Houston TX: Gulf
1990, 224 pages

A0704
Practical seismic interpretation
Bradley-Boston M E
Lancaster: D Reidel
1985, vi + 266 pages

A0705
Practical seismic interpretation
Badley M E
Englewood Cliffs NJ: Prentice Hall
1989, 266 pages

A0706
Production seismology
White J E and Sengbush R L
Oxford: Pergamon Press
1987, 386 pages
[Handbook of Geophysical Exploration 10]

A0707
Quantitative dynamic stratigraphy
Cross T A (editor)
Englewood Cliffs NJ: Prentice Hall
1990, 640 pages

A0708
Seismic and acoustic velocities in reservoir
rocks, volume 1: experimental studies
Nur A and Wang Z (editors)
Tulsa OK: SEG
1989, 420 pages
[Geophysics Reprints Series 10]

A0709
Seismic exploration fundamentals
Coffeen J A
Tulsa OK: PennWell
1986, 360 pages

A0710
Seismic methods
Lavargne M
Paris: Éditions Technip
1990, 192 pages

A0711
Seismic on screen: an introduction to
interactive interpretation
Coffeen J A
Tulsa OK: PennWell
1991, 400 pages

A0712
Seismic prospecting for sedimentary formations
Gogonenkov G N
Rotterdam: A A Balkema
1990, 228 pages [Russian Translation Series
76]

A0713
Structural styles in petroleum exploration
Lowell J D
Tulsa TX: OGCI
1987, 477 pages

A0714
Vertical seismic profiling and its exploration
potential
Galperin E I
Dordrecht: D Reidel
1985, 426 pages

A0715
Wellsite geological techniques for petroleum
exploration: methods and systems of formation
evaluation
Sahay B et al
Rotterdam: A A Balkema
1988, 513 pages

Petroleum reservoir

Reservoir geology

A0716
Carbonate reservoir characterization: a
geologic-engineering analysis, part 1
Chilingar G V et al (editors)
Amsterdam: Elsevier
in preparation
[Developments in Petroleum Science 30]

A0717
Geology in petroleum production
Dikekrs A J
Amsterdam; Elsevier
1985 xiv + 240 pages
[Developments in Petroleum Science 20]

A0718
Petroleum development geology, 3rd edition
Dickey D A
Tulsa OK: PennWell
1986, 560 pages

A0719
Reservoir characterization
Richardson TX: SPE
1989, 2 volumes
[SPE Reprint Series 27]

A0720
Reservoir characterization: 1st international
conference; Dallas, 1985
Lake L W and Carroll H B (editors)
London: Academic Press
1986, xii + 659 pages

A0721
Reservoir characterization II: 2nd international conference; Dallas, June 1989.
Lake, L W et al (editors)
London: Academic Press
1991 xviii + 726 pages

Petrophysics

A0722
Advanced well-log interpretation
Hilchie D W
Boulder CO: Hilchie Inc
1989, 378 pages

A0723
Advances in core evaluation I: accuracy and precision in reserves estimation; lst European symposium; London, 21–23 May 1990
Worthington P E (editor)
New York: Gordon & Breach for Society of Core Analysts
1990, xi + 555 pages

A0724
Advances in core evaluation II: reservoir appraisal; 2nd European symposium; London, 20–22 May 1991
Worthington P F (editor)
Reading: Gordon & Breach for Society of Core Analysts
1991, xi + 484 pages

A0725
Advances in well-test analysis
Earlougher R C
Richardson TX: SPE
1977, 264 pages
[SPE Monograph 5]

A0726
Applied geophysics
Telford W M et al
Cambridge: Cambridge University Press
1989, 540 pages

A0727
Applied open-hole interpretation
Hilchie D W
Boulder CO: Hilchie Inc.
1982, 330 pages

A0728
Applied open-hole log analysis
Brock J
Houston TX: Gulf
1986, 284 pages [Contributions in Petroleum Geology & Engineering 2]

A0729
Cores and core logging for geologists
Blackbourn G A
Latheronwheel, Caithness: Whittles Publishing Service
1990, 120 pages

A0730
Encyclopedia of well-logging
Desbrandes R
London: Graham & Trotman
1985, 608 pages

A0731
Essentials of modern open-hole log interpretation
Dewan J
Tulsa OK: PennWell
1983, 360 pages

A0732
Field geologists' training guide: an introduction to oilfield geology, mud-logging and formation evaluation
Whittaker A
Englewood Cliffs NJ: Prentice Hall
1985, 291 pages

A0733
Finding oil and gas from well-logs
Etnyre L M
New York: Van Nostrand Rheinhold
1989, 305 pages

A0734
Fundamentals of well-log interpretation, part I: the acquisition of logging data
Serra D
Amsterdam: Elsevier
1984, xii + 424 pages
[Devt in Petroleum Science 15A]

A0735
Fundamentals of well-log interpretation, part II: the interpretation of logging data
Serra D
Amsterdam: Elsevier
1987, 684 pages
[Devt in Petroleum Science 15B]

A0736
Geological applications of wireline logs
Hurst A et al (editors)
Oxford: Blackwell Scientific Publications
1990, 365 pages
[Geological Society Special Publication 48]

A0737
Geophysical well-logging
Titman J
London: Academic Press
1986, 192 pages
[Methods of Experimental Physics 24]

A0738
Handbook of well-log analysis for oil and gas
formation evaluation
Pirson S J
Englewood Cliffs NJ: Prentice Hall
1963, xv + 326 pages

A0739
Induction logging
Kaufman A A and Keller G V
Amsterdam: Elsevier
1989, xviii + 600 pages
[Methods in Geochemistry & Geophysics 27]

A0740
Log data acquisition and quality control
Theys P
Paris: Éditions Technip
1990, 352 pages

A0741
Log analysis handbook, volume 1: quantitative
log analysis methods
Crain, E R
Tulsa OK: PennWell
1986, 684 pages

A0742
Mud logging handbook
Whittaker A
Englewood Cliffs NJ: Prentice Hall
1991, 464 pages

A0743
Open-hole log analysis and formation
evaluation
Bateman R M
Englewood Cliffs NJ: Prentice Hall
1989, 319 pages

A0744
Open-hole well-logging
Richardson TX: SPE
1986, 575 pages
[SPE Reprint Series 21]

A0745
Petrophysics
Kobranova V N (Translator, Kuznetsov, V V)
Berlin: Springer Verlag
1990, 375 pages

A0746
A practical introduction to borehole geophysics
Labo J
Tulsa OK: SEG
1987, 336 pages

A0747
Principles of applied geophysics, 4th edition
Parasnis D S
London: Chapman & Hall
1986, 448 pages

A0748
Production logging
Richardson TX: SPE
1985, 375 pages
[SPE Reprint Series 19]

A0749
Production logging: theoretical and
interpretative elements
Hill A D
Richardson TX: SPE
1990, vii + 154 pages
[SPE Monograph 14]

A0750
SPWLA 30th annual logging symposium;
Denver, 11–14 June 1989
*Houston TX: Society of Professional Well-Log
Analysts*
1989, 49 papers, 2 volumes

A0751
SPWLA 31st annual logging symposium;
*Houston TX: Society of Professional Well-Log
Analysts*
1990, 52 papers, 2 volumes

A0752
SPWLA 32nd annual logging symposium;
*Houston TX: Society of Professional Well-Log
Analysts*
1991, 54 papers, 2 volumes

A0753
Stress history effects on petrophysical
properties of granular rocks
Rathore J S et al
Trondheim: IKU
1989, 60 pages

A0754
Well-logging for earth scientists
Ellis D
Amsterdam: Elsevier
1987, 450 pages

A0755
Well-logging for the non-technical person
Johnson D E and Pile K E
Tulsa OK: PennWell
1988, 200 pages

A0756
Well-logging, volume 1: rock properties,
borehole environment, mud and temperature
logging
Jorden J R amd Campbell F L
Richardson TX: SPE
1984, vii + 167 pages
[SPE Monograph 9]

A0757
Well-logging, volume 2: electric and acoustic
logging
Jorden J R and Campbell F L
Richardson TX: SPE
1986, viii + 182 pages
[SPE Monograph 10]

A0758
Vertical seismic profiling, part A
Hardage B A
Oxford: Pergamon Press
1985, 386 pages
[Handbook of Geophysical Exploration 14 Pt A]

Reservoir engineering

A0759
Application of optimal control theory to
enhanced oil recovery
Ramirez W F
Amsterdam: Elsevier
1987, vii + 244 pages
[Developments in Petroleum Science 21]

A0760
Applied petroleum reservoir engineering, 2nd
edition
Craft B C and Hawkins M (revised by Terry R E)
Englewood Cliffs NJ: Prentice Hall
1991, xv + 431 pages

A0761
Compressibility of sandstones
Zimmerman R W
Amsterdam: Elsevier
1990, 174 pages
[Developments in Petroleum Science 29]

A0762
The design engineering aspects of
waterflooding
Rose S C et al
Richardson TX: SPE
1989, viii + 358 pages

A0763
Enhanced oil recovery
Lake L W
Englewood Cliffs NJ: Prentice Hall
1989, xxv + 550 pages

A0764
Enhanced oil recovery, I: fundamentals and
analyses
Donaldson E C et al (editors)
Amsterdam: Elsevier
1985, xvi + 358 pages
[Developments in Petroleum Science 17A]

A0765
Enhanced oil recovery, II: processes and
operations
Donaldson E C et al (editors)
Amsterdam: Elsevier
1989, xiv + 228 pages
[Developments in Petroleum Science 17B]

A0766
Enhanced oil recovery, 6th SPE/DOE
symposium; Tulsa, 17–20 April 1988
Richardson TX: SPE
1988, 936 pages

A0767
EOR in the 1990s: global prospects and
perspectives of enhanced oil recovery, 7th
symposium; Tulsa, OK, 22–25 April
1990
Richardson TX: SPE/DOE
1990, 944 pages

A0768
First and second international forum on
reservoir simulation; Alpach, 12–16 September
1988 and 4–8 September 1989
P Steiner (coordinator)
Loeben: P. Steiner
1990, 632 pages

A0769
Formation damage control, 8th SPE
symposium; Bakersfield, CA, 8–9 February
1988
Richardson TX: SPE
1988, 267 pages

A0770
Formation damage control, 9th SPE
symposium; Lafayette LA, 22–23 February 1990
Richardson TX: SPE
1990, 283 pages

A0771
Fundamentals of fractured reservoir engineering
Van Golf-Racht T D
Amsterdam: Elsevier
1982, 710 pages
[Developments in Petroleum Science 12]

A0772
Fundamentals of gas reservoir engineering
Hagoort J
Amsterdam: Elsevier
1988, xii + 327 pages
[Developments in Petroleum Science 23]

A0773
Fundamentals of numerical reservoir simulation
Peaceman D W
Amsterdam: Elsevier
1977, xiii + 176 pages
[Developments in Petroleum Science 6]

A0774
Fundamentals of reservoir engineering
Dake L P
Amsterdam: Elsevier
1978 xv + 443 pages
Developments in Petroleum Science 8]

A0775
Heavy oil and gas condensates
Deighton P and Robinson G (editors)
Canterbury: SREA
1985/6, 21 pages
[Offshore Business 18]

A0776
Hydrocarbon phase behavior
Ahmed T
Houston TX: Gulf Publishing
1989, 418 pages
[Contributions in Petroleum Geology and
Engineering 7]

A0777
Hydrocarbon reservoir and well performance
Nind T E W
London: Chapman & Hall
1989, x + 368 pages

A0778
Improved oil recovery, 5th European
symposium; Budapest 25–27 April 1989
Budapest: Hungarian Hydrocarbon Institute
1989, 787 pages

A0779
Introduction to petroleum reservoir analysis
Koederitz L F et al
Houston TX: Gulf Publishing
1989, vi + 250 pages + workbook
[Contributions in Petroleum Geology and
Engineering 6]

A0780
Mathematics in oil production: Institute of
Mathematics conference; Cambridge, July 1987
Edwards S Sir and King P R (editors)
Oxford: Clarendon Press
1988, x + 375 pages

A0781
Microbial enhanced oil recovery
Donaldson E C et al (editors)
Amsterdam: Elsevier
1989, xiv + 228 pages
[Developments in Petroleum Science 22]

A0782
Microbial enhanced oil recovery
Yen T F (editor)
Boca Raton: CRC Press
1990, 192 pages

A0783
Microbial enhancement of oil recovery—recent
advances, international conference; Norman,
OK, 27 May–1 June 1990
Donaldson E C (editor)
Amsterdam: Elsevier
1991, 540 pages
[Developments in Petroleum Science 31]

A0784
Miscible displacement
Stalkup F I
Richardson TX: SPE
1983, vi + 204 pages
[SPE Monograph 8]

A0785
Miscible processes
Richardson TX; SPE
1985 2 vols
[SPE Reprint Series 18]

A0786
Multiphase flow: 4th international conference;
Nice, 1989
Fairhurst C P (editor)
Cranfield: BHRA
1989, x + 560 pages

A0787
Multiphase production: 5th international
conference; Cannes, 19–21 June 1991
Burns A P (editor)
Barking: Elsevier Applied Science
1991, 611 pages

A0788
Petroleum reservoir engineering and its
peripheral engineering
Hirakawa S
Tokyo: Uichida Rokakuho
1985, vi + 406 pages

A0789
Petroleum reservoir simulation
Aziz K and Settari A
London: Applied Science Publishers
1979, xxi + 475 pages

A0790
Polymer flooding
Littmann W
Amsterdam: Elsevier
1988, x + 212 pages
[Developments in Petroleum Science 24]

A0791
Polymer-improved oil recovery
Sorbie K S
Glasgow: Blackie
1991, xii + 359 pages

A0792
The practice of reservoir engineering
Dake L P
Amsterdam: Elsevier
in preparation

A0793
Pressure transient analysis
Stanislav J F and Kabir C S
Englewood Cliffs NJ: Prentice Hall
1990, xiv + 287 pages

A0794
Properties of oil and natural gases
Pedersen K S et al
Houston TX: Gulf Publishing
1989, 252 pages
[Contributions in Petroleum Geology and
Engineering 5]

A0795
The properties of petroleum fluids, 2nd edition
McCain W D Jr
Tulsa OK: PennWell
1990, xxxi + 548 pages

A0796
Relative permeability of petroleum reservoirs
Honarpour M et al
Boca Raton FL: CRC Press
1986, 143 pages

A0797
Reservoir compaction and surface subsidence
due to hydrocarbon extraction
Jones M E et al, for the Department of Energy
London: HMSO
1987, 175 pages
[OTH 87 276]

A0798
Reservoir engineering of fractured
reservoir–fundamental and practical aspects
Saidi A M
Paris: Total Edition Presse
1987, xix + 864 pages

A0799
Reservoir engineering and management
Robinson G (editor)
Canterbury: SREA
1990, iv + 28 pages
[Offshore Business 32]

A0800
Reservoir Stimulation, 2nd edition
Economides M J and Nolte K G
Englewood Cliffs NJ: Prentice Hall
1989, 400 pages

A0801
Surfactant/polymer chemical flooding
Richardson TX: SPE
1988, 2 vols
[SPE Reprint Series 24]

A0802
Thermal methods of oil recovery
Boberg T C
New York: Wiley
1988, xviii + 413 pages
[Exxon Monograph]

A0803
Thermal recovery of oil and bitumen
Butler R M
Englewood Cliffs NJ: Prentice Hall
1991, 608 pages

A0804
Waterflooding
Willhite G P
Richardson TX; SPE
1986 vii + 326 pages

A0805
Well test analysis for naturally fractured
reservoirs
Prat G da
Amsterdam: Elsevier
1990, 240 pages
[Developments in Petroleum Science 27]

A0806
Well testing in heterogeneous formations
Streltsova T D
New York; Wiley
1988 xviii + 413 pages [Exxon Monograph]

Field development

A0807
North Sea development guide, 3rd edition
Ledbury: OPL
1990, 1400 pages

A0808
Satellite field development
Robinson G (editor)
Canterbury: SREA
1988, iv + 46 pages [Offshore Business 24]

A0809
World field development guide, Volume 1:
Africa, Mediterranean and Middle East
Ledbury: OPL
1991, 750 pages

A0810
World field development guide, Volume 2: Asia,
Australasia and Far East
Ledbury: OPL
1991, 750 pages

Drilling engineering

A0811
Abnormal pressures while drilling
Mitchell A and Mouchet J-P
Houston: Gulf Publishing
1990, 256 pages

A0812
Advanced drilling techniques
Maurer W C
Tulsa OK: PennWell
1980, 698 pages

A0813
Applied drilling engineering
Bourgoyne, A T et al
Richardson TX: SPE
1986, 510 pages

A0814
Casing design, theory and practice
Rashman S S and Chilingar G V
Amsterdam: Elsevier
in preparation

A0815
Composition and properties of drilling and
completion fluids, 5th edition
Darley H C H and Gray G
Houston TX: Gulf Publishing
1988, 644 pages

A0816
Developments in petroleum engineering,
volume 1: Stability of tubulars, deviation control
Collected works of Arthur Lubinski: Miska S
(editor)
Houston: Gulf Publishing
1988, viii + 464 pages

A0817
Developments in petroleum engineering,
Volume 2: Offshore drilling, strength of
tubulars, drilling practices, reservoir
characterization
Collected works of Arthur Lubinski; Miska S
(editor)
Houston: Gulf
1988, viii + 408 pages

A0818
Directional drilling
Inglis T A
London: Graham & Trotman
1987, 232 pages
[Petroleum Engineering & Development
Studies 2]

A0819
Directional drilling
Richardson TX: SPE
1990, 208 pages
[SPE Reprint Series 30]

A0820
Directional drilling and deviational control
technology
French Oil & Gas Industry Association
Houston: Gulf Publishing
1991, 265 pages

A0821
Drilling and drilling fluids
Chillingarian G V and Vorabutr P
Amsterdam: Elsevier
1983, xx + 801 pages
[Developments in Petroleum Science 11]

A0822
Drilling and production safety code for onshore
operations: part 4 of the IP Model Code of Safe
Practice in the Petroleum Industry
*Chichester: John Wiley, for the Institute of
Petroleum*
1986, xi + 54

A0823
Drilling and well servicing
Robinson G (editor)
Canterbury: SREA
1990, iv + 51 pages
[Offshore Business 31]

A0824
Drilling data handbook, 6th edition
Gabolde G and Nguyen J P
Paris: Éditions Technip
1991, 552 pages

A0825
Drilling engineering–a complete well planning
approach
Adams, N J (assisted by Charrier T)
Tulsa OK: PennWell
1985, 976 pages

A0826
Drilling fluids in the oil industry, conference;
London, 3–5 September 1990
*Royal Society of Chemistry and the Society of
Chemistry Industry*
London: Royal Society of Chemistry
1990, 16 papers

A0827
Drilling fluids optimization: a practical field
approach
Lummus J L & Azar J J
Tulsa OK: PennWell
1987, 352 pages

A0828
Drilling practices manual, 2nd edition
Moore P L
Tulsa OK: PennWell
1986, 480 pages

A0829
Drilling symposium 1989: 12th annual
energy-sources technology conference;
Houston, 22–25 January 1989 (PD vol22)
Rowley J C (editor)
*New York: American Society of Mechanical
Engineers*
1988, v + 116 pages

A0830
Drilling technology symposium 1990: 13th
annual energy-sources technology conference;
New Orleans 14–18 January 1990
Weiner P D & Kastor R L (editors)
*New York: American Society of Mechanical
Engineers*
1990, vi + 184 pages
[PD Vol 27]

A0831
Drilling technology symposium 1992:
energy-sources technology conference;
Houston, 26–30 January 1992
*Fairfield NJ: American Society of Mechanical
Engineers*
1992, 158 pages [PD Vol 40]

A0832
Engineering essentials of modern drilling
*Dallas TX: Energy Publications/Harcourt Brace
Jovanovich*
1985, v + 168 pages

A0833
Fundamentals of casing design
Rabia H
London: Graham & Trotman
1987, 296 pages
[Petroleum Engineering & Development
Studies 1]

A0834
Horizontal drilling
Richardson TX: SPE
1991 231 pages
[SPE Reprint Series 33]

A0835
IADC/SPE drilling conference: Dallas TX, 28
February–2 March 1988
Richardson TX: SPE
1988, 676 pages

A0836
IADC/SPE drilling conference; New Orleans
LA, 28 February–3 March 1989
Richardson TX: SPE
1989, 874 pages

A0837
IADC/SPE drilling conference; Houston, TX, 27
February–2 March 1990
Richardson TX: SPE
1990, 764 pages

A0838
Quality and internationalism: IADC/SPE drilling
conference; Amsterdam, 11–14 March 1991
Richardson TX: SPE
1991, 892 pages

A0839
An introduction to marine drilling
MacLachlan M
Tulsa OK: PennWell
1988, 346 pages

A0840
Managing drilling operations
Fraser K et al
London: Elsevier Applied Science
1991, x + 246 pages

A0841
Offshore drilling technology, 5th annual
conference; Aberdeen, 26–27 November 1991
London: IBC Technical Services
1991, 16 papers

A0842
Oilfield mud and cementing
Robinson G (editor)
Canterbury: SREA
1990, iv + 51 pages
[Offshore Business 31]

A0843
Oilwell drilling engineering: principles and
practice
Rabia H
London: Graham & Trotman
1987, 334 pages

A0844
Oilwell fishing operations: tools and techniques
Kemp G
Houston TX: Gulf Publishing
1990, 128 pages

A0845
Recommended practice for design of control
systems for drilling well control equipment
Washington DC: API
1990, 36 pages
[API RP 16E]

A0846
Recommended practice for blowout prevention
equipment systems for drilling wells, 2nd
edition
Washington DC: API
1984   [API RP 53]

A0847
Recommended practice, standard procedure
for laboratory testing drilling fluids, 4th edition
Washington DC: American Petroleum Institute
1990, 42 pages [API RP 13I]

A0848
SCR and new technology in electric rig drilling:
a safety and efficiency handbook
McNair W L
Tulsa OK: PennWell
1991, 250 pages

A0849
Specification for drill pipe/casing protectors
(DC/CPO)
Washington DC: API
1984

A0850
Specification for drill through equipment
Washington DC: API
1986 [API Spec 16A]

A0851
Specification for materials and testing for well cements, 5th edition
Washington DC: API
1990 [API Spec 10]

A0852
Specification for rotary drilling equipment, 37th edition
Washington DC: API
1990, 100 pages
[API Spec 7]

A0853
Theory and application of drilling fluid hydraulics
Whittaker A (editor)
Boston: IHRDC
1985, xi + 203 pages
[EXLOG Series of Petroleum Geology and Engineering Handbooks]

A0854
Use of polymers in drilling and oilfield fluids
London: Plastics & Rubber Institute
1991

A0855
Well cementing
Nelson E B et al
Amsterdam: Elsevier
1990, 496 pages
[Developments in Petroleum Science 28]

A0856
Well control when drilling with oil-based mud: recent British experience in deep wells
Department of Energy, Operations & Safety Branch
London: HMSO
1987, 34 pages
[OTH 86 260]

Production engineering

A0857
Acidizing
Richardson TX: SPE
1991, 219 pages
[SPE Reprint Series 32]

A0858
Advances in operations research in the oil and gas industry, conference; Montreal, June 13–14 1991
Breton M and Zaccour G (editors)
Paris: Éditions Technip
1991, 228 pages

A0859
Artificial lift
Richardson TX: SPE
1989, 230 pages
[SPE Reprint Series 26]

A0860
Corrosion and water technology for petroleum producers
Jones L W
Tulsa OK: OGCI
1988, 202 pages

A0861
Downhole production engineering
Robinson G (editor)
Canterbury: SREA
1988, iv + 46 pages
[Offshore Business 24]

A0862
Formation damage
Richardson TX: SPE
1990, 223 pages
[SPE Reprint Series 29]

A0863
Gas production engineering
Kumar S
Houston TX: Gulf
1987, 646 pages
[Contributions in Petroleum Geology and Engineering 4]

A0864
Handbook of horizontal drilling and completion technology
World Oil Magazine
Houston TX; Gulf
1990 128 pages

A0865
Horizontal drilling, completion and production
Aquilera R
Houston TX; Gulf
1991 352 pages
[Contributions in Petroleum Geology and Engineering 9]

A0866
Horizontal well technology
Joshi S D
Tulsa OK: PennWell
1991, xiv + 535 pages

A0867
Hydraulic fracturing
Richardson TX: SPE
1990, 2 volumes
[SPE Reprint Series 28]

A0868
Hydraulic proppant fracturing and gravel packing
Mader D
Amsterdam; Elseiver
1989 xxxvi + 1240 pages
[Developments in Petroleum Science 26]

A0869
Meeting economic challenges thru technology, symposium; Oklahoma City, 12–14 March 1989
Richardson TX: SPE
1989, 664 pages

A0870
Offshore multiphase production, European conference 1989
Bedford: BHRA
1989

A0871
Oil field chemicals, symposium; Geilo, 19–21 March 1990
Oslo: Norwegian Society of Chartered Engineers
1990, 2 volumes

A0872
Oil field chemistry: enhanced recovery and production stimulation
Borchardt (editor)
Washington DC: American Chemical Society
1989

A0873
Perforating
Richardson TX: SPE
1991, 211 pages [SPE Reprint Series 31]

A0874
Producing solutions for the '90s, symposium; Oklahoma City, 7–9 April 1991
Richardson TX: SPE
1991, 976 pages

A0875
Production facilities
Richardson TX: SPE
1989, 239 pages
[SPE Reprint Series 25]

A0876
Production operations, volume 1, 3rd edition
Allen T O and Roberts A P
Tulsa OK: OGCI
1989, 310 pages

A0877
Production operations, volume 2, 3rd edition
Allen T O and Roberts A P
Tulsa OK: OGCI
1989, 350 pages

A0878
Production optimization using NODAL* analysis
Beggs D H
Tulsa OK: OGCI
1991, 398 pages

A0879
Quality assurance in the offshore oil and gas industry
Rogerson J H (editor)
London: Graham & Trotman
1988, 178 pages [Petroleum Development & Engineering Studies 3]

A0880
Recent advances in hydraulic fracturing
Gidley J L et al
Richardson TX: SPE
1990 464 pages
[SPE Monograph Series 12]

A0881
Recommended practice for offshore well completion, servicing, workover and plug and abandonment operations
Washington DC: API
1986, 23 pages
[API RP 57]

A0882
Recommended practice for training and qualification of personnel in well control ... for completion and workover ... offshore, 1st edition
Washington DC: American Petroleum Institute
1986, 9 pages [RP T-6]

A0883
Surface production operations, volume 1: design of oil-handling systems and facilities
Arnold K and Stewart M
Houston: Gulf Publishing
1986, 432 pages

A0884
Surface production operations, volume 2:
design of gas-handling systems and facilities
Arnold K and Stewart M
Houston: Gulf Publishing
1988, 525 pages

A0885
The technology of offshore drilling, completion
and production
ETA Offshore Seminars Inc
Tulsa OK: PennWell
1976, 426 pages

A0886
Thermal methods of petroleum production
Baibakov N K and Garushev A R
Amsterdam: Elsevier
1989, vi + 210 pages
[Developments in Petroleum Science 25]

A0887
Well test performance, 2nd edition
Golan M and Whitson C H
Englewood Cliffs NJ: Prentice Hall
1991, ix + 669 pages

Petroleum technology

A0888
1990 gas facts
German M I (editor)
Arlington VA: American Gas Association
1990, 217 pages

A0889
Developments in petroleum engineering, 1
Dawe R A and Wilson D C
Barking, Essex: Elsevier Applied Science
Publishers
1985, x + 297 pages

A0890
European industry outlook
Robinson G (editor)
Canterbury: SREA
1991, iv + 44 pages
[Offshore Business 37]

A0891
The European oil and gas conference: a
multidisciplinary approach to exploration and
production research and development
Imarisio G et al (editors)
London: Graham & Trotman
1991, 600 pages

A0892
A first course in petroleum technology
Donohue D A T and Lang K R
Englewood Cliffs: Prentice Hall
1986, 210 pages

A0893
A guide to professional registration of
petroleum engineers
Richardson TX: SPE
1991, 104 pages

A0894
Hydrocarbons: source of energy
Imarisio G et al (editors) for Commission of the
European Communities
London: Graham & Trotman
1989, 544 pages

A0895
In search of oil
Grossling B F and Nielsen D T
London: FT Business Information
1985, 2 volumes

A0896
Introduction to oil and gas technology
Giuliano F A (editor)
Englewood Cliffs NJ; Prentice Hall
1989 208 pages

A0897
Modern petroleum technology
Hobson G D
Chichester: Wiley (for the Institute of Petroleum)
1984, 1282 pages

A0898
Oil shale technology
Lee S and Iredell R
London: CRC Press, Wolfe Publishing
1991, 240 pages

A0899
Petroleum engineering : principles and practice
Archer J S and Wall C G
London; Graham & Trotman
1986, xi + 362 pages

A0900
Petroleum engineering handbook
Bradley H B (editor)
Richardson TX: SPE
1987, 1856 pages

A0901
Petroleum engineering handbook for the
practising engineer, volume 1
Mian M A
Tulsa OK: PennWell
1991, 450 pages

A0902
The petroleum handbook
Compiled by staff of the Royal Dutch/Shell
Group of Companies
Amsterdam: Elsevier
1983, xvii + 710 pages

A0903
SPE guide to petroleum engineering schools:
petroleum engineering and technology schools
1990–91
SPE Education & Accreditation Committee
Richardson TX: SPE
1991, vi + 134 pages

A0904
Standards, specifications, recommended
practices (comprehensive documentation)
Washington DC: American Petroleum Institute
Regularly updated
[available in Europe through Infonorme]

OFFSHORE OPERATIONS

A0905
Diesel engines in hazardous areas on offshore
installations
Sokolov A
London: Institute of Marine Engineers
1989

A0906
Dispersion of gases in offshore platforms
Lindburg B et al
London: Institute of Marine Engineers
1989, 11 pages [presented 21 February 1989]

A0907
Hook-up and commissioning conference;
Aberdeen 26–27 March 1985
Kingston-upon-Thames: Offshore Conferences
& Exhibitions
1985, 2 volumes [8 papers, 6 papers]

A0908
Managing oil and gas operations on the outer
continental shelf
Washington, DC: US Dept of the Interior,
Minerals Management Service]
1986,vi + 60 pages

A0909
Offshore and arctic operations symposium and
workshop; Dallas, 15–20
February 1987
Konuk I, editor
Fairfield, NJ: ASME Petroleum Division
1987, 352 pages

A0910
Offshore facilities and design: opportunity for
the future, conference;
London 3–4 December 1987
Rugby: Institution of Chemical Engineers
1987, 4 papers

A0911
Offshore installation managers' handbook
Lowestoft: Petroleum Training Association
North Sea
1984,

A0912
Offshore installation practice
Crawford J
Guildford: Butterworths
1988, x + 389 pages

A0913
Offshore installations (construction and survey) regulations 1974
Department of Energy
London: HMSO
1974, 21 pages [SI 1974 No 289]

A0914
Offshore installations (managers) regulations 1972
Department of Trade & Industry
London: HMSO
1972, 5 pages [SI 1972 No 703]

A0915
Offshore installations (registration) regulations 1972
Department of Trade & Industry
London: HMSO
1972, 5 pages [SI 1972 No 702]

A0916
Offshore logistics conference; Aberdeen, 28 March 1985
Kingston-upon-Thames: Offshore Conferences & Exhibitions
1985, 7 papers

A0917
Offshore operations 1986, 9th Annual Energy Technology Conference;
New Orleans, 23–27 February 1986
Kozik T J, editor
New York: American Society of Mechanical Engineers
1986, 243 pages

A0918
Offshore operations post *Piper Alpha*: WEMT '91; London, 6–8 February 1991
London: Marine Management (Holdings) Ltd
1991

A0919
Offshore safety (protection against victimisation) act 1992
UK Parliament
London: HMSO
1992, 4 pages
[Public General Acts–Elizabeth II, 1992, Ch 24]

A0920
Offshore safety act 1992
UK Parliament
London: HMSO
1992, 2, 10 pages [Public General Acts–Elizabth II, 1992, Ch 15]

A0921
A primer of offshore operations, 2nd ed
University of Texas at Austin, Petroleum Extension Service
1985, 114 pages

A0922
Pumps for the marine and offshore oil industries
Shiel A G and Manson D M
London: Institute of Marine Engineers
1988, 13 pages

A0923
Recommended practice Volume C: facilities on offshore installations, area classification and ventilation
Høvik: Det norske Veritas
1987 [RP C102]

A0924
Recommended practice for orientation program for personnel going offshore for the first time, 3rd edition
Washington DC: American Petroleum Institute
1990 [RP T-1]

A0925
Recommended practice for qualification programs for offshore production personnel who work with anti-pollution safety devices, revised edition
Washington DCL: American Petroleum Institute
1975 [RP T-2]

A0926
Recommended practice for training and qualification of personnel in well control ... for completion and workover ... offshore, 1st edition
Washington DC: American Petroleum Institute
1986, 9 pages [RP T-6]

A0927
Recommended standards of training, qualification and certification of key personnel on mobile offshore units (MOUS)
London: International Maritime Organization
1991 [Resolution A.712(17)]

A0928
Risk reduction in offshore operations, seminar;
London, 11 October 1990
London: Noble Denton International
1990

A0929
Safe operations of offshore production
installations, conference
Trondheim: SINTEF
1989, 111 PAGES [STF 75 - A89027]

A0930
Safety in offshore drilling: the role of shallow
surveys, conference; London, 23–24 January
1990
Ardus D A and Green C D (editors)
London: Graham & Trotman
1990, 312 pages [Advances in Underwater
Technology, 25]

Transportation & cranes

A0931
Aircraft and helicopters handbook
Harbison I
Chelmsford: Offshore Patrol
1986, 51 pages

A0932
Code for lifting appliances in a marine
environment
London: Lloyd's Register of Shipping
1987, looseleaf

A0933
Gas Transportation conference session T,
Offshore Northern Seas; Stavanger 26–29
August 1986
1986, 6 papers

A0934
Guide to helicopter/ship operations, 3rd edition
International Chamber of Shipping
London: Witherby, for ICS
1989

A0935
Helicopter safety study: main report
Ingstad O et al
Trondheim: SINTEF
1990 [STF75 A90008]

A0936
Helitech '86: international helicopter technology
and operations conference and exhibtion;
Aberdeen 8–10 April 1986
*Kingston-upon-Thames: Offshore Conferences
& Exhibitions*
1986, 16 papers

A0937
The infra-structure for transportation of
petroleum offshore: Offshore Northern Seas
conference, Stavanger 24–27 August 1982
Oslo: Norwegian Petroleum Society
1982, 1 volume

A0938
Inhospitable waters ... transport options open to
the shippers of dangerous goods to North Sea
platforms
Castle M
Hazardous Cargo Bulletin
1990, Volume 11(3), pages 63–65

A0939
An introduction to helicopter operations at sea
House D J
Ledbury: OPL Ltd
1992, 350 pages

A0940
Lessons in rotary drilling, unit V lesson 8:
orientation for offshore crane operations
*University of Texas at Austin, Petroleum
Extension Service*
1978, 35 pages

A0941
LNG–Teknikker øg muligheter; Tromsø, 18–19
June 1984
Oslo: Norwegian Petroleum Society
1984, 13 papers [9 in Norwegian, 4 in English]

A0942
Offshore aviation in the North Sea, London
March 1982
*London: Oyez Scientific & Technical Service
(IBC)*
1982, 154 pages

A0943
Offshore helicopter landing areas: guidance on
standards
London: Civil Aviation Authority
c 1981, 48 pages

A0944
Offshore logistics, bases, boats and aviation
Canterbury: Smith Rea Energy Analysts Ltd
1991 [Offshore Business No 38]

A0945
Offshore transport and installation: international
symposium; London 26–27 March 1985
London: Royal Institution of Naval Architects
1985, 2 volumes (15 papers)

A0946
Recommended practice for operation and
maintenance of offshore cranes, 2nd ed
Washington DC: American Petroleum Institute
1984 [API RP 2D]

A0947
Regulations for cranes on production
installations in Norwegian territorial waters
(with unofficial translation)
Stavanger: Norwegian Petroleum Directorate
1977, 20 pages

A0948
Risk analysis of crane accidents, Gulf of
Mexico OCS Region
*US Department of the Interior, Minerals
Management Service
NTIS*
1983, 29 pages [OCS Report MMS 84-0056,
NTIS PB85 - 142743]

A0949
Rules for certification and lifting appliances
Høvik: Det norske Veritas
1989

A0950
Specification for offshore cranes, 4th ed
Washington DC: American Petroleum Institute
1988, 44 + 2 pages

Ships

A0951
Collision regulations, 2nd edition
*Sturt R H B
Colchester: Lloyd's of London Press*
1984

A0952
Cumulative list of Admiralty Notices to
Mariners, July 1990
Taunton: Hydrographer of the Navy
1990, 173 pages

A0953
The damage inflicted by ships with bulbous
bows on underwater structures: report of
Working Group 8
*Permanent International Association of
Navigation Congresses
Supplement to PIANC Bulletin No 70*
1990, 17 pages

A0954
Design criteria for ships operating as offshore
installations
*Faulkner D
University of Glasgow Dept of Naval
Architecture & Ocean Engineering*
1989, 41 pages [NAOE-88-34]

A0955
Dynamic positioning of offshore vessels
*Morgan M J
Tulsa: PennWell*
1978, x + 513 pages

A0956
Dynamic positioning systems: principles,
design and applications
*Faÿ H
Paris: Éditions Technip*
1989, 208 pages [text in English]

A0957
Electronic aids to navigation: position fixing,
2nd edition
*Tetley L and Calcutt D
Sevenoaks: Edward Arnold*
1991, 400 pages

A0958
Gas carriers
*Ffooks R C
London: Fairplay*
1984, 108 pages

A0959
Guidance notes for the classification of
dynamic positioning systems
London: Lloyd's Register of Shipping
1980, 17 pages

A0960
A guide to the collision avoidance rules, 4th
edition
*Cockroft A N and Lameijer J N F
Oxford: Heinemann Professional*
1990, 240 pages

A0961
Guidelines for the design and construction of offshore supply vessels
London: International Maritime Organization
1982 [IMO-807E]

A0962
Guidelines for the specification and operation of dynamically positioned diving support vessels
UK Department of Energy (Petroleum Engineering Division) and the Norwegian Petroleum Directorate
London: Department of Energy
1983, 34 pages [Paper OT-R-82117]

A0963
The interface between supply and service vessels and offshore installations: safety study for the Department of Energy: final report, December 1982 [unclassified version of Report OT-R-8174 dated February 1982]
Camberley: Easams [OT-R-8293]

A0964
The management of safety in shipping
London: The Nautical Institute
1991, 38 chapters

A0965
The mariner's handbook, 5th edition
Taunton: Hydrographer of the Navy
1979, xii + 174 pages + supplements to date

A0966
The Nautical Almanac 1990
Her Majesty's Nautical Almanac Office
London: HMSO
1989

A0967
The Noble Denton towing vessel register 1991 edition (2nd edition)
Ledbury: OPL Ltd
1991, 350 pages

A0968
Operating experience with offshore workboats and support vessels; Newcastle upon Tyne, 30 March 1978
London: Society for Underwater Technology
1979, 23 pages

A0969
Preliminary rules for building and classing accommodation barges and hotel barges
Paramus, NJ: American Bureau of Shipping
1989, various pagings

A0970
Recommended practice for transportation of line pipe on barges and marine vessels, 1st edition
Washington, DC: American Petroleum Institute
1990 [RP 5LW]

A0971
Support craft: technical session H, Oceanology International '80; Brighton 2–7 March 1980
Brighton: BPS Exhibitions
1980, 5 papers

A0972
Support vessels for the oil activity on the Norwegian Continental Shelf seminar; Bergen, 14 April 1983
Norwegian Society of Chartered Engineers

A0973
Surface support for subsea operations
Dept of Ship & Marine Technology, University of Strathclyde
London: Department of Trade & Industry
1987, 2 volumes [Resources from the Sea Programme]

A0974
A survey of operational vessel traffic service in offshore locations
Technica, for Department of Energy
London: HMSO
1988, iv + 65 pages [Report OTI 87 503]

A0975
Tanker safety guide (liquefied gas), 2nd edition
International Chamber of Shipping
London: Witherby
1990

A0976
Tentative rules for the construction and classification of dynamic positioning systems for ships and mobile offshore units
Høvik: Det norske Veritas
1977, 40 pages

A0977
UKOOA/BOSVA code of conduct: supply vessel operations at offshore installations
London: United Kingdom Offshore Operators Association
1990

HEALTH & SAFETY

A0978
Area classification code for petroleum installations
Institute of Petroleum
Chichester: John Wiley
1990, 136 pages
[Model Code of Safe Practice in the petroleum Industry, Part 15]

A0979
Collective agreement for crews on maritime mobile offshore units
London: International Transport Workers' Federation (ITF)
no date, 17 pages

A0980
Formal safety assessments for the offshore industry, conference; London, 5–6 December 1990
London: IBC Technical Services
1990

A0981
Guide to the principles and operation of permit-to-work procedures as applied in the UK petroleum industry
Oil Industry Advisory Committee, HSE
Sheffield: Health & Safety Executive
1986

A0982
Guide to work on the North Sea oil rigs
Whyte P J
Mark Saunders
1984

A0983
Guidelines for occupational health services in the oil industry
Oil Industry Advisory Committee, HSE
Sheffield: Health & Safety Executive
1987

A0984
Guidelines for operators on naturally occurring radioactive substances on offshore installations
London: United Kingdom Offshore Operators Association
1987, 16 pages

A0985
Health, safety and environment in oil and gas exploration and production, 1st conference: The Hague, 10–14 November 1991
Richardson TX: Society of Petroleum Engineers
1991, 2 volumes [Publication 031439]

A0986
Human factors in offshore safety, specialist course; Aberdeen, 23–24 April 1991
London: IBC Technical Services
1991

A0987
ICOSH 1985: international conference on oceans–safety and health; Sydney, Nova Scotia, 16–20 June 1985
Sydney: University College of Cape Breton
1985, 30 papers

A0988
International offshore safety conference; London, 6–7 October 1987
London: IBC Technical Services
1987

A0989
The introduction of formal safety assessments for the offshore industry, seminar: London, 5–6 December 1990
London: IBC Technical Services
1990, 14 papers

A0990
IRM/ROV 90, the practical approach to offshore safety, conference: Aberdeen, 6–9 November 1990
Kingston upon Thames: Spearhead
1990, 4 volumes

A0991
Managing safety
Concawe Major Hazards Management Group
The Hague: Concawe
1989, 21 pages [Report 4/89]

A0992
Methane in Marine Sediments, conference; Edinburgh, 19–21 September 1990
Shallow Gas Group
Edinburgh: Heriot-Watt University
Abstracts of papers

A0993
NOGEPA Handboek inzake veiligheidstraining = NOGEPA policy on safety training
Den Haag: Nogepa
no date, loose leaf

A0994
Noise and vibration control offshore: volume 1,
guidance for project management
London: UEG
1984, 31 pages [Report UR25]

A0995
North Sea safety: a crucial investment
McKee R E
London: Conoco
1990, 12 pages

A0996
North Sea safety, 2nd conference; London, 12
September 1990
Wembley: Technology Forum
1990, 13 papers

A0997
North Sea safety, 3rd conference: progress
since Cullen; London, 30 October 1991
Wembley: Technology Forum
1991, 8 papers

A0998
The notification of accidents and dangerous
occurrences regulations 1980
Department of Employment
London: HMSO
1980, 14 pages [SI 1980 No 804]

A0999
Occupational health and safety in the North
Sea: a comparison between the British and the
Norwegian legislation
*Barrett B N and Howells R W L, for the Health
and Safety Executive*
London: Middlesex Polytechnic
1981, 2 volumes

A1000
Occupational safety in the North Sea: a review
of fatalities, injuries and reporting procedures in
offshore exploration and production, 1975–1981
London: E & P Forum
1983, 30 pages [Technical Review 15]

A1001
Offshore construction: health, saftey and
welfare
Health and Safety Executive
London: HMSO
1980 [HS(G)1 2]

A1002
The offshore health handbook: a practical
guide to coping with injury and illness, rev ed
Norman J Nelson and Brebner J
London: Martin Dunitz
1987, 204 pages

A1003
Offshore installation and pipeline work
Health & Safety Executive
London: HMSO
1990 [ACOP 32]

A1004
Offshore installations: formal safety
assessments: discussion document by the
Department of Energy
Petroleum Engineering Division
London: Department of Energy
1989, lv (various pagings0 [PEA 89/93/4]

A1005
Offshore installations: protection against fire
and explosion: discussion document by the
Department of Energy
Petroleum Engineering Division
London: Department of Energy
1989 [PEA 584/411/7]

A1006
Offshore medicine: medical care of employees
in the offshore oil industry, 2nd edition
Cox R A F (editor)
Berlin: Springer-Verlag
1987, 257 pages

A1007
Offshore para-medicine conference; Edinburgh,
21–22 April 1982
London: Offshore Conferences & Exhibitions
1982, 16 papers

A1008
Offshore para-medicine conference; 9–10
March 1983
London: Offshore Conferences & Exhibitions
1983, 19 papers

A1009
Offshore safety: a report of the Committee
Burgoyne J H (Chairman)
London: HMSO
1980, vii + 300 pages [Cmmd 7841]

A1010
Offshore safety: 2nd international conference;
Miami, 19–21 March 1986
Houston: International Association of Drilling
Contractors
Miami: Rosenstiel School of Marine &
Atmospheric Science
1986, 109 pages

A1011
Offshore safety and reliability, meeting; 26
March 1990
London: Institute of Marine Engineers
1990, 43 pages

A1012
Offshore safety and reliability, symposium;
Sutton Colfdield, 18–19 September 1991
Cox R F and Walter M H (editors, for Safety &
Reliability Society)
London: Elsevier Applied Science
1991, viii + 264 pages

A1013
Offshore safety and training programme
London: International Transport Workers'
Federation
1989, 18 pages

A1014
Offshore safety cases: preparation and
implementation; London, 18–19 Novembr 1991
London IBC Technical Services
1991 [Publication E2513]

A1015
Offshore safety management: report together
with proceedings of the Committee and
minutes of evidence with appendices: 7th report
House of Commons Energy Committee
(Chairman: M Clark)
London: HMSO
1991, xx + 135 pages [House of Commmons
Papers, Session 1990–91, paper 343]

A1016
Offshore safety management: Government
observations on the 7th report from the
Committee (Session 1990–1991)
House of Commons Energy Committee
London: HMSO
1991, 13 pages [House of Commons Papers,
Session 1990–91, paper 672]

A1017
The offshore safety regime
London: Department of Energy
1989, 68 pages

A1018
Offshore safety task-force report to the
EPOA/APOA Safety Committee
Calgary: Pallister Resource Management
1983, xi + 155 pages

A1019
Offshore safety: the economic implications
Canterbury: Smith Rea Energy Analysts
1989 [Offshore Business No 28]

A1020
Offshore safety: the way ahead, conference;
London 11 November 1990
London: Institute of Petroleum
1990, 138 pages

A1021
OMAE 1990: 9th offshore mechanics and arctic
engineering conference, Volume II, safety and
reliability; Houston, 18–23 February
1990
New York: American Society of Mechanical
Engineers
1990, viii + 314 pages

A1022
OMAE 1991: 10th offshore mechanics and
arctic engineering conference, Volume II,
safety and reliability; Stavanger, 23–28 June
1991
New York: American Society of Mechanical
Engineers
1991, 363 pages

A1023
Permits to work and you: an introduction for
workers in the petroleum industry
Oil Industry Advisory Committee, HSE
Sheffield, Health & Safety Executive
1986 [IND(G)39(L)]

A1024
Pipa Alpha: lessons for life-cycle safety
management
Rugby: Institution of Chemical Engineers
1990 [Symposium Series 122]

A1025
The preparation & operation of contracts in the
petroleum industry: health & safety guidelines
Oil Industry Advisory Committee, HSE
London: Health & Safety Executive
1987, 6 pages

A1026
Procedure: formal safety assessment
London: UK Offshore Operators Association
1990, 34 pages

A1027
Quality assurance and certification of safety
and pollution prevention equipment used in
offshore oil and gas operations
*New York, NY: American National Standards
Institute*
1988 [ANSI/ASME SPPE-1-1988]

A1028
Recommended practice for analysis, design,
installation and testing of basic surface safety
systems on offshore production
platforms, 4th edition
Dallas: American Petroleum Institute
1986 [API RP 14C]

A1029
Recommended practice for occupational safety
for oil and gas well drilling and servicing
operations, 2nd edition
Washington DC: American Petroleum Institute
1991 [RP 54]

A1030
Safe operation of offshore production
installations: research program final report
Bodsberg L et al
Trondheim: SINTEF
1989, 42 pages [STF75 A89029]

A1031
Safety and contingency preparedness for
petroleum activities in ice-infested arctic
waters, conference; Stavanger 19 April 1989
*Oslo: Royal Norwegian Council for Scientific &
Industrial Research*
1989, 9 papers

A1032
Safety and health in the construction of fixed
offshore installations in the petroleum industry
Geneva: International Labour Office
1981, xi + 135 pages [ILO Code of Practice]

A1033
Safety and health in the oil and gas extractive
industries, international symposium;
Luxembourg, 19–20 April 1983
*London: Graham & Trotman (for the
Commission of European Communities)*
1983, 434 pages

A1034
Safety and offshore oil: background papers of
the Committee on Assessment of Safety and
Outer Continental Shelf Activities
Washington DC: National Academy Press
1981, 212 pages

A1035
Safety and offshore oil
*Committee of Assessment of Safety of OCS
Activities, Marine Board, Assembly of
Engineering, National Research Council
Washington DC: National Academy Press*
1981, xix + 313 pages

A1036
Safety developments in the offshore industry,
conference; Glasgow 23–24 April 1991
London: Institution of Mechanical Engineers
1991, 160 pages

A1037
Safety first–the offshore approach
Hughes H W D
London: MTD Ltd
1992, 20 pages [Mike Adye Memorial Lectures,
3][MTD Ltd Publication 92/101]

A1038
Safety in exploration and production operations
Charlton R M (Shell)
London: Shell International Petroleum Co
1989, 12 pages

A1039
Safety in offshore drilling: the role of shallow
gas surveys, seminar; London 4–5 November
1987
London: Society for Underwater Technology
1987

A1040
Safety in offshore operations, training course;
London, 10–11 October 1991
London: BPP Technical Services
1991, 8 papers

A1041
Safety in the offshore petroleum industry: the
law and practice for management
Barrett B et al
London: Kogan Page
1987, 252 pages

A1042
Safety management for offshore exploration seminar; St John's, Newfoundland, 1982
St John's: Memorial University of Newfoundland
1982, 1 volume

A1043
Safety management for offshore operations on the Canadian East Coast, conference; Calgary, 1983
St John's: Memorial University of Newfoundland, Continuing Education Centre
1983, 13 papers

A1044
Safety offshore
Karstad O and Wulff E
Rogaland, Norway: Universitetsforlaget
1984, 141 pages

A1045
Safety offshore: outline of reports, final edition
Oslo: NTNF (Royal Norwegian Council for Scientific & Industrial Research)
1983, 153 pages

A1046
Safety representatives and safety committees on offshore installations: a brief guide for the workforce
London: Department of Energy
1989, folded leaflet

A1047
Safety representatives and safety committees on offshore installations: guidance notes
Department of Energy
London: HMSO
1989, 41 pages

A1048
Ship safety handbook 2
Bureau Veritas and Lloyds of London Press
Colchester: Lloyds of London Press
1989, 310 pages

A1049
Technical and human implications of automatic safety systems
Bodsberg L and Ingstad O
Trondheim: SINTEF
1989, 13 pages [STF75 A80919]

A1050
The UK Offshore Safety Conference; Eastbourne, 10–11 February 1982
Kingston-upon-Thames: Spearhead Exhibitions
1982, preprints

A1051
The UK offshore safety conference 1984; Eastbourne, 2–4 October 1984
Andover: Vortex
1984, 23 papers

A1052
Working conditions and working environment in the petroleum industry, including offshore activities
Geneva: International Labour Office
1980, 116 pages [Petroleum Committee, 9th Session]

Accidents and Risk

A1053
Abnormal events in offshore petroleum production: an analysis of daily activity reports submitted to the Norwegian Petroleum Directorate
Bodsberg L et al
Trondheim: SINTEF
1986, 46 pages [STF75 A86021]

A1054
Accidents associated with oil and gas operations, outer continental shelf 1956–86
Tracey L (compiler)
Vienna VA: US Dept of the Interior Minerals Management Service
1988, 264 pages [OCS Report MMS 88-0011]

A1055
Accidents connected with Federal oil and gas operations on the Gulf of Mexico outer continental shelf; Volume I, 1956–1979, revised
New Orleans: US Dept of the Interior Minerals Management Service
1986 [Report OCS MMS 86-0038]

A1056
Accidents connected with Federal oil & gas operation on the OCS Gulf of Mexico Region; Volume 2, January 1980–December 1986
New Orleans: US Minerals Management Service
1987, 115 pages [MMS 87-0049]

A1057
An analysis of accidents in offshore operations where hydrocarbons were lost
Gulf R & D Company for Gulf Canada
Calgary: Pallister Resource Management
1983 [Beaufort Environmental Input Statement Support Document 19]

A1058
The assessment of impact damage caused by dropped objects on concrete offshore structures
Wimpey Laboratories et al
London: HMSO
1989 [Report OTH 87-240]

A1059
Boat impact study
Lloyd's Register of Shipping, prepared for the Department of Energy
London: HMSO
1986, ix + 77 pages [Report OTH 85 224]

A1060
Capsizing and sinking of the mobile offshore unit *Rowan Gorilla I* in the North Atlantic Ocean, December 15 1988
Washington DC: National Transportation Safety Board
1989

A1061
Collisions of attendant vessels with offshore installations
National Maritime Institute, for the Department of Energy
London: HMSO
1985, 2 volumes (40 pages, 158 pages) [OTH 84 208 and OTH 84 209]

A1062
Comparison of Norwegian and UK offshore accident statistics
Institute of Offshore Engineering
London: UK Offshore Operators Association
1989

A1063
The consequences of offshore accidents
Bentham R W
Petroleum Accounting & Financial Management Journal
1990, Volume 9(1), pages 63–73

A1064
Disasters at sea: from the *Titanic* to the *Exxon Valdez*
Cahill R A
London: Century Publishing
1990, 272 pages

A1065
Dropped objects on subsea installations
Sundet I
Trondheim: SINTEF
1989, 87 pages [STF75 A89039]

A1066
Explosions in gas compressor electric motor drives
Bartels A H
London: HMSO
1990 [Report OTH 90-332]

A1067
Exposure to catastrophe in the North Sea: a re-assessment of risks in the immediate aftermath of the *Piper Alpha* disaster
Smillie J B and Ross K C L
London: Rush Johnson Associates
1989, 35 pages

A1068
Guidelines for detection and control of hydrogen sulphide during offshore drilling operations
London: UK Offshore Operators' Association
1982, 16 pages

A1069
Insurance & liability issues in offshore construction & supply contracts, seminar; London 13–14 April 1987
London: IBC Legal Studies & Services
1987

A1070
Investigation of Shell Pipeline Corporation pipeline leak, South Pass block 65, December 30 1986, Gulf of Mexico off the Louisiana coast
Shoennagel C et al
New Orleans: Minerals Management Service
1988 [Report MMS 87-0114]

A1071
Man and accidents offshore: an examination of the costs of stress among workers on oil and gas rigs
Sutherland V J and Cooper C L
Colchester: Lloyd's of London Press
1986

A1072
Marine accident report: capsizing and sinking of the US mobile offshore drilling unit, *Ocean Ranger*, February 15, 1982
US National Transportation Safety Board, Bureau of Accident Investigation
1983, iii + 100 pages

A1073
Maritime accidents: what went wrong
Gates E T
Houston: Gulf Publishing
1989, 144 pages

A1074
Norwegian Public Reports: *Alexander L Keilland* accident; from a Commission appointed by Royal decree, 28 March 1980, presented to
the Ministry of Justice, March 1981
Stavanger: Translator Service
1981, 472 pages

A1075
Offshore accident statistics: an analysis and review 1975–1988
Institute of Offshore Engineering
London: UK Offshore Operators Association
1990

A1076
Offshore hazards and their prevention, conference; London, 30 April–1 May 1990
London: IBC Technical Services
1990, 15 papers

A1077
Offshore mobile rig accidents 1955 to present
Houston: Offshore Data Services
routinely updated

A1078
Offshore reliability data handbook
OREDA participants
Tulsa: PennWell (Veritec)
1984

A1079
Offshore safety cases: preparation and implementation; London, 18–19 November 1991
London: IBC Technical Serbvices
1991, 16 papers [Publication E2513]

A1080
Piper Alpha
Lovegrove M
London: James Capel et al
1988, 8 + 10 pages

A1081
Piper Alpha: a survivor's story
Punchard E with Higgins S
London: W H Allen
1989, 212 pages

A1082
Piper Alpha technical investigation interim report
Petrie J R
London: Department of Energy, Petroleum Engineering Division
1988, 2 volumes (various pagings)

A1083
Piper Alplha technical investigation: further report
Petrie J R
London: Department of Energy, Petroleum Enginering Division
1988, 12 pages

A1084
Piper Alpha technical investigation: further report: annexes
London: Department of Energy
1988, 10 volumes

A1085
The public inquiry into the *Piper Alpha* disaster, presented to Parliament by the Secretary of State for Energy
The Hon Lord Cullen
London: HMSO
1990, 2 volumes [CM 1310]

A1086
Regulations concerning implementation and use of risk analyses in the petroleum activities
Stavanger: Norwegian Petroleum Directorate
1990, 15 pages [in Norwegian and English]

A1087
Report on the accident to Sikorsky S-61n G-BEID 29 nm north east of Sumburgh, Shetland Isles, on 13 July 1988
Air Accidents Investigation Branch
London: HMSO
1990, x + 57 pages [Aircraft Accident Report 3/90]

A1088
Report on the accident to Sikorsky S-76A helicopter G-BHYB near Fulmar platform in the North Sea on 9 December 1987
Department of Transport
London: HMSO
1988 [Aircraft Accident Report 5/88]

A1089
Report One: The loss of the semisubmersible drill rig *Ocean Ranger* and its crew
Royal Commission on the Ocean Ranger Marine Disaster
Ottawa: Canadian Government Publishing Centre
1984, x + 400 pages

A1090
Report Two: Safety Offshore Eastern Canada
Royal Commission on the Ocean Ranger Marine Disaster
Ottawa: Canadian Government Publishing Centre
1985, ix + 308 pages

A1091
Review of the applicability of predictive methods to gas explosions in offshore modules
British Gas plc, Research & Development Division
London: HMSO
1990, 175 pages

A1092
Risk analysis for offshore structures and equipment l'Association Scientifique et Technique pour l'Exploitation des Oceans (ASTEO),
prepared for the Commission of the European Communities
London: Graham & Trotman
1987, 204 pages

A1093
Risk analysis in offshore development projects
Andersen R S et al
Trondheim: SINTEF
1984, 107 pages [STF 18 A83503]

A1094
Risk analysis in the offshore industry II–after *Piper Alpha*, workshop; Aberdeen 25–27 March 1991
London: IBC Technical Services
1991

A1095
Risk assessment of emergency evacuation from offshore installations
Technica, for the UK Department of Energy
London: HMSO
1983, x + 264 pages

A1096
The risk assesssment of environmental and human health
Paustenbach D J (editor)
Chichester: John Wiley
1989, 1176 pages

A1097
The risk of ship/platform collisions in the area of the United Kingdom Continental Shelf
Technica, for the Department of Energy
London: HMSO
1986, 80 pages [OTH 86 217]

A1098
Safety and risk: models and reality, inaugural lecture; Edinburgh, 30 October 1991
Wolfram J
Edinburgh: Heriot-Watt University
1991, 19 pages

A1099
Ship impact on steel tubulars
Allan J D and Marshall J
London: HMSO
1988, lv (various pagings)

A1100
Shipping routes in the area of the United Kingdom Continental Shelf
Technica, for the Department of Energy
London: HMSO
1985, 71 pages [OTH 85 213]

A1101
Smoke hazard in offshore petroleum fires
Wighus R et al (for Norwegian Fire Research Laboratory)
Trodheim: SINTEF
1991, 43 pages [Report STF25 A91007]

A1102
Subsea safety: risk assessment, hazard analysis and systems protection, seminar; London 1 March 1990
Swindon: Knighton Enterprises
1990, 8 papers

A1103
Summary of investigations No 2/90
Marine Accidents Investigation Branch
Southampton: Department of Transport
1990

A1104
Summary of occupational injuries, illnesses and fatalities in the petroleum industry–1990
Washington DC: American Petroleum Institute
1991, 30 pages

A1105
Technical investigation of *F Ocean Ranger* accident
Russell W E et al
Government of Newfoundland & Labrador, Petroleum Directorate
1983, 2 volumes

A1106
Working offshore: the other price of Newfoundland's oil
House J D
St John's: Memorial University of Newfoundland, Institute of Social & Economics Research
1985, 50 pages [ISER Research & Policy Papers No 2]

A1107
Worldwide offshore accident databank: WOAD statistical report 1990: statistics on accidents to offshore structures engaged in oil
and gas activities in the period 1970–89
Høvik: Veritas Offshore Technology & Services
1990, various pagings

Safety, Emergency Procedures

A1108
Accident and emergency management
Theodore L et al
Chichester: Wiley
1989, 504 pages

A1109
Assessment of the suitability of stand-by vessels attending offshore installations
Department of Transport
London: HMSO
1991

A1110
Contingency planning for the offshore industry; Aberdeen, 24–25 May 1989
London: IBC Techical Services
1989, 11 papers

A1111
Department of Energy offshore immersion suits: guidance on performance criteria
Marine Technology Support Unit
London: HMSO
1989 [Report OTH 89-292]

A1112
Design and development of a fast carriage lifeboat
Hudson F D
London: Institute of Marine Engineers
1990 [paper read on 3 April 1990]

A1113
Escape II: risk assessment of emergency evacuation of offshore installations
Technica for Department of Energy
London: HMSO
1988, 80 pages [Report OTH 88-285]

A1114
Escape, survival, rescue at sea, international conference; 29–31 October 1986
London: Royal Institution of Naval Architects
1986, 2 volumes

A1115
Evacuation of personnel by sea
Steering Committee for Emergency Preparedness
Høvik: Det norske Veritas
1983, 53 pages [in English and Norwegian] [Summary Report SSB 2.2]

A1116
Experimental investigation: effect of water sprays on gas explosions
Chr Michelsen Institute for the Department of Energy
London: HMSO
1991 [Report OTH 90-316]

A1117
Fire and explosion protection offshore: possibilities and limitations (paper presented at 8th International Fire Protection Seminar, 1990)
Pedersen K C
Trondheim: SINTEF
1990 [Report STF25 A90021]

A1118
Fire engineering in petrochemical and offshore applications, conference: Stratford-upon-Avon, 23–24 June 1987
Bedford: BHRA
1987, 19 papers

A1119
Fire hazard offshore: management, evaluation and protection, seminar; London 22 June 1988
Bedford: BHRA
1988, 7 papers

A1120
Fire prevention and control on open type offshore production platforms, 2nd ed
Washington DC: American Petroleum Instiute
1985, 26 pages [API RP14G]

A1121
Fire safety engineering: 2nd international conference on fire engineering and loss prevention in offshore petrochemical and other hazardous applications; Brighton, 27-29 June 1989
Smith D N (editor)
Bedford: BHRA
1989, 276 pages

A1122
First aid on offshore installations and pipeline works: Offshore Installations & Pipeline Works (First-aid) Regulations 1989 ...
guidance
Health & Safety Commission
London: HMSO
1990, iv + 40 pages [Approved Code of Practice 32]

A1123
Further in-water performance assessment of lifejacket and immersion suit combinations
Light I M and Slater P
London: HMSO
1991, viii + 74 pages [OTI 91 550]

A1124
Guidelines for offshore emergency drill and exercises on installations
London: UK Offshore Operators' Association
1982, 20 pages

A1125
Guidelines for offshore emergency safety training on installations
London: UK Offshore Operators' Association
1980, 12 pages

A1126
Helicopter rescue from offshore survival craft
Techword Services and Marine Technology Support Unit
London: HMSO
1990, vi + 41 pages [Report OTH 90-319]

A1127
Hybrid personal flotation devices
New York, NY: American National Standards Institute
1988 [ANSI/UL 1517-1988]

A1128
International convention for the Safety of Life at Sea (SOLAS)
London: International Maritime Organization
1986, xii + 439 pages (consolidated text)
1990/1991 amendments (1991)

A1129
Lessons in rotary drilling, unit V lesson 7: helicopter safety and survival procedures, revised edition
University of Texas at Austin, Petroleum Extension Services
1980, 40 pages

A1130
Marine survival and rescue systems
House D J
London: E & F Spon
1988, xiv + 256 pages

A1131
Marine survival craft: liferafts, lifeboats, survival systems; international conference, London 14–16 November 1983
London: Royal Institution of Naval Architects
1983, 22 papers

A1132
Offshore conference on safety and maritime training; Haugesund, 21–22 August 1985
Haugesund: Rogaland Regional Council of Higher Education
1985, 15 loose papers

A1133
The offshore health handbook: a practical guide to coping with injury and illness
Norman J Nelson and Brebner J
London: Martin Dunitz
1987, 208 pages

A1134
Offshore installations (fire-fighting equipment) regulations 1978
Department of Energy
London: HMSO
1978, 10 pages [SI 1978 No 611]

A1135
Offshore installations: guidance on fire fighting appliances
Department of Energy
London: HMSO
1980, 102 pages

A1136
Offshore installations: guidance on life-saving appliances, 3rd ed
Department of Energy
London: HMSO
1984

A1137
Offshore installations (life-saving appliances) regulations 1977
Department of Energy
London: HMSO
1977, 7 pages [SI 1977 No 486]

A1138
Offshore installations (operational safety, health and welfare) regulations 1976
Department of Energy
London: HMSO
1978, 37 pages [SI 1976 No 1019]

A1139
Offshore passive fire protection seminar; London 31 March–1 April 1987
London: Plastics & Rubber Institute
1987, 15 papers

A1140
6th Offshore search & rescue, communications & marine safety conference; Edinburgh, 27–30 October 1987
London: Institution of Electronic & Radio Engineers
1987, 12 papers

A1141
7th Offshore search & rescue, marine communications, marine & aviation safety conference; Edinburgh, 1–4 November 1988
Edinburgh: Leith International Converences
1988, 18 papers

A1142
8th Offshore search & rescue, marine communications, marine and aviation safety conference; Edinburgh, 1–3 November 1989
Edinburgh: Leith International Conferences
1989, 15 papers

A1143
9th Offshore search & rescue, marine communications, marine and aviation safety conference; Edinburgh 30 October–2 November 1990
Edinburgh: Leith International Conferences
1990, 15 papers

A1144
10th Offshore search & rescue, marine communications, marine & aviation safety conference; Edinburgh, 28–31 October 1991
Edinburgh: Leith International Conferences
1991, 19 papers

A1145
Recommended practice for training of personnel in rescue of persons in water, 1st edition
Washington, DC: American Petroleum Institute
1990, 11 pages [RP T-7]

A1146
Recommended practice for training of offshore personnel in non-operating emergencies, 1st edition
Washington DC: American Petroleum Institute
1986, 5 pages [RP T-4]

A1147
Recommended practice Volume C, facilities on offshore installations: emergency support systems
Høvik: Det Norsk Veritas
1981 [RP C106]

A1148
Recommended practice Volume C, facilities on offshore installations: escape routes
Høvik: Det norske Veritas
1988 [RP C103]

A1149
Recommended practices/guidelines for evaluation of mobile offshore drilling unit emergency power and fire protection systems, 1st edition
Washington DC: American Petroleum Institute
1988 [API RP 62]

A1150
Regulations concerning stand-by vessels in use on the Norwegian continental shelf
Oslo: Ministry of Local Government & Labour
1983, amended 1984, 12 pages [unofficial translation from the Norwegian]

A1151
Safety/standby vessels: the new requirements, conference; London, 26 November 1991
London: Institute of Petroleum
1990

A1152
SASMEX '88: safety at sea and marine electronics, conference; London, 27–28 April 1988
Redhill: Safety at Sea
1988, 7 papers

A1153
The stand-by boat service on the Norwegian Continental Shelf Report (summary and main conclusions) from working group appointed by the Norwegian Petroleum Directorate
Oslo: NPD
1982

A1154
Survey of life-saving appliances: volume 2, instructions for the guidance of surveyors: testing of life-saving appliances
Department of Transport
London: HMSO
1990

A1155
Testing and evaluation of life-saving appliances
London: International Maritime Organization
1985, 63 pages

A1156
Training for survival and rescue at sea, conference; Aberdeen, 21–22 May 1986
International Association for Sea Survival Training
1986, 12 papers

A1157
Training for survival and rescue at sea, 3rd IASST conference; St John's, 2–5 October 1991
St John's: Marine Institute
1991

A1158
United Kingdom search and rescue handbook, revised edition
Department of Transport
London: HMSO
1991, viii + 28 pages

ENVIRONMENTAL PROTECTION

A1159
Acute toxicity of drilling muds containing hydrocarbon additives and their fate and partitioning among liquid, suspended and solid phases: final report
Broteler R J et al (Battelle New England Marine Research Laboratory)
Washington, DC: API
1985, xiv + 133 pages

A1160
Annual book of ASTM standards: water and environmental technology: water (I)
Philadelphia, PA: American Society for Testing & Materials
1991, 594 pages [Volume 11.01]

A1161
Annual book of ASTM standards: water and environmental technology: water (II)
Philadelphia, PA: American Society for Testing & Materials
1991, 594 pages [Volume 11.02]

A1162
13th annual report on the activities of the Oslo Commission; June 1987–June 1988)
London: Oslo and Paris Commissions
1989

A1163
10th annual report on the activities of the Paris Commission; June 1987–June 1988
London: Paris Commission
1989

A1164
10th annual report on the activities of the Paris Commission; June 1987–June 1988

A1165
Cleaning of drilled cuttings
Wiig O
Trondheim: SINTEF
1985, 108 pages + 13 appendices
[Report STF21 A85073]

A1166
The control of oil pollution, 2nd editon
Wardley-Smith J (editor)
London: Graham & Trotman
1983, xi + 285 pages

A1167
Controls over the use and associated
discharges of oil base drilling muds in the UK,
Norway, the Netherlands and USA
Edinburgh: Institute of Offshore Engineering
1989, 1 volume [IOE Report 89/839]

A1168
Developing oil & gas resources: environmental
impact and responses
Cairns W J (editor)
Barking: Elsevier Applied Science
due 1992

A1169
Discharges of oil from platforms on the
Norwegian Shelf during 1987
Oslo: State Pollution Control Authority
1989, 9 pages

A1170
Drilling wastes: 1988 international conference;
Calgary, 5–8 April 1988
Englehardt F R et al (editors)
Barking: Elsevier Applied Science
1989, xvii + 867 pages

A1171
Environment Northern Seas
Lange R et al (editors)
Oslo: Norwegian University Press
1991, 63 pages

A1172
Environment Northern Seas, conference;
Stavanger, 26–30 August 1991
Stavanger: ENS Secretariat
1991, 7 volumes

A1173
ENS 91: Executive summary and proposals for
action
Stavanger: ENS Secretariat
1991, 1 volume

A1174
Environmental code of practice for treatment
and disposal of waste discharges from offshore
oil and gas operations
*Industrial Programs Branch, Environmental
Protection, Conservation and Protection*
Ottawa: Environment Canada
1990, x + 85 pages [Report EPS 1/PN/2]

A1175
Environmental considerations in the offshore
use of oil-based drilling muds; workshop,
Ottawa, 23–24 May 1984
Greene G and Engelhardt F R
*Ottawa: Canada Oil & Gas Lands
Administration*

A1176
Environmental cooperation between the North
Sea States: success or failure?
London: Belhaven Press
1988

A1177
Environmental impact analysis: theory and
practice
Walthem P
London: Unwin Hyman
1988, xx + 332 pages

A1178
The environmental impact of North Sea
oil-related developments on Scotland
Robertson J G
Portree: Habitat Scotland
1984, 88 pages

A1179
Environmental issues in future offshore developments, seminar; London 20–21 May 1991
London: IBC Technical Services
1991

A1180
Environmental protection in the 1990s:
Concawe senior management seminar;
8–9 October 1987
The Hague: Concawe
1987, 59 pages [Report 9/87]

A1181
Environmental protection in the European
Community: the role of Concawe
Brussels: Concawe
1990, 12 pages

A1182
Environmental protection of the North Sea;
international conference, London 24–27 March
1987
Newman P J and Agg A R (editors)
London: Heinemann
1988, xxviii + 886 pages

A1183
Forties environmental review, 2nd edition
Brummage K G et al
Dyce: BP Petroleum Development
1982, 89 pages

A1184
Guidelines for monitoring methods to be used
in the vicinity of platforms in the North Sea
London: Oslo and Paris Commissions
1991

A1185
ICOE '84, 3rd international conference on oil
and the environment; Bergen, 14–16 May 1984
Oslo: Norwegian Petroleum Society
1984, 16 papers

A1186
IMAS '90, marine technology and the
environment; London 23–25 May 1990
London: Institute of Marine Engineers
1990 [Trans IMarE (C) Vol 102, Conference 2]

A1187
The implementation of the ministerial
declaration of the 2nd international conference
on the protection of the North Sea
*The Hague: Ministry of Transport & Public
Works*
1990, 553 pages

A1188
Important issues in enhancing scientific
understanding of the North Sea environment
London: Oslo and Paris Commissions
1990 [North Sea Environment Report No 2]

A1189
International conference on the protection of
the North Sea; Bremen, 31 October–1 November 1984
Declaration
1984, 7 + 22 + 15 pages

A1190
3rd international conference on the protection
of the North Sea: the implementation;
ministerial declaration
*The Hague: Ministry of Transport & Public
Works*
1990

A1191
3rd international conference on the protection
of the North Sea: interim report on the quality
status of the North Sea
*The Hague: Ministry of sTransport & Public
Works*
1990, 48 pages

A1192
International convention for the prevention of
pollution of the sea by oil; 1954, as amended in
1962 and 1969
London: International Maritime Organization
1981, 1 volume + supplement

A1193
Land and water issues related to energy devel-
opment: 4th annual meeting of the International
Society of Petroleum Industry Biologists;
Denver, 22–25 September 1981
Rand P J (editor)
Ann Arbor: Ann Arbor Science
1982, 469 pages

A1194
Managing troubled waters: the role of marine
environmental monitoring
*Committee on a Systems Assessment of
Marine Monitoring, Marine Board, Commission
on Engineering and Technical Systems,
National Research Council*
Washington, DC: National Academy Press
1990, x + 125 pages

A1195
Manual on oil pollution: Section 1, prevention
London: International Maritime Organization
1983, 68 pages

A1196
Marine ecology: a comprehensive, integrated
treatise on life in oceans and coastal waters.
Volume V, Ocean Management: Part 3,
Pollution and protection of the seas: radioactive
materials, metals and oil
Kinne O (editor)
Chichester; John Wiley & Sons
1984, pages 1019–1618

A1197
The marine environment of the Moray Firth;
symposium, Aberdeen, 27–28 March 1985
*Edinburgh: Proceedings of the Royal Society of
Edinburgh, Section B*
1986, Volume 91, 358 pages

A1198
Ministerial declaration of the 3rd international conference on the protection of the North Sea
The Hague: Ministry of Transport & Public Works
1990, 198 pages

A1199
Monitoring in the marine environment; Texel, Netherlands, 4–6 June 1984
Kramer C J M and Hekstra G P (editors)
Environmental Monitoring & Assessment
1986, Volume 7 (2 and 3), pages 115–304

A1200
The North Sea Forum report
London: Council for Environmental Conservation (for The North Sea Forum)
1987, vi + 198 pages

A1201
North Sea oil and the environment: developing oi and gas resources: environmental impacts and response
Cairns W J (editor)
Barking: Elsevier Applied Science
due 1992

A1202
1990 North Sea Report
London: The Marine Forum for Environmental Issues
1990, 164 pages

A1203
2nd North Sea Seminar: the status of the North Sea environment: reasons for concern; Rotterdam, 1–3 October 1986
Amsterdam: Werkgroep Noordzee
1987,1988, 2 volumes

A1204
North Sea Task Force five year plan 1989–93
London: Oslo and Paris Commissions
1989 [North Sea Environment Report No 1]

A1205
Northern hydrocarbon development: environmental problem solving: 8th annual meeting, International Society of Petroleum Industry Biologists; Banff, Canada, 24–26 September 1985
Calgary: SPIB
1986, 374 pages

A1206
Offshore environmental studies program: final reports, publications and presentations, 2nd edition
Herndon VA: Minerals Management Service
1990 [OCS Report MMS 89-0018, NTIS Report PB 89-207815]

A1207
The offshore industry and the environment, conference; London, 10 September–1 October 1991
London: IBC Technical Services
1991, 13 papers

A1208
Offshore oil and gas: environmental studies program meets most user needs but changes needed
Resources, Community and Economic Development Division
Washington DC: United States General Accounting Office
1988, 88 pages

A1209
Oil based drilling fluids: cleaning and environmental effects of oil contaminated drill cuttings
SFT/Statfjord Unit Joint Research Project
1986, 20 pages

A1210
Oil based drilling fluids: cleaning and environmental effects of oil contaminated drill cuttings; conference, Trondheim, 24–26 February 1986
Statfjord Unit Owners (Mobil, SFT et al)
1986, 136 pages

A1211
Oil based drilling muds: off structure monitoring
Seakem Oceanography Ltd
Ottawa: Environmental Studies Revolving Funds
1988, 192 pages [ESRF Report 101]

A1212
Oil based muds: continuing education seminar: Aberdeen, 20 April 1989
Society of Petroleum Engineers, Aberdeen Section
1989, 7 papers

A1213
Oilspill risk analysis for the St George Basin
(December 1984) and North Aleutian Basin
(April 1985) Outer Continental Shelf lease
offerings
US Minerals Management Service
Reston VA: NTIS
1984, 75 pages [PB84-158310]

A1214
Options for treatment and disposal of oil-based
mud cuttings in the Canadian Arctic
Drinnan R W (Dobrocky Seatech) et al
Ottawa: Environmental Studies Revolving
Funds
1987, 81 pages + appendices [ESRF Report
No 063]

A1215
Outer Continental Shelf environmental
assessment program: comprehensive
bibliography
Anchorage: NOAA
1988, 746 pages

A1216
The Oslo and Paris Commissions: the first
decade (1974–1984)
London: Oslo and Paris Commissions
1984, 377 pages

A1217
Oslo Commission: procedure and decisions
anual
London: Oslo and Paris Commissions
1991

A1218
Paris Commission: procedure and decisions
manual
London: Oslo and Paris Commissions
1991

A1219
Petroleum and the environment, conference;
Oslo, 9–10 November 1990
Oslo: Norwegian Petroleum Society
1990, 5 papers

A1220
PetroPisces: lst international conference on
fisheries and offshore petroleum exploitation;
Bergen, 23–25 October 1989
Bergen: SCANEWS
1989, loose papers

A1221
The pollution control policy of the European
Communities, 2nd edition
Johnson S P
London: Graham & Trotman
1983, xvii + 244 pages

A1222
Procedure for minimising oil-based mud
discharges to the marine environment
London: UK Offshore Operators Association
1990, 8 pages

A1223
Progress report on the activities of the Oslo
and Paris Commissions (covering .. November
1987–March 1990)
London: Oslo and Paris Commissions
1990

A1224
Protection of the environment in offshore oil
developments located in inshore areas
Brocklehurst M S (BP Exploration)
London: Institute of Marine Engineers
1991, 16 pages

A1225
Recommended practice standard procedure for
field testing oil-based drilling fluids, 2nd edition
Washington DC: API
1991 [RP 13B-2]

A1226
Recommended practice standard procedure for
field testing water-based drilling fluids, 1st
edition
Washington DC: API
1990 [RP 13B-1]

A1227
Regulations for the prevention of pollution by
oil: articles of the International Convention for
the prevention of pollution from ships, 1973,
and the protocol of 1978 relating thereto;
Annex 1 of MARPOL 73/78
London: International Maritime Organisation
1982, 165 pages

A1228
A review of treatment and disposal options in
the Canadian arctic for cuttings contaminated
with oil-based drilling muds
Drinnan R W, Davies S R H et al, for
Environmental Studies Revolving Funds
Sidney BC: Dobrocky Seatech
1986

A1229
Simpilified toxicity testing of drilling muds I:
marine phytoplankton *Skeletonema costatum
grown in test tubes*
Østgaard K and Bonaunet K
Trondheim: SINTEF
1990, 22 pages [Report STF21 A90050]

A1230
Strategy for the protection of the marine
environment
London: International Maritime Organization
1989, 40 pages

A1231
Treatment of oil and oily wastes
Halmo G
Trondheim: SINTEF
1987, 28 pages [Report STF21 A87117]

A1232
United Kingdom marine pollution law
Bates J H
London: Lloyd's of London Press
1985, xxxii + 461 page

Spill Control

A1233
Accreditation of testing laboratories for safety
and pollution prevention equipment used in
offshore oil and gas operations
*New York, NY: American National Standards
Institute*
1988 [ANSI.ASME SPPE-2-1988]

A1234
Action against oil pollution: a guide to the main
intergovernmental and industry organisations
concerned with oil pollution in the marine
environment
London: Witherby & Co
1981, 20 pages

A1235
The agreement for cooperation in dealing with
pollution of the North Sea by oil (Bonn
Agreement); position paper on dispersants
London: Oslo and Paris Commissions
1991

A1236
Annual book of ASTM standards: pesticides;
resource recovery; hazardous substances and
oil spill responses; waste management;
biological effects
*Philadelphia PA: American Society for Testing
& Materials*
1992, 1444 pages [Volume 11.04]

A1237
Application and environmental effects of oil spill
chemicals
*London: International Petroleum Industry
Environmental Conservation Association*
1980, 19 pages

A1238
10th Arctic and Marine Oilspill Program
technical seminar; Edmonton, 9–11 June 1987
Ottawa: Environment Canada
1987, 467 pages

A1239
11th Arctic & Marine Oilspill Program technical
seminar; Vancouver, 7–9 June 1988
Ottawa:Environment Canada
1989, 514 pages

A1240
12th Arctic & Marine Oilspill Program technical
seminar; Calgary, 7–9 June 1989
Ottawa: Environment Canada
1989, 334 pages

A1241
13th Arctic and Marine Oilspill Program
technical seminar; Edmonton, 608 June 1990
Ottawa: Environment Canada
1990, vi + 485 pages

A1242
14th Arctic and Marine Oilspill Program technical seminar; Vancouver, 12–14 June 1991
Ottawa: Environment Canada
1991, x + 673 pages

A1243
Arctic oil spill response guide for the Alaskan
Beaufort Seas
Robert J Myers & Associates Inc
Springfield VA: NTIS
1988, 335 pages [US Coastguard report
CG-D-18-88 NITIS AD-A-204-788]

A1244
ASTM standards on hazardous substances and oil spill response
Philadelphia PA: American Society for Testing & Materials
1990, 119 pages

A1245
Biotechnology and oil spills, conference paper; Leeds, September 1991
Morgan P
London: Shell International Petroleum Co
1991, 7 pages [Shell Selected Papers, November 1991]

A1246
The Bridgeness incident
London: Department of Transport Marine Pollution Control Uniit
1985, 16 pages

A1247
A catalogue of oil spill containment barriers
Solsberg L B
Ottawa: Conservation & Protection, Environment Canada
1986, vii + 207 pages [Report EPS9/SP/1]

A1248
Characterization of spilled oil samples: purpose, sampling, analysis and interpretation
Butt J A (editor)
Chichester: John Wiley (for Institute of Petroleum)
1985, ix + 95 pages

A1249
Combatting oil spills: some economic aspects
Paris: OECD (HMSO)
1982, 140 pages

A1250
Committee of effectiveness of oil spill dispersants
US National Research Council
Washington DC: National Academy Press
1989

A1251
Control of pollution encyclopaedia
Garner J F (editor)
London: Butterworths
1977, 2 looseleaf binders + updating service

A1252
Definitions of terms relating to spill response barriers
Philadelphia PA: ASTM
1986 [ASTM F818-86]

A1253
A discussion of limitations on dispersant application
Koops W (Netherlands North Sea Directorate)
Oil & Chemical Pollution
1988, Volume 4(2) pages 139-153

A1254
Dispersants in the marine environment (selected papers from the international seminar on chemical and natural dispersion of oil on sea; Trondheim, 10–12 November 1986)
Oil & Chemical Pollution
1986/7, Volume 3(6), pages 400–503

A1255
Environmental atlas for Beaufort Sea oil spill response
Dickins D et al for Environment Canada Environmental Protection Service
Vancouver: D F Dickins Associates
1987, 186 pages + 5 appendices

A1256
An evaluation of the methodology used for shipboard monitoring of oil spills
Carter J A and MacGregor C D R
Oil & Chemical Pollution
1989, Volume 5(1), pages 47–63

A1257
A field guide to the application of dispersants to oil spills
Concawe Oil Spill Clean-up Technology Special Task Force No 19
Den Haag: Concawe
1988, 64 pages

A1258
Guidelines on the use and acceptability of oil spill dispersants, 2nd edition
Environmental Protection Service
Ottawa: Environment Canada
1984, 31 pages [Report EPS 1-EP-84-1]

A1259
Guidelines on the use of oil spill dispersants, 2nd edition
Institute of Petroleum
Chichester: John Wiley
1987, 52 pages

A1260
UNI/UNEP guidelines on oil spill dispersant
application and environmental considerations
London: International Maritime Organization
1982, 43 pages

A1261
Improved arrangements to combat pollution at
sea
London: Department of Trade & Industry
1979, 64 pages

A1262
International oil spill control directory, 8th editon
Oil Spill Intelligence Report
Arlington MA: Cutter Information
1988, 102 pages

A1263
International oilspill conference; San Diego,
4–7 March 1991
Washington, DC: American Petroleum Insrtitute
1991

A1264
Manual on oil pollution: Section II, contingency
planning
London: International Maritime Organization
1978, 52 pages

A1265
Manual on oil pollution: Section III, salvage
London: International Maritime Organization
1983, 45 pages

A1266
Manual on oil pollution: Section IV, practical
information on dealing with oil spillages
London: International Maritime Organization
1980, 143 pages

A1267
Members' handbook and red alert directory, 4th
edition
London: British Oil Spill Control Association
1986, 63 pages

A1268
Oil dispersants: new ecological approaches,
symposium; Williamsburg VA, 12–14 October
1987
Flaherty L M (editor)
*Philadelphia PA: American Society for Testing
& Materials*
1989, 302 pages [STP 1018]

A1269
Oil pollution control
Pritchard S Z
Beckenham: Croom Helm
1987, x + 231 pages

A1270
Oil spill cleanup of the coastline: a technical
manual
Stevenage: Warren Spring Laboratory
1982

A1271
1985 oil spill conference (prevention, behavior,
control, cleanup); Los Angeles, 25–28 February
1985
Washington DC: American Petroleum Institute
1985, xl + 651 pages

A1272
10th oil spill conference (prevention, behavior,
control, cleanup); Baltimore, 6–9 April 1987
Washington DC: American Petroleum Institute
1987, xxx + 634 pages

A1273
1989 20th anniversary oil spill conference
(prevention, behavior, control, cleanup); San
Antonio, 13–16 February 1989
Washington DC: American Petroleum Institute
1989, xxviii + 587 pages

A1274
1991 oil spill conference (prevention, behavior,
control, cleanup); San Diego, 4–7 March 1991
Washington DC: American Petroleum Institute
1991, xxviii + 739 pages

A1275
Oil spill contingency planning: a brief guide
*London: Oil Companies International Marine
Forum*
1979, 17 pages

A1276
Oil spill response guide
R J Meyers & Associates
Park Ridge, NJ: Noyes Data
1989

A1277
Oil spill response: options for minimizing
adverse ecological impacts
Washington DC: American Petroleum Institute
1985, vi + 98 pages [API Publication No 4398]

A1278
Petroleum spills in the marine environment: the chemistry and formation of water-in-oil emulsions and tar balls
Payne J R and Phillips C R
Chelsea Michigan: Lewis
1985, x + 148 pages

A1279
Pollution control instrumentation for oil and effluents
Parker H D and Pitt G D
London: Graham & Trotman
1987, xviii + 304 pages

A1280
The remote sensing of oil slicks, meeting; London 17–18 May 1989
Institute of Petroleum
Chichester: John Wiley
1989, ix + 165 pages

A1281
Remote sensing of pollution of the sea, colloquium; Oldenburg, 31 March–3 April 1987
Reuter R and Gillot R H (editors)
Oldenburg, W Germany: University of Oldenburg
1987, 555 pages

A1282
Response to oil and chemical marine pollution
Cormack D
Barking: Elsevier Applied Science
1983, xxv + 531 pages

A1283
Response to marine oil spills
International Tanker Owners Pollution Federation Ltd
London: Witherby
1987, 150 pages

A1284
A survey of chemical spill countermeasures
Solsberg I B and Parent R D
Ottawa: Environment Canada, Conservation & Protection, Technology Department and Technical Services Branch, Environmental Emergencies Technology Division
1986, xxviii + 392 pages [Report EPS 9/SP/2]

A1285
Using oil spill dispersants on the sea
Committee of Effectiveness of Oil Spill Dispers-ants, Marine Board, National Research Council
Washington DC: National Academy Press
1989, xv + 35 pages

A1286
World catalog of oil spill response products
Schulze R (editor)
Elkridge MD: World Catalog
1987, 500 pages

Marine pollution

A1287
ACOPS yearbook 1987–8
London: Advisory Committee on Pollution of the Sea
1988, 141 pages

A1288
ACOPS yearbook 1990
Dent P (editor)
Oxford: Pergamon
1990, viii + 180 pages

A1289
Amoco Cadiz: fates and effects of the oil spill; international symposium, Brest, 1981
Brest: Centre Océanologique de Bretagne
1981, 881 pages

A1290
Aquatic toxicology and environmental fate, 11th symposium; Cincinnati, 10–12 May 1987
Philadelphia PA: American Society for Testing & Materials
1989, 605 pages [STP 1007]

A1291
Aquatic toxicology and risk assesssment, 13th symposium; Atlanta GA, 16–18 April 1989
Philadelphia PA: American Society for Testing & Materials
1990, 378 pages [STP 1096]

A1292
Assessment of environmental fate and effects of discharges from offshore and oil and gas operations
Dalton Dalton Newport Inc, amended by Technical Resources Inc
Washington DC: US Environmental Protection Agency
1985 [EPA 440/4-85/002]

A1293
Assessment of pollution loads to the North Sea
Norton R L
Medmenham: Water Research Centre
1982, 28 pages [Technical Report no 182]

A1294
Assessment of sublethal effects of pollutants in the sea, discussion; London, 24–25 May 1978
Cole H A (organiser)
London: The Royal Society
1979, viii + 235 pages

A1295
Behaviour and impact assessment of heavy metals in estuarine and coastal zones
Salomons W et al
Delft: Delft Hydraulics Laboratory
1987, 47 pages [Communication No 375]

A1296
Characteristics of petroleum and its behaviour at sea
CONCAWE's Special Task Force
Den Haag: CONCAWE
1983, 47 pages [Report No 8/83]

A1297
Characterisation and prediction of the weathering properties of oils at sea: a manual for the oils investigated in the DIWO project
Daling P S et al
Trondheim: SINTEF
1991, 140 pages [DIWO Report No 16]

A1298
Characterization of spilled oil samples: purpose, sampling, analysis and interpretation
Butt J A, Duckworth D F and Perry S G
Chichester: John Wiley & Sons
1986, 70 pages (updating of Marine Pollution by Oil, 1974)

A1299
Chemical dispersability testing of fresh and weathered oils: an extended study with eight oil types
Brandvik P J et al
Trondheim: SINTEF
1991, 78 pages [DIWO Report No 12]

A1300
A coastal zone oil spill model: development and sensitivity studies
Reed M et al
Oil & Chemical Pollution
1989, Volume 5(6), pages 411–449

A1301
Cold water oil spills
Etkin D S
Arlington: Cutter Information Corp
19l90, 63 pages

A1302
Conference on the establishment of an international compensation fund for oil pollution damage, 1971; final act, including the text of the adopted convention
London: International Maritime Organization
1971, 1 volume plus supplement
[in English and French]

A1303
Contaminated marine sediments: assessment and remediation
Committee on Contaminated Marine Sediments, US National Research Council
Washington DC: National Academy Press
1990, 508 pages

A1304
Contaminated sediments
Förstner U
Berlin: Springer Verlag
1989, 157 [Lectures Notes in Earth Sciences 21]

A1305
Control of marine pollution in international law
Soni R
Colchester: Lloyd's of London Press
1985

A1306
The cost of oil spills: expert studies; OECD seminar
Paris: OECD (HMSO)
1982, 252 pages

A1307
Diffusion of contaminants in the ocean
Ozmidov R V [translation]
Dordrecht, London: Kluwer
1990, xii + 283 pages [Oceanographic Sciences Library Vol 4]

A1308
Drilling discharges in the marine environment
Panel on assessment of fates and effects of drilling fluids and cuttings in the marine environment, Marine Board, Commission Engineering & Technical Systems, National Research Council
Washington DC: National Academy Press
1983, xii + 180 pages

A1309
Ecological impacts of the oil industry,
symposium; London, 4–5 November 1987
B Dicks (editor)
Chichester: Wiley, for Institute of Petroleum
1989, ix + 316 pages

A1310
Effects of offshore petroleum operations on
cold water marine minerals: a literature review
Washington DC: American Petroleum Institute
1989 [Publication 4485]

A1311
Effects of oil pollution on major municipal and
industrial sea water abstractions: final report to
the Commission of the European Communities
Directorate-General for the Environment
Sir M MacDonald & Partners
*London: International Tanker Owners Pollution
Federation*
1987

A1312
The effects of pollution on marine ecosystems,
FAO/UNEP meting; Blanes, Spain, 7–11
October 1985
*Rome: Food & Agriculture Organization of the
United Nations*
1987, vii + 279 pages [FAO Fisheries Report
No 352 supplement]

A1313
Environmental ecology: the impacts of pollution
and other stresses on ecosystem structures
and function
Freedman B
London: Academic Press
1989, 399 pages

A1314
Environmental effects of North Sea offshore oil
operations; London 19–20 February 1986
London: The Royal Society
1987, 217 pages [Phil Trans Roy Soc B, 316]

A1315
Estuarine and coastal pollution: detection,
research and control; IAWPRC/NERC
conference,, Plymouth, 16–19 July 1985
Moulder D S and Williamson P (editors)
Oxford: Pergamon
1986, 364 pages [Water Science & Technology
volume 18 (4/5]

A1316
An exploratory study of the buoyancy
behaviour of weathered oils in water
Wilson D et al
Ottawa: Environmental Canada
1986, v + 50 pages [Report EE-85]

A1317
Fate and effects of oil dispersants and chem-
ically dispersed oil in the marine environment
Tetra Tech
*Los Angeles: US Department of the Interior,
Minerals Management Service, Pacific OCS
Region*
1985, xx + 114 pages

A1318
Fate and effects of oil in marine ecosystems,
IAWPRC oil pollution conference; Amsterdam,
23–27 February 1987
Dordrecht: Martinus Nijhof for TNO
1987, xiii + 338 pages

A1319
Fate and effects of produced water discharges
in nearshore marine waters
*API Health & Environmental Sciences
Department*
Washington DC: American Petroleum Institute
'1989, xi + 300 pages [Publication No 4472]

A1320
Fate and weathering of petroleum spills in the
marine environment
Jordan R E and Payne J R
Ann Arbor: Ann Arbor Science (Butterworth)
1980, 174 pages

A1321
Fate of hydrocarbons discharged at sea
Mackay D and McAuliffe C D
Oil & Chemical Pollution
1989, Volume 5(1), pages 1–20

A1322
Final act of international conference on marine
pollution, 1973
London: HMSO
1974 [Cmnd 5748]

A1323
Greeenhouse gas emissions from offshore oil
and gas production on the Norwegian
continental shelf
Lindeberg Erik G B et al
Trondheim: IKU
1990 [Report I90.145]

A1324
Guidelines for monitoring methods to be used in the vicinity of platforms in the North Sea
Norwegian State Pollution Control Authority (STF)
London: Paris Commission
1988, 28 pages

A1325
Hydrocarbons with water and seawater, part 1: hydrocarbons C5–C7
Shaw D G (editor)
Oxford: Pergamon
1988, 548 pages

A1326
Hydrocarbons with water and seawater, part 2: hardrocarbons C8–C36
Shaw D G (editor)
Oxford: Pergamon
1988, 584 pages

A1327
Impacts of exploratory drilling for oil and gas on the benthic environment of Georges Bank
Neff J M et al
Marine Environmental Research
1989, Volume 27(2), pages 77–114

A1328
Input of contaminants to the North Sea from the United Kingdom
Grogan W C
Edinburgh: Institute of Offshore Engineering
1985, viii + 203 pages

A1329
Integrated approaches to water pollution problems: SISIPPA 89; IISBON, 19–23 June 1989
Bau J et al (editors)
London: Eslevier Applied Science
1991, ix + 340 pages

A1330
International convention on civil liability for oil pollution damage (1969) and protocol (1976)
London: International Maritime Organization
1977, 20 pages

A1331
International Ocean Pollution, Ist symposium; Puerto Rico, 28 April–3 May 1991
Melbourne FL: Florida Institute of Technology
1991

A1332
Law relating to compensable damage caused by marine oil pollution with special reference to environmental damage
Advisory Committee on Pollution of the Sea
London: Graham & Trotman (for the Commission of the European Communities)
1984, 350 pages

A1333
Long-term consequences of low-level marine contamination
IMO/FAO/UNESCO .. Joint Group of Experts
Rome: GESAMP
1990, 14 pages [Reports & Studies 40]

A1334
The long-term effects of oil polllution on marine populations, communities and ecosystems: meeting, 28–29 October 1981
Clark R B (editor)
London: The Royal Society
1982, viii + 259 pages

A1335
Long term environmental effects of offshore oil and gas developments
Boesch D F and Rabalais N N (editors)
Barking: Elsevier Applied Science
1987, x + 708 pages

A1336
Marine environmental pollution
Geyer R A (editor)
Amsterdam, Oxford: Elsevier Science Publishers
Volume 1: hydrocarbons
1980, xxii + 592 pages
Volume 2: dumping and mining
1981, xxii + 574 pages
[Elsevier Oceanography Series 27a/b]

A1337
Marine pollutants: impacts on human health and the environment
World Health Organization & United Nations Environment Programme
Houston: Gulf
1990, 286 pages
[Advances in Applied Biotechnology Vol 5]

A1338
Marine pollution, 3rd edition
Clark R B
Oxford: Clarendon Press
1992, vii + 172 pages

A1339
Metal pollution in the aquatic environment, 2nd edition
Fürstner U et al
Berlin: Springer-Verlag
1983, xviii + 486 pages

A1340
Marine pollution, proceedings; Arab School on Science & Technology
Albaiges J (editor)
Washington, DC: Hemisphere
1989, 15 + 365 pages

A1341
Methods for predicting the physical changes in oil spilt at sea
Buchanan I
Stevenage: Warren Spring Laboratory
1987, 38 pages [Report LR 609 (OP) M]

A1342
Methods for predicting the physical changes in oil spilt at sea
Buchanan I and Hurford N (Warren Spring Laboratory)
Oil & Chemical Pollution
1988, Volume 4(4) pages 311–328

A1343
Natural dispersion of oil
Delvigne G A L and Sweeney C E
Oil & Chemical Pollution
1988, Volume 4(4), pages 281–310

A1344
Natural recovery of cold water marine environments after an oil spill (presented at 13th AMOP Technical Seminar, 1990)
Baker J M et al
Edinburgh: Institute of Offshore Engineering
1990, 111 pages

A1345
The Nordic Council's International Conference on the Pollution of the Seas; Copenhagen, 16–18 October 1989
Stockholm: The Nordic Council
1989, 108 pages

A1346
A North Sea computational framework for environmental and management studies: an application for eutrophication and nutrient cycles (presented at SISIPPA 1989)
Glas P C G and Nauta T A
Delft: Delft Hydraulics
1990, 16 pages [Publication 432]

A1347
1990 North Sea Report
London: The Marine Forum for Environmental Issues
1990, 164 pages

A1348
OCES: Offshore Clean Seas & Emergency Services
London: UK Offshore Operators Association
1988, 1 folded sheet [Fact sheet on offshore oil & gas activities]

A1349
The offshore ecology investigation: effects of oil drilling and production in a coastal environment
Ward C H (editor)
Houston: Rice University
1979, 589 pages [Rice University Studies Volume 65 (4 and 5)]

A1350
Oil in the sea: inputs, fates and effects
Steering Committee for the Petroleum in the Marine Environment Update, Board on Ocean Science & Policy et al
Washington DC: National Academy Press
1985, xix + 601 pages

A1351
Oil pollution: claims, liability and environment, conference; London, 23 October 1991
London: IBC Legal Studies & Services
1991

A1352
Oil pollution: environmental and commercial concerns, conference; London, 23 November 1991
London: IBC Legal Studies & Services
1991

A1353
Oil pollution is seas and estuaries, seminar proceedings
Somers C (editor)
Dublin: Institution of Engineers of Ireland
1991, 150 pages

A1354
Oil pollution of the sea: 8th report of the Royal Commission on Environmental Pollution
Kornberg H (Chairman)
London: HMSO
1981, xiv + 307 pages [Cmnd 8358]

A1355
Oil pollution of the sea: the Government response to the 8th report of the Royal Commission on Environmental Pollution
Department of the Environment
London: HMSO
1982, v + 39 pages [Pollution Paper No 20]

A1356
Oil pollution sensitivity atlas
Trondheim: SINTEF, Division of Applied Chemistry
1985, 7 reports

A1357
Petroleum effects in the arctic environment
Engelhardt F R (editor)
London: Elsevier Aplied Science
1985, xiv + 281 pages

A1358
Petroleum in the marine environment; symposium, Miami Beach, 13–14 September 1978
Petrakis L AND Weiss F T (editors)
Washington DC: American Chemical Society
1980, 371 pages [Advances in Chemistry 185]

A1359
Pollution in the North Atlantic Ocean; conference, Halifax, 19–23 October 1981
Farrington J W (editor)
Canadian Journal of Fisheries & Aquatic Sciences
1983, Volume 40 Supplement 2, 363 pages

A1360
Pollution of the marine environment, international conference; Venice, 27–30 October 1987
London: ACOPS

A1361
Pollution of the North Sea: an assessment
Salomons W and Bayne B L et al (editors)
Berlin, London: Springer
1988, xi + 687 pages

A1362
Pollution of the North Sea from the atmosphere
van Aalst R M et al
Delft: TNO
1982

A1363
Port and ocean engineering under arctic conditions: Volume II, symposium on noise and marine mammals; Fairbanks, 17–21 August 1987
Fairbanks: University of Alaska
1988, xii + 111 pages

A1364
Practical handbook of marine science
Kennish M J (editor)
Boca Baton FL: CRC Press
1988, 720 pages

A1365
Progress report on the activities of the Oslo and Paris Commissions, November 1987– March 1990
London: Secretariat of the Oslo and Paris Commissions
1990, 51 pages

A1366
Provisions concerning the reporting of incidents involving harmful substances under MARPOL 73/78
London: International Maritime Organization
1986, 24 pages [Publication 516-86-14E]

A1367
Quality status of the North Sea: 2nd International Conference on the Protection of the North Sea, Scientific & Technical Working Group
London: Department of the Environment
1987, vii + 88 pages

A1368
Remote sensing of pollution of the sea; inter- national colloquium, Oldenburg, W Germany, 31 March–3 April 1987
Reuter R and Gillot R H (editors)
University of Oldenburg
1987, 555 pages

A1369
Report of the ICES Advisory Committee on Marine Pollution 1988
Copenhagen:International Council for the Exploration of the Sea
1989 [Cooperative Research Report 160]

A1370
Report of the ICES Advisory Committee on
Marine Pollution 1989
*Copenhagen:International Council for the
Exploration of the Sea*
1989, 173 pages [Cooperative Research
Report 167]

A1371
Research on environmental fate and effects of
drilling fluids and cuttings; symposium, Lake
Buena Vista, 21–24 January 1980
Washington DC: American Petroleum Institute
2 volumes

A1372
Resource damage assessment in the marine
environment
Cormack D (editor)
Oil & Chemical Pollution (special issue)
1989, Volume 5(2+3), pages 81–238

A1373
Response of marine animals to petroleum and
specific petroleum hydrocarbons
Neff J M and Anderson J W
Barking: Elsevier Applied Science
1981, x + 182 pages

A1374
Responses of marine organisms to pollutants
Roesijadi G et al
Marine Environmental Research
1989, Voilume 28, 545 pages

A1375
Sea mammals and oil: confronting the risks
Geraci J R and St Aubin D J (editors)
San Diego, London: Academic Press
1990, xvi + 282 pages

A1376
Seabirds in the North Sea: final report of phase
2 of the Nature Conservancy Council Seabirds
at Sea project, November 1983 –October 1986
Tasker M L et al
Peterborough: Nature Conservancy Council
1987, 336 pages

A1377
The state of the marine environment
GESAMP
*Nairobi: United Nations Environment
Programme*
1990 [UNEP Regional Seas Reports & Studies,
115]

A1378
The state of the marine environment
GESAMP
Oxford: Blackwell Scientific
1991, 152 pages

A1379
Strategies for the assessment of the biological
impacts of large coastal oil spills: European
coasts
Dicks B, with CONCAWE's Special Task Force
Den Haag: CONCAWE
1985, 62 pages [Report No 5/85]

A1380
Stream, lake, estuary and ocean pollution, 2nd
edition
Nemerow N L
Wokingham: Van Nostrand Reinhold
1991, 464 pages

A1381
Tidal and dispersant project: fate and effects of
chemically dispersed oil in the nearshore
benthic environment
Gilfillan E S et al
Washington DC: American Petroleum Institute
1986, 17 + 215 pages [API Publication No
4440]

A1382
Two years after the spill: environmental
recovery in Prince William Sound and the Gulf
of Alaska
Baker J M et al
Edinburgh: Institute of Offshore Engineering
1991, 10 pages

A1383
The UK marine oil pollution legislation
Bates J H
Colchester: Lloyd's of London Press
1987 [LLP Annotated Acts]

A1384
Vulnerable concentrations of birds in the North
Sea
*Tasker M L and Pienkowsi M W (Seabirds at
Sea Team)*
Peterborough: Nature Conservancy Council
1987, 38 pages

A1385
Wastes in the ocean: volume 4, energy wastes
in the ocean
Dredall I, Kester D R and Ketchum B H
Chichester: John Wiley & Sons
1985, 818 pages

A1386
Water pollution law
Howarth W
London: Shaw & Sons
1988, xxxvii + 608 pages

A1387
Water pollution: modelling, measuring and prediction, 1st conference; Southampton, 3–5 September 1991
Wrobel L C et al (editors)
Southampton: Computational Mechanics
1991, 750 pages

A1388
Water quality management plan North Sea: framework for analysis
Koudstaal R
Rotterdam: A A Balkema for International Federation of Institutes for Advanced Study and Delft Hydraulics
1987 141 pages [Coastal Water series]

DECOMMISSIONING & ABANDONMENT

A1389
Abandonment: current legal issues of interest to oil company lawyers
Beazley R
Oil & Gas Law & Taxation Review
1986/7, Volume 5 (8) pages 202–208

A1390
The abandonment of offshore installations and pipelines
London: UK Offshore Operators Association
1988, 6 pages
[Fact sheet on offshore oil and gas activities]

A1391
Abandonment of United Kingdom offshore oil and gas installations, fiscal, security and accounting issues
London: Ernst & Whinney
1989, 66 pages

A1392
Alternative uses of offshore installations: final report on SERC funded study (1983–1985)
Johnston C S (Principal Investigator)
Side J C (report author)
Edinburgh: Institute of Offshore Enginering
1985, 84 pages

A1393
3rd International Artificial Reef conference; Newport Beach, California, 3–5 November 1983
Bulletin of Marine Science
1985, volume 37 (1) pages 1–402

A1394
Artificial reefs: marine and freshwater applications
d'Itri F M (editor)
Chelsea MI: Lewis Publishers
1985, 589 pages

A1395
Decommissioning, 1988, lst international conference on decommissioning offshore, onshore and nuclear works; Manchester, 23–24 March 1988
Whyte I (editor)
London: Thomas Telford
1988, 327 pages

A1396
Decommissioning and demoliltion, 2nd conference on decommissioning offshore, onshore demolition and nuclear works; Manchester, 24–26 April 1990
Whyte I L (editor)
London: Thomas Telford
1990, 254 pages

A1397
Decommissioning and removal of North Sea structures: conference; London, 7–8 April 1987
London: IBC Technical Services
1987, 12 papers

A1398
Decommissioning and removal of offshore structures, conference; London, 19–20 April 1989
London: IBC Technical Services
1989, 1 volume

A1399
The decommissioning of offshore installations: a world-wide survey of timing, technology and anticipated costs
London: Oil Industry International Exploration & Production Forum
1984 [Report 10.5/108]

A1400
Decommissioning of oil and gas fields: report with appendices together with proceedings of the Committee: 4th Report
House of Commons Energy Committee (Chairman, M Clark)
London: HMSO
1991, xxii + 142 pages [House of Commons Papers, Session 1990–91, paper 33]

A1401
Decommissioning of oil and gas fields: Government observations on the 4th Report from the Committee (Session 1990–91); 6th Special Report
House of Commons Energy Committee (Chairman, M Clark)
London: HMSO
1991, xii pages [House of Commons Papers, Session 1990–91, paper 585]

A1402
Economic and fiscal aspects of field abandonment with special reference to the UK continental shelf
Kemp A G and Reading D
Aberdeen: University of Aberdeen Dept of Economics
1991 [North Sea Study Occasional Paper 34]

A1403
Economics of field abandonment in the UKCS
Kemp A G
Aberdeen: University of Aberdeen Dept of Economics
1989, 28 pages [North Sea Study Occasional Paper 29]

A1404
Engineering aspects of platform removal and the selection of crane vessels
Penney P W
Underwater Technology
1988, Volume 14(4), pages 16–25, 28

A1405
Field abandonment and decommissioning
Robinson G (editor)
Canterbury: Smith Rea Energy Associates
1987, 34 pages [Offshore Business No 19]

A1406
Fiscal aspects of field abandonment in the UKCS
Kemp A G et al
Aberdeen: University of Aberdeen Dept of Economics
1985 [North Sea Study Occasional Paper 22]

A1407
Fishing offshore platforms Central Gulf of Mexico: an analysis of receational and commercial fishing use at 164 major offshore petroleum structures
US Minerals Management Service
Reston VA: NTIS
1984, xii + 159 pages [PB84-216605]

A1408
The future of redundant offshore installations
London: Department of Energy
1982, 7 pages

A1409
Guidelines and standards for the removal of offshore installations and structures on the contintental shelf and in the exclusive economic zone
London: International Maritime Organization
1988, 6 pages [MSC/CIRC 490]

A1410
OAR 90, offshore abandonment and removal, conference: Aberdeen, 27–29 March 1990
Kingston-upon-Thames: Offshore Conference & Exhibitions
1990, 3 volumes

A1411
Offshore de-commissioning conference '86; Heathrow, 25–26 November 1986
Kingston-upon-Thames: Offshore Conferences & Exhibtions
1986, 15 papers

A1412
Offshore pipeline abandonment
Sharp W R
Solihull: UK Offshore Ooperators Association
1989, 1 volume

A1413
Planning for abandonment: conference; London, 9 July 1985
Rutland: European Study Conferences
1985, 5 papers

A1414
Platform abandonment conference; London, 1 June 1984
Rutland: European Study conference
1984, loose papers

A1415
Platform removal demands complex explosive designs
Al-Hassani S T S
Oil & Gas Journal
1988, Volume 86(20), pages 35–39

A1416
Programmatic environmental assessment: structural removal activities, central and western Gulf of Mexico planning areas
New Orleans: US Minerals Management Service
1987, 84 pages [MMS 87-0002]

A1417
Removal of offshore platforms and the development of international standards
Kasoulides G C
Marine Policy
1989, Volume 13(3) pages 249–265

A1418
Rigs-to-reefs: the use of obsolete petroleum structures as artificial reefs
Reggio V, Jr
New Orleans: US Minerals Management Service
1987, 17 pages [MMS 87-0015]

A1419
Structure removal activities, central and western Gulf of Mexico planning areas
Washington DC: US Dept Interior Minerals Management Service
1987, v + 84 pages [OCS EIS/EA MMS 87-0002]

A1420
UK abandonment policy: development and implementation
Butler W C F
London: Department of Energy
1990, 8 pages

MARINE RESOURCES
OTHER THAN PETROLEUM

GENERAL

A1421
Advances in the science and technology of
ocean management, Challenger Society/SUT
conference; Cardiff, 12–13 April 1988
Smith H D
London: Routledge
1992, xvi + 240 pages
[Ocean management and policy series]

A 1422
Annual review of ocean affairs: law and policy:
main documents 1985–1987
*United Nations Office for Ocean Affairs and the
Law of the Sea*
Colchester: Lloyd's of London Press
1989, 2 volumes

A1423
Cable laying: guidelines to exposures and
insurances
Bashford A S
London: Witherby
1991, 31 pages

A1424
Case studies in oceanography and marine
affairs
Open University Course Team
Oxford: Pergamon
1991, 248 pages

A1425
4th Circum-Pacific energy and mineral
resources conference; Singapore, 17–22
August 1986
Horn M K, (editor)
*Circum-Pacific Council for Energy and Mineral
Resources*
1987, 636 pages

A1426
Conflicts and cooperation in managing
environmental resources
Pethig R (editor)
Berlin, London, New York: Springer-Verlag
1992, 338 pages

A1427
COSU 89, coastal ocean space utilization; New
York, 8–10 May 1989
Halsey S D and Abel R B
Barking: Elsevier Applied Science
1990, 408 pages

A1428
The development of integrated sea use
management
Smith H and Vallega A
London: Routledge
1991, xv + 283 pages

A1429
The economics of coastal zone management: a
manual of assessment techniques: the 'Yellow
Manual'
Penning-Rowsell E (editor)
London: Belhaven Press
1991, 368 pages

A1430
Energy, resources and environment
*Blunden J and Reddish A (for Open University
Course Team)*
London: Hodder & Stoughton
1991, 339 pages

A1431
Europe and the Sea: marine sciences and
technology in the 1990s, conference; Hamburg,
27–29 September 1988
Hoefeld J et al (editors)
*Hamburg: Deutsches Kommittee für
Meeresforschung und Meerestechnik e V*
1989, 321 pages

A1432
The exclusive economic zone: an exciting new frontier
McGregor B A and Oldfield T W
Washington DC: US Government Printing Office for US Geological Survey
c 1985, 20 pages

A1433
Exclusive economic zones: British Dependencies and non-living resources
Smith A J and Docherty J I C for Marine Technology Directorate
Bedford: BHRA for Marine Technology Directorate
1989, 88 pages [MTD Publication 89/100]

A1434
Exclusive economic zones: resources, opportunities and the legal regime
London: Graham & Trotman (for the Society for Underwater Technology)
1986, 104 pages
[Advances in Underwater Technology, Vol 8]

A1435
Report of the NOAA Exclusive Economic Zone bathymetric and geophysical survey workshop ; Washington DC, 11–12 December 1984
Lockwood M (compiler)
Rockville: US Department of Commerce, NOAA National Ocean Service
1985, 63 pages

A1436
Exploring the new ocean frontier: the exclusive economic zone symposium; Washington DC, 2–3 October 1985
Lockwood M and Hill G (editors)
Rockville: US Department of Commerce, NOAA National Ocean Service
1985, vii + 270 pages

A1437
Georges Bank
Backus R H (editor)
Cambridge MA; London: MIT Press
1987, 632 pages

A1438
Gold, silver and uranium from seas and oceans: the emerging technology
Burk M
Los Angeles: Ardor Publishing
1989, 252 pages

A1439
Handbook on ocean politics and law
Wang J
Greenwood Press
1992, 592 pages

A1440
Harmonization of Netherlands North Sea policy 1989–1992
Smit-Kroes N et al
s'Gravenhage: SDU Uitgeverij
1989, 75 pages

A1441
The Indian ocean: exploitable mineral and petroleum resources
Toonwal G S
Berlin, London: Springer Verlag
1986, xvi + 198 pages

A1442
International maritime boundaries
Charney J I and Alexander L M (editors)
Dordrecht: Martinus Nijhoff
due 1992, 2 looseleaf volumes + updates

A1443
International organizations and the Law of the Sea documentary yearbook 1985
Netherlands Institute for the Law of the Sea
London: Graham & Trotman
1987, viii + 645 pages

A1444
International organizations and the Law of the Sea: documentary yearbook 1986, volume 2
Netherlands Institute of the Law of the Sea
London: Graham & Trotman
1988

A1445
International organizations and the Law of the Sea: documentary yearbook 1987, volume 3
Netherlands Institute for the Law of the Sea
London: Graham & Trotman/Martinus Nijhoff
1989

A1446
International resource management
Anderson S and Ostreng W (editors)
London: Belhaven Press
1989, 288 pages

A1447
The Law of the Sea, 2nd ed
Churchill R Rand Lowe A V
Manchester: Manchester University Press
1988, 370 pages

A1448
Law of the Sea: Caracas and beyond
Anand R P (editor)
Dordrecht: Martinus Nijhoff
1981, 411 pages

A1449
The Law of the Sea: documents 1983–1989:
preparatory commission for the International
Sea-bed Authority and for the International
Tribunal for the Law of the Sea
Platzoder R (editor)
Dobbs Ferry, NY: Oceana
1990, 10 volumes

A1450
The Law of the Sea: marine scientific research:
a guide to the implementation of the relevant
provisions of the United Nations convention
*United Nations Office for Ocean Affairs and the
Law of the Sea*
1991, xii + 38 pages [E.91.V.3]

A1451
The Law of the Sea: maritime boundary
agreements (1942–1969)
*United Natioins Office for Ocean Affairs and the
Law of the Sea*
1991, x + 96 pages [E.91.V.II]

A1452
The Law of the Sea: national legislation,
regulations and supplementary documents on
marine scientific research in areas under
national jurisdiction
*New York: UN Office for Ocean Affairs & the
Law of the Sea*
1989

A1453
The Law of the Sea: official text of the United
Nations Convention
London: Croom Helm
1983, xxxvii + 224 pages

A1454
The Law of the Sea: where we are–where to
go, conference; London 7 November 1989
London: Greenwich Forum
1989, 53 pages

A1455
Leasing energy resources on the outer
continental shelf
*Washington DC: US Dept of the Interior,
Minerals Management Service*
1987, vi + 49 pages

A1456
Managing marine environments
Kenchington R A
New York, London: Taylor & Francis
1990, iv + 248 pages

A1457
Managing the ocean: resources, research, law
Richardson J G
Mt Airy: Lomond Publications
1985, xv + 407 pages

A1458
Managing the Outer Continental Shelf lands:
oceans of controversy
Farrow R S et al
New York, London: Taylor & Francis
1990, viii + 168 pages

A1459
Marine policy for America
Mangone G J
London; Washington DC: Taylor & Francis
1989, 400 pages

A1460
Marine scientific research boundaries and the
Law of the Sea: discussion and inventory of
national claims
Ross D A and Landry T W
*Woods Hole: Woods Hole Oceanographic
Institution*
1987, 173 pages

A1461
Maritime boundaries and ocean resources
Blake G (editor)
Beckenham: Croom Helm
1987, 15 + 284 pages

A1462
Mediterranean continental shelf delimitations
and regimes: international and national legal
sources
Leanza U and Sico L
Dobbs Ferry, NY: Oceana
1988, 4 volumes

A1463
Mineral resource exploration in the Irish Sea
Breathnach P
Geographical Society of Ireland
1989, Special Publication No 3, pages 58–64

A1464
Neptune's Domain: a political geography of the sea
Glassner M I
London: Unwin Hyman
1990, 160 pages

A1465
New directions in Law of the Sea
Dobbs Ferry, New York: Oceana
11 volumes 1973–81, continued looseleaf, 1981–

A1466
A new handbook on the Law of the Sea, Volume 1
Dupuy R-J and Vignes D (editors)
Dordrecht: Martinus Nijhoff
1991, 900 pages
[Recueil des Cours, Colloque 13]

A1467
A new handbook on the Law of the Sea, Volume 2
Dupuy R-J and Vignes D (editors)
Dordrecht: Martinus Nijhoff
1992, 882 pages
[Recueil des Cours, Colloque 14]

A1468
The North Sea: basic legal documents on regional environmental cooperation
Freestone D and Ijstra T (editors)
Dordrecht, London: Kluwer
1991, 468 pages [Basic Legal Documents on Regioinal Environmental Cooperation 1]

A1469
The North Sea: perspectives on regional environmental cooperation
Freestone D and Ijstra T (editors)
Dordrecht, London: Kluwer
1990, 356 pages

A1470
The ocean basins: their structure and evolution
Open University Course Team
Oxford: Pergamon
1989, 171 pages

A1471
The ocean in human affairs
Singer S F (editor)
New York: Paragon House
1990, 374 pages

A1472
Ocean management: the way forward
London: Royal Institute of Chartered Surveyors
1990, 9 leaflets in folder

A1473
Ocean Resources: Ist international ocean technology congress; Honolulu, January 1989; Volume I, assessment and utilization
Ardus D A and Champ M A (editors)
Dordrecht: Kluwer
1990

A1474
Ocean Resources: Ist international ocean technology congress; Honolulu, January 1989; Volume II, subsea work systems and technologies
Ardus D A and Champ M A (editors)
Dordrecht: Kluwer
1990

A1475
Ocean technology, development, training and transfer; proceedings of Pacem in Maribus XVI; Halifax, Canada
Vandermuelen J and Walker S (editors)
Oxford: Pergamon
1990

A1476
Ocean Yearbook 6
Borgese E M and Ginsburg N (editors)
Chicago, London: University of Chicago Press
1987, 686 pages

A1477
Ocean Yearbook 7
Borgese E M et al (editors)
Chicago, London: University of Chicago Press
1989, 650 pages

A1478
Ocean Yearbook 8
Borgese E M et al (editors)
Chicago, London: University of Chicago Press
1990, xv + 685 pages

A1479
Ocean Yearbook 9
Borgese E M et al (editors)
Chicago, London: University of Chicago Press
1992, 508 pages

A1480
Oceans under threat
Neal P
London: Dryad
1990, 64 pages

A1481
Our seabed frontier: challenges & choices
Silva A J et al
Washington, DC: National Academy Press
1989, 138 pages

A1482
PACON '84: Pacific congress on marine
technology; Honolulu, 24–27 April 1984
*Manoa: Marine Technology Society (Hawaii
Section)*
1984, 95 papers or abstracts

A1483
PACON '86: 2nd Pacific conference of marine
technology: Honolulu, 24–28 March 1986
*Manoa: Marine Technology Society (Hawaii
Section)*
1986, 146 papers

A1484
Plan for National Ocean Service Exclusive
Economic Zone bathymetric and geophysical
program
*Rockville: US Department of Commerce,
NOAA National Ocean Service*
1984, viii + 51 pages

A1485
Preliminary assessment of offshore platform
conversion to geothermal power station
Manchester University (Turner M J)
London: Department of Energy
1989, lv (various pagings) [ETSU G145]

A1486
Productivity of the ocean, past and present
Berger W H et al (editors)
Chichester: Wiley
1989, 416 pages
[Dahlem Life Sciences Research Report 44]

A1487
Rights to oceanic resources: deciding and
drawing maritime boundaries symposium;
Athens, Georgia, 1–2 May 1987
Dallmeyer D G and DeVorsey L (editors)
Dordrecht, London: Martinus Nijhoff
1989, xiii + 107 pages
[Publications on Ocean Development Vol 13]

A1488
Seabed energy and mineral resources and the
Law of the Sea: volume 1, the areas within
national jurisdiction
Brown E D
London: Graham & Trotman
1984, 300 pages
[looseleaf with updating service]

A1489
Seabed energy and mineral resources and the
Law of the Sea: volume 2, the area beyond the
limits of national jurisdiction
Brown E D
London: Graham & Trotman
1986, 430 pages
[hardback or looseleaf with updating service]

A1490
Seabed energy and mineral resources and the
Law of the Sea: volume 3, selected documents,
tables and bibliography
Brown E D
London: Graham & Trotman
1986, 460 pages
[hardback or looseleaf with updating service]

A1491
Technologies for exploitation of novel marine
resources
General Technology Systems Ltd
London: Dept of Trade & Industry
1987, 2 volumes
[Resources from the Sea Programme]

A1492
Tourist oceanology international '90
conference; Monaco, 9–11 October 1990
Kingston upon Thames: Spearhead
1990, 2 volumes

A1493
Undersea resources: the Mike Adye Memorial
Lecture
Patten T D
London: Marine Tehcnology Directorate
1990, 17 pages [MTD Publication 90/100]

A1494
UnderSea World '91, conference; San Diego,
16–18 April 1991
Spring Valley, CA: West Star Productions
1991, 20 papers

A1495
United Nations convention on the Law of the
Sea 1982: a commentary
Volume 2: Second Committee: articles 1–132,
annexes I and II, final act annex II
Nordquist M H (editor)
Dordrecht: Kluwer
in preparation

A1496
United Nations convention on the Law of the
Sea 1982: a commentary
Volume 3: First Committee: articles 133–191,
annexces III and IV, final act annex I
Nordquist M H (editor)
Dordrecht: Kluwer
1990

A1497
United Nations convention on the Law of the
Sea 1982: a commentary
Volume 4: Third Committee: articles 192–278
and final act annex VI
Rosenne S and Yankov A (editors)
Dordrecht: Kluwer
1990

A1498
United Nations convention on the Law of the
Sea 1982: a commentary
Volume 5: Settlement of disputes, general and
final provisions
Rosenne S and Sohn L B (editors)
Dordrecht: Kluwer
1989

A1499
United Nations convention on the Law of the
Sea 1982: a commentary
Volume 6: comprehensive index to series,
consolidated lists of treaties, cases ...
Nordquist M H (editor)
Dordrecht: Kluwer
in preparation

A1500
The wasted ocean: the ominous crisis of
marine pollution and how to stop it
Bullough D K
New York: Lyons & Burford
1989

A1501
Wealth from the oceans: an advanced
technology programme forming part of the
DTI's Research & Technology Initiative: Ocean
Technology—international programmes and
markets
Smith Rea Energy Analysts Ltd
London: Department of Trade & Industry
1989, 154 pages

A1502
World ocean management
Smith H D and Lalwani C S
London: Routledge
due 1992
[Ocean management and policy series]

MINING

A1503
Analysis of exploration and mining technology
for manganese nodules
*London: Graham & Trotman (with United
Nations)*
1984, xi + 140 pages
[Seabed Minerals Series Vol 2]

A1504
Analysis of processing technology for
manganese nodules
*United Nations Ocean Economics &
Technology Branch*
London: Graham & Trotman
1986, x + l97 pages
[Seabed Minerals Series Volume 3]

A1505
Assessment of manganese nodule resources
London: Graham & Trotman
1982, 79 pages
[Seabed Minerals Series, Volume 1]

A1506
A bibliography of mineral law
Spencer D
*Dundee: Dundee University Centre for
Petroleum & Mineral Law Studies*
1988, 197 pages

A1507
Deepsea mining and the Law of the Sea
Post A M
The Hague: Martinus Nijhoff
1983, xxiii + 358 pages

A1508
Delineation of mine-sites and potential in
different sea areas
United Nations Ocean Economics &
Technology Branch
London: Graham & Trotman
1987, 79 pages

A1509
Handbook of geophysical exploration at sea,
volume 2, 2nd edition
Geyer R A
Boca Raton: CRC Press
due 1992, 432 pages

A1510
Marine mineral exploitation
Kunzendorf H (editor)
Amsterdam: Elsevier Science Publishers
1986 [Oceanography Series volume 41]

A1511
Marine minerals: advances in research and
resources assessment
Telek P G (editor) et al
Dordrecht: D Reidel
1987, 588 pages [NATO ASI Series C: Vol 194]

A1512
Marine minerals: exploring our new ocean
frontier
US Congress, Office of Technology
Assessment
Washington, DC: US Government Printing
Office
1987, 349 pages [OTA-O-342]

A1513
Marine mineral resources
Earney F C F
London: Routledge
1990, 387 pages

A1514
Marine mineral resources, short course;
London, 12–13 March 1990
Royal School of Mines and the London Centre
for Marine Technology, with the Centre for
Continuing Education at Imperial College
London: Imperial College
1990, lv (various pagings)

A1515
Marine minerals in exclusive economic zones
Cronan D
London: Chapman & Hall
1992, xi + 209 pages
[Topics in the Earth Sciences, 5]

A1516
Marine mining: a new beginning: conference;
Hilo, Hawaii, 18–21 July 1982
Humphrey P B (editor)
Honolulu: State of Hawaii Department of
Planning & Economic Development
1985, 319 pages

A1517
Marine mining on the Outer Continental Shelf,
environment effects overview
Cruikshank M et al
Vienna VA: US Minerals Management Service
1987, 66 pages [MMS 87-0035]

A1518
Offshore mineral resources, 2nd international
seminar; Brest, 19–23 March 1984
Orleans: Germinal
1984, 733 pages

A1519
Polymetallic sulfides and oxides of the US EEZ:
metallurgical extraction and related aspects of
possible future development
Harvey W W
Arlington MA: Arlington Technical Services
1987, 62 pages

A1520
Selection of sites for seabed manganese
nodule processing plants
London: Graham & Trotman for the United
Nations
1989, 104 pages [Seabed Minerals Series, 5]

A1521
Subsea mineral resources
McKelvey V E
UK Geological Survey Bulletin 1689-A
1986, 106 pages

A1522
Technology assessment and research program
for offshore minerals operations, 1986 report
Gregory J B and Smith C E
Reston, VA: US Dept of the Interior, Minerals
Management Services
1986, 204 pages [OCS Study MMS 86-00083]

DREGDING

A1523
Dredging: a handbook for engineers, 2nd
edition
Bray R N et al
London: Edward Arnold (Publishers)
due 1992, 352 pages

A1524
Dredging technology: 3rd international
symposium; Bordeaux, March 1980
Cranfield: BHRA Fluid Engineering
1980, 450 pages

A1525
Environmental effects of deep sea dredging
Speiss F N et al
Washington DC: NTIS
1986, 119 pages [PB87-138319]

A1526
Marine dredging for sand and gravel: minerals
planning research project No PECD
7/1/163-99/84
Nunny R S and Chillingworth P C H
*London: HMSO (for Department of the
Environment Minerals Division)*
1986, 192 pages plus appendices

A1527
Minerals and mineraloids in marine sediments:
an optical identification guide
Rothwell R G
London: Elsevier Applied Science
1989, xvi + 279 pages

A1528
Seabed dredging over considerable distances
Newman R and Cookson E W
Underwater Systems Design
1988, Volume 10(6), pages 26–29

A1529
Specification and dredging technology,
workshop; Amsterdam, 21 April 1989
*Rijswijk: Hydrographic Society, Netherlands
Branch*
1989, 5 papers

A1530
World Dredging Congress 1983, combining
WODCON X, 4th International symposium on
Dreging Technology, and SEATEC IV;
Singapore, 19–22 April 1983
Cranfield: BHRA Fluid Engineering
1983, 729 pages

SALVAGING

A1531
IMSC '88, international marine salvage
conference; Heathrow, 18–19 October 1988
*Kingston upon Thames: Offshore Conferences
& Exhibitions*
1988, 15 papers

A1532
International tug convenion and marine salvage
symposium; Halifax, Nova Scotia, 24–28
September 1990
Troup K D (editor)
*East Molesey, Surry: Thomas Reed
Publications*
1991, xv + 264 pages

A1533
Marine salvage in the United States
National Research Council, Marine Board
Washington DC: National Academic Press
1982, 141 pages

A1534
Marine salvage: international symposium; New
York, 1–3 October 1979
Searle, W F (editor)
Washington DC: Marine Technology Society
1980, 323 pages

A1535
Marine salvage: international symposium; New
York, 1–3 October 1979
New York: The Maritime Exchange
1980

A1536
Maritime law of salvage
Brice G
London: Stevens & Sons
1983

A1537
Peril at sea and salvage: a guide for masters,
3rd edition
*International Chamber of Shipping and Oil
Companies International Marine Forum*
London: Witherby, for ICS and OCIMF
1988

A1538
Reed's commercial salvage practice
Hancox D
New Malden: Thomas Reed
1987, 2 volumes + annual supplements

A1539
Salvage from the sea
Forsberg G
London: Routledge & Kegan Paul
1977, 169 pages

A1540
Salvage operations
Baptist C N T
London: Stanford Maritime Press
1979, 155 pages

WASTE DISPOSAL

A1541
Assessing the impact of deep sea disposal of
low level radioactive waste on living marine
resources
Vienna: International Atomic Energy Agency
1988, 127 pages [Technical Report 288]

A1542
Bibliography for a data base for a review of the
scientific and technical considerations related
to the dumping at sea of radioactive waste:
part II, complete bibliography
Vienna: International Atomic Energy Agency
1984, 148 pages

A1543
Biological processes and wastes in the ocean
Capuzzo J M and Kester D R (editors)
Melbourne, Florida: Kreiger
1987, 280 pages
[Oceanic processes in marine pollution, I]

A1544
Co-ordinated research and environmental
surveillance programme related to sea disposal
of radioactive waste: progress report at end of
1983
Paris: OECD Nuclear Energy Agency
1984, 55 pages

A1545
Deep sea waste disposal [papers presented at
5th international ocean disposal symposium]
Somerset NJ, Chichester: John Wiley
1985, 432 pages [Wastes in the Ocean, 5]

A1546
Disposal of chemical waste in the marine
environment: implicatins of the international
dumping conventions: international symposium;
Rotterdam 11–12 September 1980
Hueck van der Plas, E H (editor)
Oxford: Pergamon Press
1981 [Chemosphere, Volume 10 (6) pages
505–715]

A1547
Disposal of nuclear waste at sea: a review of
development and industrial opportunities
Spagni D A et al
London: UEG
1986, 106 pages [UEG Publication UR29]

A1548
Disposal of radioactive and other hazardous
wastes, workshop; Stockholm, January 1989
*Stockholm: Swedish National Institute of
Radiation Protection*
1989

A1549
Disposal of radioactive waste in seabed
sediments, conference; Oxford, 20–21
September 1988
Freeman T J (editor)
*London: Graham & Trotman for the Society of
Underwater Technology*
1989, 333 pages
[Advances in Underwater Technology, Ocean
Science & Offshore Engineering, Volume 18

A1550
Dumping at Sea Act 1974
UK Parliament
London: HMSO
1974, 13 pages [Chapter 20]

A1551
Dumping sewage sludge at sea
Macquitty M
London: Greenpeace UK
1987, 44 pages [Research Report 8]

A1552
Dumping of waste at sea
*House of Lords Select Committee on the
European Communities*
London: HMSO
1986, 234 pages
[HL 219, Session 1985–86 17th Report]

A1553
Energy wastes in the ocean [papers presented at 4th international ocean disposal symposium]
Somerset NJ, Chichester: John Wiley
1985, 862 pages [Wastes in the ocean, 4]

A1554
The field assessment of effects of dumping wastes at sea: 12–the disposal of sewage sludge, industrial wastes and dredged spoils in Liverpool Bay
Norton M G et al
Lowestoft: MAFF Directorate of Fisheries Research
1984, 50 pages [Fisheries Research Technical Report No 76]

A1555
Geotechnical engineering of ocean waste disposal, symposium; Orlando FL, 26 January 1989
Demars K R and Chaney R C (editors)
Philadelphia, PA: American Society for Testing & Materials
1990, 310 pages [STP 1087]

A1556
Handling and management of hazardous materials and waste
Allegri T H
London: Chapman & Hall
1986, 458 pages

A1557
Inter-governmental conference on the dumping of wastes at sea; London, 30 October–13 November 1982: final act, including the convention on the prevention of marine pollution by dumping of wastes and other matter, 1982 edition
London: International Maritime Organization
1982, 32 pages

A1558
Integrated modelling as a tool for assessment of environmental impacts from dumping of polluted dredging sludge (presented at the International Seminar on the Environmental Aspects of Dredging Activities, Nantes, France, 27 November–1 December 1989)
van Pagee J A and Winterwerp J C
Delft: Delft Hydraulics
1990, 28 pages [Publication No 434]

A1559
7th international ocean disposal symposium; Wolfville, Nova Scotia, 21–25 September 1987
Melbourne FL: Florida Institute of Technology
1987, 111 pages [program & abstracts]

A1560
Legislation and testing requirements relating to the use and disposal of oil and chemical products in the North Sea offshore industry: 1985 environmental seminar; Brixham, 24–27 June 1985
Brixham: ICI
1985, 106 pages

A1561
Marine waste management: science & policy
Champ M A and Park, R Kilho (editors)
Malabar FL: Krieger
1989, 341 pages [Oceanic processes in marine pollution, 3]

A1562
8th Ocean disposal international symposium; Dubrovnik, 9–13 October 1989 (program and abstracts)
Melbourne FL: Florida Institute of Technology
1989, vii + 134 pages

A1563
Ocean disposal of radioactive waste by penetration emplacement
Ove Arup & Partners, editors
London: Graham & Trotman
1986, 256 pages

A1564
Ocean dumping and marine pollution: geological aspects of waste disposal at sea
Palmer H D and Gross M G (editors)
New York, London: Academic Press
1979, 268 pages

A1565
Ocean dumping of industrial wastes: based on 1st international ocean dumping symposium, University of Rhode Island
Ketchum B H (editor)
New York, London: Plenum Publihsing
1981, ix + 525 pages

A1566
An oceanographic model for the dispersion of wastes disposed of in the deep sea
Joint Group of Experts on the Scientific Aspects of Marine Pollution–GESAMP
Vienna: International Atomic Energy Agency
1983, 182 pages [Report & Studies No 19]

A1567
Offshore oil and gas production waste
characteristics, treatment methods, biological
effects and their applications to Canadian
regions
Arctic Laboratories et al
Ottawa: Environment Canada Environmental
Protection Service
1984, 1 volume

A1568
Offshore ship and platform incineration of
hazardous wastes
Hooper G V (editor)
Park Ridge, NJ: Noyes Data
1981, 468 pages

A1569
Oily water discharges: regulatory, technical and
scientific considerations: IOE seminar;
Edinburgh 28–29 June 1978
Johnston C S and Morris R J (editors)
Barking: Elsevier Applied Science
1980,xv + 225 pages

A1570
Physicochemical processes and wastes in the
ocean
O'Connor T P (editor) et al
Melbourne FL: Krieger
1987, 17 chapters [Oceanic processes in
marine pollution, 2]

A1571
Public waste management and the ocean
choice: ocean disposal of public wastes,
symposium; technology & policy for the future,
April 1985
Stolzenbach K D and Kildow J T (editors)
Cambridge MA: MIT Sea Grant Program
1986, 280 pages [MITSG 85-36]

A1572
Report on disposal of waste at sea 1986 + 1987
Ministry of Agriculture, Fisheries & Food
London: HMSO
1989

A1573
Review on sewage sludge disposal at sea
London: Oslo Commission
1989, 84 pages

A1574
Scientific monitoring strategies for ocean waste
disposal
Hood D W (editor) et al
Malabar, FL: Krieger
1989, xv + 286 pages
[Oceanic processes in marine pollution, 4]

A1575
Seabed stability of debris
Wimpey Offshore
London: HMSO
1990 [Report OTH 89/313]

A1576
Thermal discharges in the marine environment
Joint Group of Experts on the Scientific
Aspects of Marine Pollution (GESAMP)
Rome: Food & Agriculture Organization of the
UN (HMSO)
1984, 44 pages [Reports & Studies No 74]

A1577
Use of the ocean for man's wastes:
engineering and scientific aspects
National Research Council, Marine Board
Washington DC: National Academy Press
1982, 300 pages

A1578
Waste discharge into the marine environment:
principles and guidelines for the Mediterranean
Action Plan
Oxford: Pergamon Press (with Institute of
Sanitary Engineering, Polytechnic of Milan,
Italy)
1982, xv + 422 pages

A1579
Waste disposal in the oceans: minimising
impact, maximising benefits
Soule D and Walsh D (editors)
Guildford: Westview Press
1984, 175 pages

A1580
Wastes in marine environments
Office of Technology Assessment Task Force
Cambridge MA, London: Hemisphere
1988, xi + 312 pages

RENEWABLE ENERGY RESOURCES

A1581

Assessment of wave power available at key United Kingdom sites: a description of work undertake in the Department of Energy's Wave Energy Programme
Crabb J A
Wormley: Institute of Oceanographic Sciences
1984, 113 pages [Report 186]

A1582

Developments in tidal energy, proceedings of 3rd conference on tidal power; London 28–29 November 1989
Institution of Civil Engineers
London: Thomas Telford
1990, 334 pages

A1583

The development of wave power: a techno-economic study
East Kilbride: National Engineering Laboratory
1975

A1584

Emerging energy technology symposium; Houston, 26–30 January 1992
Fairfield NJ: American Society of Mechanical Engineers
1991, 128 pages [PD Vol 41]

A1585

Energy for islands, conference; London, 10 December 1987
Society for Underwater Technology
London: Graham & Trotman
1988, vii + 135 pages [Advances in Underwater Technology, Vol 17]

A1586

Energy from ocean waves, colloquium; Bristol, 26–28 September 1988
Bristol: University of Bristol
1988 [Euromechanics Colloquium 243]

A1587

Energy from the waves, 2nd edition
Ross D
Oxford: Pergamon
1981, xix + 14 pages

A1588

Energy resources and the environment
Blunden J and Reddish A (editors)
London: Hodder & Stoughton
1991, 320 pages

A1589

Hydrodynamics of ocean wave-energy utilization: IUTAM symposium; Lisbon, Portugal, 1985
Evans D V and Falco A F de O (editors)
Berlin: Springer Verlag
1986, 452 pages

A1590

8th Miami international conference on alternative energy sources, proceedings
Vezirolgu T N (editor)
Washington, DC: Hemisphere
1989, 2 volumes

A1591

Ocean energies: resources for the future
Charlier R H and Justus J R
Amsterdam: Elsevier Science Publishers
due 1992 [Elsevier Oceanography Series]

A1592

Ocean energy recovery, proceedings of ICOER '89
Krock Hans-Jurgen (editor)
New York: American Society of Civil Engineers
1990, 367 pages

A1593

Ocean engineering for ocean thermal energy conversion
National Research Council, Marine Board
Washington DC: National Academy Press
1982, 69 pages

A1594

Ocean thermal energy conversion: legal, political and institutional aspects
Knight H G et al
Lexington Books (Gower Publishing)
1977, 272 pages

A1595

Ocean thermal energy conversion
Lavi A (editor)
Oxford: Pergamon Press
1980, 80 pages
[special issue, Energy, Volume 5 (6)]

A1596
Ocean wave energy conversion
McCormick M E
New York, Chichester: John Wiley & Sons
1981, 300 pages

A1597
Offshore mean wind profile
Wills J A B, Grant A & Boyack C F (for Dept of Energy)
London: HMSO
1986, 48 pages

A1598
Offshore wind-energy resource around the United Kingdom
White P A
Bracknell: Meteorological Office
1987, 88 pages [ETSU-WN-05621/2]

A1599
Renewable energy sources
Laughton M A
London: Elsevier Applied Science
1990, vii + 168 pages
[Watt Committee on Energy Report No 22]

A1600
Renewable energy technologies: their applications in developing countries a study by the Beijer Institute, The Royal Swedish Academy of Sciences
Kristoferson L A and Bokalders V
Oxford: Pergamon
1986, xviii + 319 pages

A1601
Report on offshore wind energy assessment: phase IIB study; volume 1, executive summary
Taywood Engineering et al
London: Department of Energy
1985, lv

A1602
Report on the West Sole wind structure project 1986
Wills J A B
British Maritime Technology
1986, l volume

A1603
Strategic review of the renewable energy technologies: an economic assessment
Energy Technology Support Unit
London: HMSO
1982, 2 volumes

A1604
Taking power from water: water energy technology in Britain
London: Department of Energy
1989, 12 pages

A1605
Tidal power
Baker A C
Hitchin: Institution of Electrical Engineers (Peter Peregrinus)
1991, 260 pages

A1606
Water for energy: 3rd international symposium on wave, tidal, OTEC, and small scale hydro energy; Brighton 14–16 May 1986
Cranfield: BHRA
1986, 409 pages

A1607
Wave and tidal energy: 2nd international symposium; Cambridge, 23–25 September 1981
Bedford: BHRA Fluid Engineering
1981, 449 pages

A1608
Wave energy: a design challenge
Shaw R
Chichester: Ellis Horwood
1982, 202 pages

A1609
Wave energy devices, conference; Coventry, 30 November 1989
Duckers L J (editor)
London: Solar Energy Society
1990, 80 pages

A1610
Wave energy utilisation, 2nd international symposium; Trondheim, 22–24 June 1982
Berge H (editor)
Trondheim: TAPIR
1982, 461 pages + 1 annexe

A1611
Wave power: the use of water as an alternative source of energy
Simeons C
Oxford: Pergamon Press
1980, 200 pages

OCEANOGRAPHY AND METEOROLOGY

General

A1612
Acoustics and the seabed: conference; Bath, 6–8 April 1983
Pace N G (editor)
Bath: Bath University Press
1983, 436 pages

A1613
Acoustic remote sensing of the atmosphere and oceans
Weill A (editor)
Amsterdam: Elsevier Science Publishers
1986, li + 232 pages [Atmospheric Research, Volume 20 number 2/3]

A1614
Advances in meteorological services to the offshore and shipping industries (paper presented 27 March 1990 to the Joint Offshore Group of IME and RINA)
Hopkins J S
London: Institute of Marine Engineers
1990, 13 pages

A1615
The application of remote sensing techniques to the determination of marine environmental parameters
Seaconsult Ltd for the Department of Energy
London: HMSO
1988, ix + 162 pages [Report OTI 88 531]

A1616
Applied dynamics of ocean surface waves
Mei C C
Singapore: World Scientific
1989

A1617
Atlas of tidal elevations and currents around the British Isles
Howarth M J
London: HMSO
1990, lv (various pagings)

A1618
The British seas: an introduction to the oceanography and resources of the north-west European continental shelf
Hardisty J
London: Routledge
1990, xiv + 272 pages

A1619
Case studies in oceanography and marine affairs
Open University Course Team
Oxford: Pergamon
1991

A1620
Civil marine applications of satellite date: final report
Fleet, Surrey: Earth Observation Sciences Ltd
1991, 136 pages

A1621
Comparisons of near-surface currents measured by HF radar and current meters
Griffiths C R et al
London: HMSO
1990 [Report OTH 90 330]

A1622
Current measurements offshore: conference; London, 17 May 1984
London: Society for Underwater Technology
1984, ten papers

A1623
The deep sea bed: its physics, chemistry and biology, meeting; London, 5–6 April 1989
Charnock H (editor) et al
London: The Royal Society
1990

A1624
Directional wave measurements: Southern North Sea
Clayson C H
Wormley: Institute of Oceanographic Sciences Deacon Laboratory
1989, 192 pages [Report No 266]

A1625
Directional ocean wave spectra a record of the
Labrador Sea extreme waves experiment,
symposium; Baltimore, 18–20 April 1989
Beal R C (editor)
Baltimore MD: Johns Hopkins University Press
1991, xv + 218 pages

A1626
Dynamics of ocean tides
Marchuk G I and Kagan B A
Dordrecht: Kluwer
1989, 340 pages
[Oceanographic Sciences Library]

A1627
Dynamics of the coupled atmosphere and
ocean
Charnock H and Philander S G H (editors)
London: The Royal Society
1989, 163 pages [Phil Trans A 329]

A1628
Energy industry Metocean data around the UK
Metocean Consulting
London: HMSO
1986, 68 pages [OTH 86 227]

A1629
Environmental data atlas and catalogue for the
Norwegian Shelf and adjacent waters
Saetra H J et al
Trondheim: OTTER Group
1981, 940 pages [Report STF88 - F81041]

A1630
Environmental data for the Barents Sea
Ø Johansen et al
Trondheim: SINTEF
1989, 33 pages [Report STF 60 A89094]

A1631
Environmental parameters on the UK
Continental Shelf
Noble Denton & Associates
London: HMSO
1985, 31 pages, 8 maps [OTH 84 201]

A1632
Estimating wave climate parameters for
engineering applications
Carter D J T et al
London: HMSO
1986, vii + 130 [OTH 86 228]

A1633
Feasibility study on marine environmental
measurement
Metocean Consultancy
London: Department of Trade & Industry
1989, various pagings
[Resources from the Sea Programme]

A1634
Fernerkundung des Ozeans: probleme ünd
Moglichkeiten der Fernkundung des Ozeans
mit optischen Mitteln
Zimmermann G
Berlin: Akademie Verlag
1991, 419 pages

A1635
Global wave statistics
British Maritime Technology
Woking: Unwin Brothers
1987, 656 pages

A1636
IOC Committee on Ocean Processes and
Climate, 3rd session; Paris, 27–29 June 1989
Paris: UNESCO
1990, 43 pages

A1637
An introduction to marine science, 2nd edition
Meadows P S and Campbell J I
Glasgow: Blackie & Son
1987, ix + 285 pages

A1638
Introductory oceanography, 4th edition
Thurman H V
Columbus, Ohio: Charles E Merrill
1985, viii + 503 pages

A1639
JOSNMOD '82: workshop of the Joint North
Sea Modelling Group on mathematical models
of the North Sea and surrounding Continental
shelf areas; Edinburgh 1982
Dyke P P G and Heeps H S (editors)
Oxford: Pergamon Press
1984, 136 pages

A1640
List of sea state parameters
International Association for Hydraulic Research
Rotterdam: A A Balkema
1986, 50 pages

A1641
Making MAREP work: MAREP national
conference; Seattle, June 1985
*Baton Rouge: Louisiana Sea Grant College
Program for National Sea Grant College
Program, National Weather Service*
1986, 103 pages

A1642
A manual of chemical and biological methods
for seawater analysis
Parsons T R et al
Oxford: Pergamon Press
1984, xii + 173 pages

A1643
MDS '86: Marine data systems, international
symposium; New Orleans, 30 April–2 May 1986
Washington DC: Marine Technology Society
1986, xiv + 611 pages

A1644
Marine climate, weather and environment
effects of weather and changing climates on
the ocean and its fishery resources
Laevastu T
Oxford: Fishing News Books
due 1992

A1645
Marine environmental measurement, feasibility
study
Metocean Consultancy
London: Dept of Trade & Industry
1987, vaious pagings
[Resources from the Sea Programme]

A1646
Marine geochemistry
Chester R
London: Unwin Hyman
1990, 720 pages

A1647
Marine physics
Dera J
Amsterdam: Elsevier Science Publishers
1991, 520 pages

A1648
Meteorological effects on currents northwest of
the United Kingdom
London: HMSO
1990 [Report OTH 89 306]

A1649
Methods of seawater analysis, 2nd edition
Grasshoff K (editor)
Basel: VCH
1983, xxviii + 419 pages

A1650
Metocean parameters: parameters other than
waves
Techword Services for Department of Energy
London: HMSO
1990, vi + 151 pages [Report OTH 89 229]

A1651
Metocean parameters: wave parameters
Department of Energy
London: HMSO
1990 [Report OTH 89 300]

A1652
Modelling the offshore environment
conference; London 1–2 April 1987
*London: Graham & Trotman (for Society for
Underwater Technology)*
1987, 20 papers
[Advances in Underwater Technology, 11]

A1653
Near-surface current measurements in the
North Sea
Ellett D J et al
London: HMSO
1991, iv + 13 pages [Report OTR 91 331]

A1654
New technology for ocean science
Woods J
London: MTD Ltd
1991, 14 pages
[Mike Adye Memorial Lectures 2] [MTD 91/100]

A1655
North Sea climate, based on observations from
ships and lightvessels
Korevaar C G
London, Dordrecht: Kluwer
1990, xi + 137 pages

A1656
North Sea–estuaries interactions, 18th EBSA
symposium; Newcastle upon Tyne, 29
August–2 September 1988
McLusky D S et al
*Dordrecht, London: Kluwer Academic
Publishers*
1990, ix + 221 pages [Reprinted from
Hydrobiologia Volume 195, 1990]

A1657
North Sea dynamics: international symposium;
Hamburg, 31 August–4 September 1981
Lenz W and Sundermann J (editors)
Berlin: Springer-Verlag
1982, xvii + 693 pages

A1658
The North Sea environmental guide
Ledbury: Oilfield Publications
1984, 68 pages

A1659
North Sea seismicity: summary report
Principia Mechanica
London: HMSO
1986, 5 + 73 pages [OTH 86 219]

A1660
Ocean chemistry and deep-sea sediments
Open University Course Team
Oxford: Pergamon
1989, 134 pages

A1661
Ocean circulation
Open University Course Team
Oxford: Pergamon
1989, 238 pages

A1662
Ocean science, 2nd edition
Stowe K
Chichester: John Wiley & Son
1983, 673 pages

A1663
Ocean variability and acoustic propagation,
workshop; La Spezia, 4–8 June 1990
Potter J and Warn-Varnas A (editors)
Dordrecht: Kluwer
1991, 620 pages

A1664
Ocean waves mechanics, computational fluid
dynamics and mathematical modelling
Rahman M
Southampton: Computational Mechanics
1990, 986 pages

A1665
Oceanography
Ingmanson D E and Wallace W J
Belmont, CA: Wadsworth Publishing
1989

A1666
Oceanography: a view of the earth, 4th edition
Gross M Grant
Englewood Cliffs: Prentice-Hall
1987, xi + 406 pages

A1667
Oceanography: perspectives on a fluid earth
Neshyba S
New York, Chichester: John Wiley & Sons
1987, 506 pages

A1668
Offshore and coastal modelling
Dyke P P G (editor) et al
Berlin: Springer Verlag
1985 [Coastal & Estuarine Studies, 12]

A1669
Offshore mean wind profile
Wills J A B et al
London: HMSO
1986, 43 pages [OTH 86 226]

A1670
Operational analysis and prediction of ocean
wind waves
Khandekar M L
Berlin, London: Springer
1989, viii + 214 pages
[Coastal & Estuarine Studies, 33]

A1671
Physical optics of ocean water
Shifrin K S
New York: American Institute of Physics
1988, 280 pages

A1672
Principles of underwater sound
Urick R
New York, Maidenhead: McGraw-Hill
1983

A1673
Regulations on environmental data in the
petroleum activities
Stavanger: Norwegian Petroleum Directorate
1990 [in Norwegian and English]

A1674
Scientific Committee on Oceanic Research,
proceedings; Halifax, July 1990
*Halifax: Dalhousie University Dept of
Oceanography*
1990, 78 pages

A1675
Seabed pockmarks and seepages: impact on geology, biology and the marine environment
Horland M and Judd A G
London: Graham & Trotman
1988, xii + 293 pages

A1676
Seawater: its composition, properties and behaviour
Open University Course Team
Oxford: Pergamon
1989, 165 pages

A1677
Seismic monitoring of the North Sea
Marrow P C (British Geological Survey)
London: HMSO
1992, vi + 53 pages [OTH 90 323]

A1678
Seismicity and seismic risk in the offshore North Sea area: NATO Advanced Research Workshop; Utrecht, 1–4 June 1982
Dordrecht: D Reidel (Kluwer)
1983, xxv + 420 pages

A1679
Seismicity of the British Isles and the North Sea
Ambraseys N and Melville C
London: Imperial College, Marine Technology Centre
1983, xxv + 132 pages

A1680
Standard and reference materials for marine science
Paris: UNESCO
1990, various pagings [Intergovernmental Oceanographic Commission, Manuals & Guides No 21]

A1681
Statistical analysis of measured directional wave and current data
Karunakaran D and Spidsøe N
Trondheim: SINTEF
1986, 48 pages [Report No 416]

A1682
Three-dimensional models of marine and estuarine dynamics: 18th International Liège Colloquium on Ocean Hydrodynamics
Nihoul J C J and Jamart B M (editors)
Amsterdam: Elsevier Science Publications
1987,xii + 630 pages
[Elsevier Oceanography Series 45]

A1683
Tidal hydrodynamics [based on papers presented at the conference on tidal hydrodynamics; Gaithersburg MD, 15–18 November 1988]
Parker B B (editor)
New York, Chichester: John Wiley
1991, xv + 883 pages

A1684
Tides, surges & mean sea-level: a handbook for engineers & scientists
Pugh D T
Chichester: Wiley
1987, 472 pages

A1685
Time series analysis of ocean waves
Dommermuth D G
Cambridge MA: Massachusetts Institute of Technology, Sea Grant College Program
1986, 1v [Report MITSG 87-8TN]

A1686
Transport atlas of the southern North Sea
de Ruijter W P M et al
Delft: Delft Hydraulics
1987, 33 pages + floppy disk

A1687
Water mass transformations in the Nordic Seas: the ICES deep water project, meeting; Copenhagen, October 1986
Clarke R Allyn (editor)
Deep Sea Research
1990, Volume 37(9A) pages 1383-1511

A1688
Water wave kinematics, workshop; Molde, Norway, 22–25 May 1989
Torum A and Gudmestad O T (editors)
Dordrecht, London: Kluwer
1990, xiv + 771 pages
[Nato Advanced Sciences Institutes Series E: Applied Sciences 178]

A1689
Wave climate atlas of the British Isles
Draper L
London: HMSO
1991, ii + 22 pages [OTH 89-303]

A1690
Wave predictions by an empirical ray tracing model
Elliott A J
London: HMSO
1988, 53 pages [OTI 87-502]

A1691
Wave prediction in deep water and at the
coastline
Southgate H N
Wallingford: Hydraulics Research
1987 [Research Report SR114]

A1692
Waves in the North East Atlantic
Barratt M J (BMT Fluid Mechanics Ltd)
London: HMSO
1991, 56 pages [OTI 90 545]

A1693
Waves in ocean engineering: measurement,
analysis and interpretation
Tucker M
Chichester: Ellis Horwood
1991, 431 pages

A1694
Waves recorded at ocean weather station LIMA
Institute of Oceanographic Sciences, Taunton
London: HMSO
1985, 75 pages [OTH 84 204]

A1695
Waves, tides and shallow-water processes
Open University Course Team
Oxford: Pergamon
1989, 187 pages

MARINE TECHNOLOGY

GENERAL

A1696
Advanced ocean technologies and the future of maritime industries, symposium; Cambridge, Mass , 5–6 October 1987
Cambridge MA: MIT Industrial Liaison Program

A1697
Arctic energy resources: Comité Arctique International conference; Oslo, 22–24; September 1982
Rey L (editor) and Behrens C
Amsterdam, Oxford: Elsevier Science Publishers
1983, x + 366 pages
[Energy Research Volume 2]

A1698
Arctic Offshore Technology Conference; Calgary, 6–9 October 1984
Calgary: AOTC
1984, 26 papers

A1699
Arctic Offshore Technology Conference; Anchorage, 3–5 September 1985
Calgary: AOTC
1985, 28 papers

A1700
10th Australasian conference on coastal and ocean engineering; Auckland, 2–6 December 1991
Auckland: University of Auckland
1991, 571 pages

A1701
CADMO '86; Washington DC, September 1986
Murthy T D S (editor)
Southampton: Computational Mechanics
1986, 657 pages

A1702
CADMO '88; marine and offshore computer applications, 2nd conference; Southampton, September 1988
Murthy T K S et al (editors)
Southampton: Computational Mechanics
1988, 812 pages

A1703
CADMO '91: 3rd conference on computer-aided design, manufacture and operation in the marine and offshore industries; Florida, 1991
Southampton: Computational Mechanics
1991, 419 pages

A1704
Civil engineering in the arctic offshore: conference; San Francisco, 25–27 March 1985
Bennett F L and Machemehl (editors)
New York: American Society of Civil Engineers
1985, ix + 1259 pages

A1705
Civil/Navy ocean engineering technology workshop; September 1982
Washington DC: National Research Council, Marine Board
1982, 184 pages

A1706
ConOff: IRO/MaTS international construction offshore conference; Amsterdam, 20–21 February 1986
IRO-Journal 10c jaargang (9), 76 pages
[special issue]

A1707
Computer modelling in ocean engineering: problems and solutions in coastal and offshore systems, conference; Venice, 19–21 September 1988
Schrefler B A and Zienkiewicz O C (editors)
Rotterdam: A A Balkema
1988, 800 pages

A1708
Computer models in offshore engineering, specialty conference; Halifax, Nova Scotia, 1984
Montreal: Canadian Society for Civil Engineering
1984, 488 pages

A1709
A course in ocean engineering
Gran S
Amsterdam: Elsevier Science Publishers
due 1992

A1710
Current practices and new technololgies in ocean engineering
McGuinness T and Smith H H
Fairfield, NJ: American Society of Mechanical Engineers
1986, [OED Volume 11]

A1711
Current practices and new technologies in ocean engineering, 1987: presented at the 10th Energy Sources Technology Conference; Dallas, 15–18 February 1987
Wolfe G K (editor)
Fairfield NJ: American Society of Mechanical Engineers
1987 v + 162 pages [OED Volume 12]

A1712
Current practices and new technologies in ocean engineering, 1988: presented at the 11th Energy Sources Technology Conference; New Orleans, 10–13 January 1987
Wolfe G K and Chang P Y (editors)
Fairfield NJ: American Society of Mechanical Engineers
1988, v + 177 pages [OED Volume 13]

A1713
Decade of the oceans: 16th annual conference of the Marine Technology Society; Washington DC, 6–8 October 1980
Washington DC: Marine Technology Society
1980, 607 pages

A1714
Deepwater mooring and drilling: winter annual meeting of ASME, New York, 1979
Lou Y K (editor)
New York: ASME
1979, 212 pages [OED Volume 7]

A1715
Defence oceanology international 91: extending the reach; Brighton, 6–8 March 1991
Kingston-upon-Thames: Spearhead Exhibitions Ltd
1991, 91 papers

A1716
Development of arctic offshore technology
Enkvist E and Eranti E
Finland: Technology Development Centre (TEKES)
1990, 95 pages

A1717
European offshore mechanics, lst symposium; Trondheim, 20–22 August 1990
Chung J S et al (editors)
Golden CO: International Society of Offshore & Polar Engineers
1990, xi + 564 pages

A1718
International offshore and polar engineering, 1st conference; Edinburgh, 11–16 August 1991
Chung J S et al (editors)
Golden CO: International Society of Offshore & Polar Engineers
1991, 4 volumes

A1719
Handbook of coastal and ocean engineering, volume 2: offshore structures, marine foundations, sediment processes and modelling
Herbich J B (editor)
Houston, London: Gulf Publishing Company
1991

A1720
Harsh environment and deep water handbook
Tulsa: PennWell Publishing
[supplement to the November1985 issue of OFFSHORE]

A1721
INCOE '81: 1st Indian conference on ocean engineering; Madras, 18–20 February 1981
Madras: Indian Institute of Technology
1981, various pagings, 2 volumes

A1722
3rd Indian conference on ocean engineering; Bombay, 11–13 December 1986
Bombay: Indian Institute of Technology
1986,2 volumes

A1723
Marine safety conference; Glasgow, 7–9
September 1983
*Glasgow: University of Glasgow Department of
Naval Architecture and Ocean Engineering*
1983, 31 papers

A1724
Marine technology in the 1990s: meeting,
17–18 March 1982
London: The Royal Society
1982, 202 pages

A1725
Marine technology reference book
Morgan N (editor)
London: Butterworth
1990, various pagings

A1726
Marinetech China '85 conference; English
language edition: Volume 3, offshore
technology; Shanghai,4–10 December 1985
Hong Kong: Lloyd's of London Press (Far East)
1985

A1727
Modelling the offshore environment:
conference; London, 1–2 April 1987
London: Society for Underwater Technology
1987, 208 pages

A1728
MTS '91: an ocean cooperative: industry,
government and academia; New Orleans,
10–14 November 1991
Washington DC: Marine Technology Society
1991

A1729
North Sea horizons: technology in the quest for
oil and gas
Offshore Supplies Office
London, Glasgow: Department of Energy
1981, 57 pages

A1730
Numerical methods in offshore engineering
Zienkiewicz O C (editor)
Chichester: John Wiley & Sons
1978, xii + 582 pages

A1731
Ocean energy: 15th annual conference of the
Marine Technology Society; New Orleans,
10–12 October 1979
Washington DC: Marine Technology Society
1979, xiii + 419 pages

A1732
Oceanology International '84; Brighton, 6–9
March 1984
Kingston-upon-Thames: Spearhead Exhibitions
1984, 47 papers

A1733
Oceanology [Oceanology International '86;
Brighton, 4–7 March 1986]
London: Graham & Trotman
1986, 486 pages
[Advances in Underwater Technology, Ocean
Science and Offshore Engineering, Volume 6]

A1734
Oceanology '88 [Oceanology International '88;
Brighton, 8–11 March 1988]
London: Graham & Trotman
1988, 304 pages
[Advances in Underwater Technology, Ocean
Science and Offshore Engineering, Volume 16]

A1735
Oceanology '90: Brighton, 6–9 March 1990
Kingston-upon Thames: Spearhead Exhibitions
1990, 4 volumes

A1736
Oceans '87: the oceans, an international
workplace; Halifax, Nova Scotia, 28
September–1 October 1987
Piscataway NJ: IEEE
1987, 5 volumes

A1737
Oceans '88: a partnership of marine interests;
Baltimore, 31 October–2 November 1988
Piscataway NJ: IEEE
1989, 4 volumes

A1738
Oceans '89: addressing methods for
understanding the global ocean; Seattle, 18–21
September 1989
New York: IEEE
1989, 6 volumes

A1739
Oceans '90; Washington DC, 24–26
September 1990
Piscataway NJ: IEEE
1990 XX + 604 pages

A1740
Oceans '91; Honolulu, Hawaii, November 1991
Piscataway NJ: IEEE
1991

A1741
Offshore and onshore engineering practices
compared: Subject Group for Oil and Natural
Gas Production (SONG) seminar; Glasgow,
21–22 November 1984
Rugby: Institution of Chemical Engineers
1984, 12 papers + loose paper

A1742
OMAE 1987; offshore mechanics and arctic
engineering, 6th symposium; Houston, 1–6
March 1987
Chung J S (editor)
New York: ASME
1987, 4 volumes

A1743
OMAE 1988; offshore mechanics and arctic
engineering, 7th international conference;
Houston, 7–12 February 1988
New York: ASME
1988, 6 volumes

A1744
OMAE 1989, offshore mechanics and arctic
engineering, 8th international conference; The
Hague, 19–23 March 1989
New York: ASME
1989, 6 volumes

A1745
OMAE 1990; offshore mechanics and arctic
engineering, 9th international conference;
Volume IV, arctic/polar technology; Houston,
18–23 February 1990
New York: ASME
1990, ix + 339 pages

A1746
OMAE 1991; offshore mechanics and arctic
engineering; 10th international conference;
Volume IV, arctic/polar technology; Stavanger,
23–28 June 1991
New York: ASME
1991, 287 pages

A1747
OMAE lst speciality symposium on offshore
and arctic frontiers: 9th annual energy sources
technology conference; New Orleans, 23–27
February 1986
Salama M M (editor)
New York: ASME
1986, ix + 492 pages

A1748
Offshore mechanics and cold ocean
engineering symposium; St John's, 13–15 June
1983
*St John's: Memorial University of
Newfoundland, Continuing Education Centre*
1983, 12 papers

A1749
Offshore technology conference, 19th;
Houston, 27–30 April 1987
Richardson TX: OTC
1987, 4 volumes

A1750
Offshore technology conference, 20th;
Houston, 2–5 May 1988
Richardson TX: OTC
1988, 4 volumes

A1751
Offshore technology conference, 21st;
Houston, 1–4 May 1989
Richardson TX: OTC
1989, 4 volumes

A1752
Offshore technology conference, 22nd;
Houston, 7–10 May 1990
Richardson TX: OTC
1990, 4 volumes

A1753
Offshore technology conference, 23rd;
Houston, 6–9 May 1991
Richardson TX: OTC
1991, 4 volumes

A1754
Offshore technology conferences: alphabetical
subject index: 1969–1988 inclusive, cumulated
Richardson, TX: OTC
1988, 432 pages

A1755
Pacem in Maribus XVI: ocean technology,
development, training and transfer conference;
Halifax, Canada, August 1988
Vandermuelen J and Walker S
Oxford: Pergamon
1990: 522 pages

A1756
Pacific/Asia offshore mechanics, 1st
symposium; Seoul, 24–28 June 1990
Chung J S et al (editors)
Golden CO: International Society of Offshore &
Polar Engineers
1990, 3 volumes

A1757
Pioneering projects in the arctic: conference;
Stavanger, 15–16 November 1983
Stavanger: ONS Foundation
1983, 5 papers

A1758
POAC '87: 9th international conference on port
and ocean engineering under arctic conditions;
Fairbanks, Alaska, 17–21 August 1987
Fairbanks AK: Geophysical Institute, University
of Alaska
1988, 2 volumes

A1759
POAC '89: 10th international conference on
port and ocean engineering under arctic
conditions; Lulea, 12–16 June 1989
Axelesson K B E and Fransson L A (editors)
Lulea: Lulea University of Technology
1989, 2 volumes

A1760
POAC '91: 11th international conference on
port and ocean engineering under arctic
conditions; St John's, Canada, 24–28
September 1991
Muggeridge D B et al (editors)
St John's, Newfoundland: Memorial University
1991, 2 volumes

A1761
Polartech '86: international offshore and
navigation conference; Helsinki, 27–30 October
1986
Espoo: Technical Research Centre of Finland
(VTT)
1986, 3 volumes + 2 loose papers

A1762
Polartech '88: international conference on
technology for polar areas; Trondheim, 15–17
June 1988
Hansen A and Storm J F (editors)
Trondheim: TAPIR
1988, 2 volumes

A1763
Polartech '90: international conference on
development and commercial utilization of
technologies in polar regions; Copenhagen,
14–16 August 1990
Horsholm: Danish Hydraulic Institute
1990, xiii + 759 pages

A1764
Responding to changes in sea level:
engineering implications
US National Research Council
Chichester: Wiley
1987, 160 pages

A1765
Risers, arctic design criteria, equipment
reliability in hydrocarbon processing: workbook
for engineers presented at 37th Petroleum
Mechanical Engineering Workshop and
conference; Dallas, 13–15 September 1981
New York: ASME
1981, 242 pages

A1766
Science and technology for a new oceans
decade, proceedings; Washington DC, 26–28
September 1990
Washington DC: Marine Technology Society
1990, 3 volumes

A1767
Subsea defence '85; London, 17–18 June 1985
Chipping Norton: Subsea Conferences
1985, 14 pages

A1768
Subsea Defence '86; Bristol, 28–29 May 1986
Chipping Norton: Subsea Conferences
1986, 12 papers

A1769
Subsea defence '87; Bournemouth, 20–21
October 1987
Chipping Norton: Hawkedon Partnership

A 1770
Techno-Ocean '88, symposium: Kobe, 16–18
November 1988
Tokyo: Techno-Ocean '88 Secretariat
1988, 2 volumes

A1771
Technology common to aero and marine
engineering, proceedings; London, 26–28
January 1988
*London: Graham & Trotman for Society for
Underwater Technology*
1988, xvi + 304 pages
[Advances in Underwater Technology, 15]

A1772
Technology for the arctic: special conference
topic A, Offshore Gøteborg '83; Sweden, 1–4
March 1983
Gøteborg: Swedish Trade Fair Foundation
1983, 5 volumes + 2 loose papers

A1773
UnderSeas Defense '87 conference; San
Diego, 2–5 November 1987
Spring Valley CA: West Star Productions
1987, 33 papers

A1774
UnderSeas defense '88: advanced technology
for undersea defense; San Diego, 3–6 October
1988
Spring Valley CA: West Star Productions
1988, 200 pages + 2 loose papers

A1775
UnderSeas defense '89: defense in depth; San
Diego 23–26 October 1989
Spring Valley CA: West Star Productions
1989, 143 pages

HYDRODYNAMICS

A1776
Action de l'environnement sur les ouvrages en
mer
ARAE
Paris: Éditions Technip
1989
[Guides Practiques sue les Ouvrages en Mer]

A1777
Advanced dynamics of marine structures
Hooft J P
New York, Chichester: John Wiley & Sons
1982, x + 345 pages

A1778
Analysis of uncertainties in environmental
loading on offshore structures: a summary
report
Carr P
London: HMSO
1989, iv + 30 pages

A1779
The applied dynamics of ocean surface waves
Mei C C
Chichester: John Wiley & Sons
1983, 760 pages

A1780
Breaking wave loads on immersed members of
offshore structures
Rainey R C T
London: HMSO
1991, ii + 28 pages [OTH 89 311]

A1781
The compilation of non-linear waves on a
current of arbitrary non-uniform profile
London: HMSO
1990 [OTH 90 327]

A1782
Development of a method to make use of
sensitivity studies and its application to analysis
of uncertainties in environmental loading on
offshore structures
Department of Energy
London: HMSO
1989 [OTI 88 537]

A1783
Dynamic loads on offshore structures,
bibliography
*Stuttgart: Information Centre for Regional
Planning and Building Construction (IRB)*
1989, 94 pages [ICONDA Bibliography No 46]

A1784
Dynamics of marine vehicles and structures in
waves; IUTAM symposium; Uxbridge, 24–27
June 1990
Price W G et al (editors)
Amsterdam, Oxford: Elsevier Science
1991, xiv + 338 pages

A1785
Environmental forces on offshore structures: a
state-of-the-art review
Soding H et al
Marine Structures
1990, Vol 3 (1), pages 59–81

A1786
Environmental forces on offshore structures
and their prediction: conference; London,
28–29 November 1990
Society for Underwater Technology
Dordrecht, London: Kluwer
1990, viii + 429 pages
[Advances in Underwater Technology, Ocean
Science.. Volume 26]

A1787
Estimation of fluid loading on offshore structures
Hogben N, et al
Feltham: National Maritime Institute [British
Maritime Technology Ltd]
1977, [NMI R11 OT-R-7614]

A1788
Field measurements of directional wave loads
on coastal structures
Heleran J van et al
Applied Ocean Research
1989, Volume 11(2) pages 58–74

A1789
Flow-induced vibration: proceedings of Ist
international conference; Bowness-on-
Windermere, 12–14 May 1987
King R (editor)
Berlin, London: Springer, for BHRA
1987, 600 pages

A1790
Fluid loading on fixed offshore structures
Barltrop N D P et al
London: HMSO
1990, two volumes [OTH 90 322]

A1791
Handbook of coastal and ocean engineering;
Volume 1: wave phenomena and coastal
structures
Herbich J B (editor)
Houston: Gulf Publishing
1990, 912 pages

A1792
Hydrodynamic forces
Naudascher E
Rotterdam: A A Balkema
1990, 300 pages
[Hydraulic Structures Design Manual 3]

A1793
Hydrodynamics in ocean engineering:
international symposium; Trondheim, 24–28
August 1981
Trondheim: Norwegian Hydrodynamic
Laboratories
1981, 2 volumes

A1794
Hydrodynamics of offshore structures:
mathematical theory and its applications in
structures
Chakrabarti S K
Southampton: CML Publications
1987, 480 pages

A1795
The hydrodynamics of waves and tides with
some applications
Rahman M
Southamtpon: Computational Mechanics
1989, 322 pages

A1796
An inventory of cylinder hydrodynamic
elements under wave conditions: a state of the
art review
Hauguel A et al
International Association for Hydraulic Research
1980, 33 pages [in English and French]

A1797
Marine hydrodynamics
Newman J N
Cambridge Mass, London: The MIT Press
1977, 402 pages

A1798
Mechanics of wave forces on offshore
structures
Sarpkaya T and Isaacson M
New York, Wokingham: Van Nostrand Reinhold
1981, xv + 651 pages

A1799
Mechanics of wave-induced forces on cylinders
Shaw T L (editor)
London: Pitman Publishing
1979, 750 pages

A1800
The nature of air flows over offshore platforms
Davies M E et al
Feltham: National Maritime Institute [British
Maritime Technology Ltd]
1977, 46 pages [NMI R14, OT-R-7726]

A1801
Ocean engineering wave mechanics
McCormick M E
New York, Chichester: John Wiley & Sons
1973, xix + 179 pages [Interscience]

A1802
Ocean waves mechanics, computational fluid
dynamics and mathematical modelling; 11th
conference
*Rahman M (editor, for Canadian Applied
Mathematical Society)*
Southampton: Computational Mechanics
1990, 968 pages

A1803
Principles of fluid flow and surface waves in
rivers, estuaries, seas and oceans
Rijn, L C van
Amsterdam: Aqua Publications
1991, 400 pages

A1804
Probabilistic offshore mechanics
Spanos P D (editor)
Southampton: Computational Mechanics
1985, 109 pages [reprinted from Journal of
Applied Ocean Research]

A1805
Probability distributions for environmental load
on offshore structures
*Lloyds Register of Shipping for Department of
Energy*
London: HMSO
1989 [OTI 88 504]

A1806
Sea loads on ships and offshore structures
Faltinsen O M
Cambridge: Cambridge University Press
1991 [Cambridge Ocean Technology Series]

A1807
Sensitivity of shallow water jacket structrues to
uncertainties in environmental loading
Atkins Oil & Gas Engineering Ltd
London: HMSO
1988, 74 pages [OTH 88 273]

A1808
Sensitivity study of member stresses to current
profile variations
London: HMSO
1989 [OTH 89 302]

A1809
Separated flow around marine structures:
international symposium; Trondheim, 26–28
June 1985
Trondheim: IAHR
1985, 398 pages

A1810
Ship and planar motions: international
workshop; 16–18 October 1983
Yeung R W (editor)
Berkeley: University of California
1984, 605 pages

A1811
Spectral analysis for engineers
Hearn G E and Metcalf A W
Sevenoaks: Edward Arnold
due 1992, 256 pages

A1812
Standardization load sequence for offshore
structures: wave action standard history
(WASH 1) final report
*Sonsino C M et al (Industrieanlagen
Betriebsgesellschaft)*
*Darmstadt: Fraunhofer Institut für
Betriebsfestigkeit*
1988, 91 pages [IABG TF-2347]

A1813
Summary of literature study on wave-current
interaction and their influence on fluid loading
on slender offshore structures
Karunakaren D
Trondheim: SINTEF
1991, 75 pages [STF15 A91009]

A1814
Tidal hydrodynamics
Parker B
New York, Chichester: Wiley
due January 1992, 912 pages

A1815
Uncertainties in the estimation of fluid loading
on offshore structures with special emphasis on
wind forces
Singh S
London: Institute of Marine Engineers
1989, 15 pages

A1816
Variability of environmental loads
Maes M A et al
Ocean Engineering
1988, Volume 15(2) pages 171–187

A1817
Water wave mechanics for engineers and scientists, 2nd edition
Dean R G and Dalrymple R A
Singapore, London: World Scientific
1991, xiii + 353 pages
[Advanced Series on Ocean Engineering Volume 2]

A1818
Waves and wave forms
Gran S
Amsterdam: Elsevier Science Publishers
due 1991 [Lectures in Ocean Engineering]

A1819
Wave force experiments at the Christchurch Bay Tower with simulated hard marine fouling
Bishop J R
London: HMSO
1988, 9 pages [OTH 89 541]

A1820
Wave forces on fixed offshore structures in short crested seas
Mitwally H and Novak M
Journal of Engineering Mechanics
1989, Volume 115(3) pages 636–655

A1821
Wave loads on offshore structure
Faltinsen O M
Annual Review of Fluid Mechanics
1990, Volume 22 pages 35–36

A1822
Wind loading of offshore structures: a summary of wind tunnel studies
Miller B L and Davies M E
Feltham: National Maritime Institute [now British Martitime Technology]
1982, 68 pages [NMI R136 OT-R-8225]

SURVEYING

A1823
Acquisition of marine surveying technologies; report of the UN Expert Group meeting; Bangkok, 28 October–1 November 1986
New York: United Nations
1987, 269 pages [UN Publication II.A.18]

A1824
Admiralty manual of hydrographic surveying
London: Hydrographer of the Navy
1965, 1969–1982, 2 volumes, 671 pages, 6 fascicles respectively, + 1 overlay and 1 supplement

A1825
Advances in hydrographic surveying: seminar; London, 24 January 1980
London: Society for Underwater Technology
1980, 76 pages

A1826
Hydro '86: 5th biennial international symposium; Southampton, 16–18 December 1986
Dagenham: The Hydrographic Society
1987, 27 papers

A1827
Hydro '88: 6th biennial international symposium; Amsterdam, 15–17 November 1988
Dagenham: The Hydrographic Society
1988 [Special Publication 23]

A1828
Hydro '90: 7th biennial international symposium; Southampton, 18–20 December 1990
Dagenham: The Hydrographic Society
1990, 1 volume [Special Publication 26]

A1829
Hydrography for the surveyor and engineer, 2nd edition
Ingham A E
London: Granada Technical Books
1984, xiii + 132 pages

A1830
2nd International hydrographic technical conference; Plymouth, 3–7 Septmber 1984
Dagenham: The Hydrographic Society
1985, 1 volume + 2 loose papers

A1831
Mapping and research in the exclusive zone
McGregor B A and Lockwood M
Washington DC: US Department of the Interior and Department of Commerce
1986, 40 pages

A1832
Marine positioning, symposium; Reston, VA, 14–17 October 1986
Dordrecht: Reidel, for Marine Technology Society
1987, xii + 474 pages

A1833
Marine surveying and consultancy
Guy J
Coulsdon: Fairplay Publications
1989

A1834
Offshore surveying for the civil engineering
industry
*London: Royal Institution of Chartered
Surveyors*
1977, 116 pages

A1835
Positioning and hydrographic survey
Canterbury: Smith Rea Energy Analysts Ltd
1991 [Offshore Business No 36]

A1836
Seabed surveying & exploration
Institute of Offshore Engineering
London: Dept of Trade & Industry
1987 [Resources from the Sea Programme]

A1837
Surveying and charting of the seas
Langeraar W
Amsterdam: Elsevier Science Publishers
1984, xviii + 612 pages [Elsevier
Oceanography Series 37]

A1838
Underwater engineering surveys
Milne P H
*London: E & F N Spon / Routledge, Chapman
Hall*
1980, 366 pages

UNDERWATER
CONSTRUCTION

A1839
An assessment of undersea teleoperators
Sofyanos T N and Sheridan T B
Cambridge, Mass: MIT
1980, 316 pages
[MITSG 80-11 NTIS PB81-102535]

A1840
Concise methods for predicting the effects of
underwater explosions on marine life
Young G A
*Silver Spring MD: Naval Surface Warfare
Center*
1991, 21 pages
[NAVSWC-MP-91-220] [NTIS AD-A241 310/2]

A1841
Diverless and deepwater technology:
conference; London 22–23 February 1989
Society for Underwater Technology
London: Graham & Trotman / Kluwer
1989, viii + 151 pages
[Advances in Underwater Technology, Ocean
Science & Offshore Engineering, 19]

A1842
Effects of explosives in the marine
environment: workshop on effects of explosives
use in the marine environment; Halifax, 29–31
January 1985
Greene G D et al
Canada Oil & Gas Lands Administration
1985, xii + 399 pages [Technical Report No 5]

A1843
Electricity in water: safety of divers: part 2,
practical applications
Diesen A et al
Høvik: Det norske Veritas
1980, 102 pages [Veritas Report No 80-0238]

A1844
Intelligent uses of explosives in engineering
offshore: seminar; London, 27 November 1986
London: Institution of Mechanical Engineers
1986, 6 papers

A1845
International advanced robotics programme:
1st workshop on mobile robots for subsea
environments; Monterey CA, 1990
Lee M J and McGhee R B
*Pacific Cove, CA: Monterey Bay Aquarium
Research Institute*
1991, 200 pages

A1846
The market for underwater construction:
international conference; London, 5–6 March
1987
London: Society for Underwater Technology
1987, 12 papers

A1847
Mortality of fish subjected to explosive shock as
applied to oil well severance on Georges Bank
Baxter L and Hays E E
*Woods Hole, MA: Woods Hole Oceanographic
Institution*
1982, 69 pages

A1848
Pénétration sous-marine, underwater
operations and techniques; international
conference, Paris, 6–8 December 1982
*Paris: Éditions Technip, for Association
Technique Maritime et Aéronautique*
1982, 498 pages
[20 papers in English, 11 papers in French]

A1849
Rules and regulations for the construction and
classification of submersibles and diving
systems
Crawley: Lloyd's Register of Shipping
1980, looseleaf

A1850
Rules for building and classing underwater
systems and vehicles
New York: American Bureau of Shipping
1979

A1851
Safe underwater electrical power
*Tucker L W and Nelson F E, for Civil
Engineering Lab, US Navy
Springfield, Virginia: NTIS*
1980, v + 23 pages [AD A091 672]

A1852
Specification for underwater welding
*New York: American National Standards
Institute*
1989 [ANSI/AWS D3.6-89]

A1853
A study of methods of measuring the
metacentric heights of semi-submersibles
London: HMSO (for the Department of Energy)
1985, 48 pages

A1854
The subsea construction forecast, 2nd edition
*Gerard Engineering Ltd
Canterbury: Smith Rea Energy Analysts Ltd*
1991

A1855
Subsea manned engineering
*Haux G F K
London: Baillière Tindall / Harcourt Brace
Jovanovich*
1981, 500 pages

A1856
Subtech '89: fitness for purpose; Aberdeen,
7–9 November 1989
*Society for Underwater Technology
London: Graham & Trotman / Kluwer*
1990 [Advances in Underwater Technology
and Offshore Engineering, 23]

A1857
Surface support for subsea operations
*Dept of Ship & Marine Technology, University
of Strathclyde
London: Dept of Trade & Industry*
1987, 2 volumes [Resources for the Sea
Programme]

A1858
Undersea teleopeators & intelligent
autonomous vehicles: papers presented at a
conference; Cambridge MA, October 1986
*Doelling N & Harding E T (editors)
Cambridge MA: MIT Sea Grant Program*
1987, 233 pages [MITSG 87-1]

A1859
Undersea work systems
*Talkington H R
New York: Marcel Dekker*
1981, 184 pages [Ocean Engineering Series
Volume 1]

A1860
UnderSea World '91 conference; San Diego,
16–18 April 1991
Spring Valley CA: West Star Productions
1991, 122 pages

A1861
Underwater acoustic positioning systems
*Milne P H
London: E & F N Spon / Routledge Chapman
Hall*
1983, x + 284 pages

A1862
Underwater blast effects from explosive
severance of offshore platform legs and well
conductors
*Connor J G
Silver Spring MD: Naval Surface Warfare
Center*
1991, 146 pages
[NAVSWC-TR-90-532] [NTIS AD-A235 964/4]

A1863
Underwater construction: development &
potential: papers of the market for underwater
construction international conference; London,
5–6 March 1987
Society for Underwater Technology
London: Graham & Trotman / Kluwer
1987, 184 pages
[Advances in Underwater Technology 1]

A1864
Underwater engineering symposium: some
challenges facing the industry now; Aberdeen,
2–3 November 1983
*Kingston-upon-Thames: Spearhead Exhibitions
(for Association of Offshore Diving Contractors*
1983, 23 papers + 3 loose papers

A1865
6th Underwater engineering symposium;
Aberdeen, 7–8 November 1984 (Association of
Offshore Diving Contractors)
*Kingston-upon-Thames: Offshore Conferences
& Exhibitions and Spearhead Exhibitions*
1984, 1 volume

A1866
Underwater explosives
London: MTD Ltd
1992, 100 pages

A1867
Underwater technology, 30th anniversary
symposium; The Hague, 24 May 1984
Delft: TNO-IWECO
1984, 4 papers

A1868
Underwater technology conference 86: deep
water technology; Bergen, 14–16 April 1986
Mellingen T (editor)

A1869
Underwater technology: 15th WEGEMT school;
Helsinki, 14–18 October 1991
*West European Graduate Education Marine
Technology*
Helsinki: Helsinki University of Technology
1991, 2 volumes, looseleaf

A1870
Underwater technology, 4th congress, with
Underwater Canada '85; Toronto, 27–28 March
1985
Cox F E (editor)
186 pages

A1871
Underwater technology and diving: special
conference topic B, Offshore; Gøteborg, 104
March 1983
Gøteborg: Swedish Trade Fair Foundation
1983, 5 volumes plus 2 loose papers

A1872
Underwater welding and cutting: international
symposium; Geesthacht, 23–24 June 1983
Geesthacht: GKSS
1983, 132 pages

A1873
Underwater welding: international conference;
Trondheim, 27–28 June 1983
*Oxford: Pergamon Press (for International
Institute of Welding)*
1983, xiv + 394 pages

A1874
Underwater welding of offshore platforms and
pipelines: proceedings of a conference; New
Orleans, 5–6 November 1980
Miami: American Welding Society
1981, ix + 189 pages

A1875
Underwater work programmes at 300m and
beyond: 4th underwater engineering
symposium; Aberdeen, 24–25 November 1982
*Kingston-upon-Thames: Spearhead Exhibitions
(for Society of Underwater Technology)*
1982, 14 papers

DIVING

A1876
XIVth Annual meeting on diving and hyperbaric
medicine and Diving Medical Advisory
Committee workshop; Aberdeen, 5–9
September 1988
European Undersea Biomedical Society
1988, 36 papers

A1877
Arctic underwater operations: medical and
operational aspects of diving activities in arctic
conditions: Icedive 1984 conference;
Stockholm,3–6 June 1984
Rey L (editor)
London: Graham & Trotman
1985, 356 pages

A1878
Aseptic bone necrosis in commercial divers
London: Underwater Engineering Group
1981, 28 pages [Technical Note UTN/25]

A1879
Atmospheric diving systems symposium;
Washington DC, September 1982
Marine Technology Society Journal
1983, Volume 17 (3), pages 3–60, 5 papers

A1880
Code of safety for diving systems
London: International Maritime Organization
1985, 20 pages

A1881
A colour atlas of dangerous marine animals
Halstead B W et al
London: Wolfe
1989, 192 pages

A1882
Commercial diving: reference and operations
handbook
Freitag M and Woods A
Chichester: John Wiley & Sons
1983, xv + 414 pages

A1883
Decompression: decompression sickness
Bühlmann A A
Berlin: Springer
1984, 87 pages

A1884
Department of Energy diving safety
memoranda: January 1974–January 1989
Submex Ltd
London: HMSO
1990, 99 pages

A1885
Developments in diving technology: Divetech
1984 conference (for Society for Underwater
Technology); London, 14–15 November 1984
London: Graham & Trotman
1985,157 pages
[Advances in Underwater Technology &
Offshore Engineering, Volume 1]

A1886
Diver's handbook of underwater calculations
Tucker W C
Centerville, Maryland: Cornell Maritime
1980, 182 pages

A1887
Diver training standards: Part I, basic air diving;
Part II, mixed gas diving; Part III, air diving
where no surface compression chamber is
required on site; Part IV, air diving with
self-contained equipment where no surface
compression chamber is required on site
Sheffield: Health & Safety Executive
1986

A1888
Divetech '81: the way ahead in diving
technology: international conference; London
24–26 November 1981
London: Society for Underwater Technology
1981, 4 volumes

A1889
Diving and life at high pressures: Royal Society
discussion meeting; London, 12–13 May 1983
Paton W (editor and organiser)
London: The Royal Society
1984 [Phil Trans R Soc Lond B304 (3–4)]

A1890
Diving communications
*Hipwell Baume Associates for Department of
Energy*
London: HMSO
1988, iv + 49 pages [OTH 88 279]

A1891
Diving operations at work regulations 1981 as
amended by the Diving Operations at Work
(Amendment) Regulations 1990: guidance on
regulations
Health & Safety Executive
London: HMSO
1991, 56 pages

A1892
The diving operations at work (amendment)
regulations 1992
London: HMSO
1992, 4 pages
[Statutory Instrument 1992 No 608]

A1893
Diving operations: technical session 6,
Oceanology International '80; Brighton, 2–7
March 1980
Brighton: BPS Exhibitions
1980, 15 papers

A1894
Diving, 13th annual international symposium;
New Orleans, 7–9 February 1983
Gretna, Louisiana: Association of Diving Contractors
1983, 169 pages

A1895
Emergency, contingency and safety in diving
Onarheim J et al
Ytre Lakesvag: Norwegian Underwater Technology Centre
1983, 34 pages [Report 2-83]

A1896
Emergency diving bell recovery, guidance notes (code of practice on certain aspects of the design and operation of diving bells)
London: Association of Offshore Diving Contractors
1980

A1897
Guidance notes for safe diving
European Diving Technology Committee
Luxembourg: Commission of the European Communities (HMSO)
1977, 58 pages

A1898
Guidelines on qualifications for personnel engaged in manned underwater operations in the petroleum activity
Stavanger: Norwegian Petroleum Directorate
1990 [in Norwegian and English]

A1899
Improvements in diver communications
PA Technology
London: HMSO
1988, 72 pages [OTI 87 500]

A1900
Joint meeting on diving and hyperbaric medicine; Amsterdam, 11–18 August 1990
Bethesda MD: Undersea Medical Society
1990, vi + 184 pages
[Undersea Biomedical Research, supplement to Volume 17]

A1901
Man underwater: 20th symposium of the Underwater Association for Scientific Research; London, 14–15 March 1986
Margate: The Underwater Association
1987, 251 pages
[Progress in Underwater Science Volume 12 New Series]

A1902
Measurement of diver resistance and the electric fields generated near the diver by impressed current anodes and by welding
Dyson R J et al
London: HMSO
1988, iii + 60 pages [OTH 88-281]

A1903
NOAA Diving manual for science and technology, 2nd edition
Miller J W (editor)
US National Oceanic & Atmospheric Administration
1979 (various pagings)

A1904
Offshore installations (diving operations) regulations 1974
Department of Energy
London: HMSO
1974, 18 pages [SI 1974 No 1229]

A1905
Offshore medicine: medical care of employees in the offshore oil industry, 2nd edition
Cox R A F (editor)
Berlin: Springer-Verlag
1987, xix + 258 pages

A1906
Oilfield diving: the diver's story
Jacobsen N G
Westwego LA: Norman H Jacobsen
1983, xii + 113 pages

A1907
The physician's guide to diving medicine
Shilling C W et al (Undersea Medical Society)
New York, London: Plenum
1984, xxxii + 736 pages

A1908
The physiology and medicine of diving, 3rd edition
Bennet P B and Elliott D H (editors)
London: Baillière Tindall
1982, 704 pages

A1909
The principles of safe diving practice
Underwater Engineering Group in association with Association of Offshore Diving Contractors and the UK Department of Energy
London: UEG
1984, 210 pages [UR23]

A1910
Procedures and language for underwater communication
London: UEG
1982, 32 pages [UTN 26]

A1911
The professional diver's handbook, 2nd edition
Sisman D (editor)
London: Submex Ltd [distributed by ASR Books]
1985, 306 pages

A1912
Registration and notification of diving operations: proposals for Diving Operations at Work (Amendment) (No 2) Regulations: consultative document
Health & Safety Commission
London: Health & Safety Executive
1990, 18 pages

A1913
Regulations concerning manned underwater operations in the petroleum activities
Stavanger: Norwegian Petroleum Directorate
1990 [in Norwegian and English]

A1914
Requiem for a diver: fifteen years of diving fatalities
Warner J and Park F
Glasgow: Brown Son & Ferguson
1990, 115 pages

A1915
Review of diving research programme: Part II, underwater electrical safety
Moulton R J
London: UEG (for Offshore Energy Technology Board)
1983, 37 pages [OTP 13]

A1916
Risk assessment of diver contingency
Jacobsen E and Nygaard A
Ytre Laksevag: Norwegian Underwater Institute
1980, 46 pages NUTEC Report 1-80]

A1917
Safety and rescue for divers
The British Sub-Aqua Club
London: Stanley Paul
1987, 144 pages

A1918
Safety in manned diving
Jakobsen E et al
Stavanger: Universitetsforlaget
1984, 89 pages

A1919
Safety of diving operations: symposium; Luxembourg, 7–8 May 1985 (European Diving Technology Committee)
London: Graham & Trotman (for Commission of the European Communities)
1986, xi + 343 pages

A1920
Scientific diving: a general code of practice
N C Flemming and Max M D (editors)
Paris: UNESCO
1990, 254 pages

A1921
Seamanship for divers
The British Sub-Aqua Club
London: Stanley Paul
1986, 159 pages

A1922
Tables for saturation and excursion diving on nitrogen-oxygen mixtures
London: UEG
1985, 50 pages [Publication UR31]

A1923
Technical and human aspects of diving and diving safety symposium; Luxembourg, 9–10 October 1980
Commission of the European Communities, Safety and Health for the Mining and Extractive Industries
1981, 320 pages

A1924
Thermal stress on divers in oxy-helium environments
London: Underwater Engineering Group
1983, 51 pages [Technical Note UTN 28]

A1925
Underwater association code of practice for scientific diving, 4th edition
Gamble J C et al (editors)
Swindon: Natural Environmental Research Council
1989, looseleaf

A1926
The underwater handbook
Shilling C W (editor)
New York, London: Plenum Publishing
1976, xxvii + 912 pages

A1927
Underwater photography: scientific and
engineering applications symposium; Woods
Hole, 21–24 April 1980
Smith P F (editor)
New York: Van Nostrand Reinhold
1984, x + 422 pages

A1928
Underwater safety: entry-level SCUBA
certification: minimum course content
*New York NY: American National Standards
Institute*
1989 [ANSI Z86.3-1989]

A1929
Underwater safety in the offshore oil industry:
seminar; London, 16 October 1986
London: Submex Ltd
1986, 79 pages

A1930
Underwater technology: salvaging underwater
work; a manual of SCUBA commercial, salvage
and construction operations, 2nd edition
Cayford J E
Centreville MD: Cornell Maritime Press
1966, xiii + 258

A1931
US Navy air decompression table handbook
and recompression chamber operator's
handbook
Supervisor of Diving, United States Navy
San Pedro: Best Publishing
1989, 211 pages

A1932
US Navy diving manual, volume 1: air diving,
revision 2
Naval Sea Systsems Command
Washington DC: US Government Printing Office
1988 [Navsea 0994-LP-001-9010]

A1933
US Navy diving manual, volume 2: mixed-gas
diving, revision 2
Naval Sea Systems Command
Washington DC: US Government Printing Office
1987 [Navsea 0994-LP-001-9020]

A1934
A work study of diver operations in the North
Sea
James G W et al
*Harwell: Department of Energy, Offshore
Energy Technology Board (HMSO)*
1985, 2 volumes
[OETB Report No OT-N-85-130]

Diving equipment

A1935
Air range diving support vessel guidance
Chief Inspector of Diving, Department of Energy
London: HMSO
1991 [OTH 90 336]

A1936
Care and maintenance of underwater breathing
apparatus: part 1, compressed air open circuit
type
London: British Standards Institution
1981 [BS 4001]

A1937
Control and monitoring of carbon dioxide in
diving bells
London: UEG (HMSO)
1986,53 pages [Publication UR34]

A1938
Diving chamber fire response times
Technica Ltd
London: HMSO
1991 [OTH 90 334]

A1939
Guidelines for evaluation of breathing
apparatus for use in manned underwater
operations in the petroleum activities
Stavanger: Norwegian Petroleum Directorate
1991

A1940
Handbook of underwater tools, 3rd edition
*London: Underwater Engineering Group [MTD
Ltd]*
1988, various pagings [UEG UR18]

A1941
Hyperbaric chamber gas contaminant
monitoring
*Murray B G and Marr I (Development
Engineering International)*
Boston Spa: BLDSC
1990, 34 pages

A1942
Immersion suits
*New York NY: American National Standards
Institute*
1989 [ANSI/UL 1197-1989]

A1943
Offshore photography
Strickland C L
San Diego: Photosea Systsems Inc
1984, 47 pages

A1944
Preliminary risk assessment for two operating
scenarios with the ... transfer under pressure
diving system
Milo (Consultant Engineers) Ltd
Boston Spa: BLDSC
1990, 106 pages [OTO 90 008]

A1945
Review of the diving research programme: part
2, underwater electrical safety
*London: Underwater Engineering Group (for
Offshore Energy Technology Board)*
1983 [OTP 13]

A1946
Rules for certification of diving systems
Oslo: Det norske Veritas
1988

A1947
Suggested limits for contaminants in hyperbaric
chambers
DEA Ltd
London: Department of Energy (HMSO)
1987, 45 pages
[Offshore Technology Report OTH 86 262]

A1948
Underwater lighting fixtures
*New York NY: American National Standards
Institute*
1985 [ANSI/UL 676-1985]

A1949
Underwater safety: low pressure hose
assemblies connecting first and second stages
on SCUBA diving
*New York NY: American National Standards
Institute*
1987 [ANSI Z86.7.2-1987]

A1950
Underwater safety: low pressure hose
assemblies on submersible pressure gauges
for SCUBA diving: design and minimum
performance requirements
*New York NY: American National Standards
Institute*
1987 [ANSI Z86.7.2-1987]

A1951
Underwater tools
Hackman D J and Caudy D W
Columbus, Ohio: Battelle Press
1981, vii + 152 pages

A1952
Underwater video acceptance standards
London: MTD Ltd
due 1992, 32 pages

Underwater vehicles

A1953
Advanced systems for underwater vehicles:
seminar; London, 25 February 1982
London: Society for Underwater Technology
1983, 110 pages

A1954
Diving and subsea vehicles: past, present and
future trends, meeting; London 18 March 1989
London: Institute of Marine Engineers
1989, 33 pages

A1955
International safety standard guidelines for the
operation of undersea vehicles
Pritzlaff J A (editor)
Washington DC: Marine Technology Society
London: Society for Underwater Technology
1979

A1956
An inventory of some force producers for use in
marine vehicle control
Wilson M B et al
*Bethesda, Maryland: David W Taylor Naval
Ship Reserach & Development Centre*
1979, 409 pages [NTIS AD-A078 423]

A1957
Launch, operation and recovery of underwater
vehicles: current practice and problems and
potential improvement: UEG seminar, 16
September 1983
London: UEG
1985, 30 pages [UEG Publication UE 30]

A1958
Motion dynamics of subsea vehicles
Kalske S
Technical Research Centre of Finland ?place
1989, 44 pages [Research Notes 924]

A1959
Power systems for small underwater vehicles,
workshops; Cambridge 5–6 October 1988
Cambridge MA: MIT Sea Grant Program
1988, 33 pages [Collegium Opportunity Brief
51, Report MITSG 88-11]

A1960
Regulations for the classification and
construction of submersibles
Hamburg: Germanischer Lloyd
1979, 14 pages

A1961
Rules and regulations for the construction and
classification of submersibles
Paris: Bureau Veritas
1976, 199 pages

A1962
Submersibles and their use in oceanography
and ocean engineering
Geyer R A (editor)
Amsterdam: Elsevier Science Publishers
1977, xii + 384 pages

A1963
Submersibles: Technical session J,
Oceanology International '80; Brighton 2–7
March 1980
Brighton: BPS Exhibtions
1980, 9 papers

A1964
Subtech '83: the design and operation of
underwater vehicles; London, 15–17 November
1983
London: Society for Underwater Technology
1983, 38 papers

A1965
Subtech '85: Submersible technology,
designing for intervention; Aberdeen, 29–31
October 1985
*London: Graham & Trotman (for Society for
Underwater Technology)*
1986, 400 pages
[Advances in Underwater Technology 5]

A1966
Subtech '87: Submersible technology, adapting
to change; Aberdeen 10–12 November 1987
*London: Graham & Trotman (for Society for
Underwater Technology*
1988, 336 pages
[Advances in Underwater Technology 14]

A1967
Undersea vehicles directory 1985
Arlington VA: Busby Associates
1985, 430 pages + appendices

Manned submersibles & habitats

A1968
Deep-sea research using the submersible
SHINKAI 2000 system, 5th Symposium;
technical reports
*Yokosuka: Japan Marine Science &
Technology Center*
1989, 342 pages [text in Japanese with
abstracts and captions in English]

A1969
Code of practice for the operation of manned
submersible craft
*London: International Association of Offshore
Diving Contractors*
1984, 139 pages [Guidance Note 022]

A1970
Deepwater oil production and manned
underwater structures
Jones M E W
London: Graham & Trotman
1981, xx + 245 pages

A1971
Manned submersibles
Busby R F
USA: Office of the Oceanographer of the Navy
1976, xii + 764 pages

A1972
Manned subsea operations
Underwater Engineering Group
London: Dept of Trade & Industry (HMSO)
1987, 113 pages
[Resources from the Sea Programme]

A1973
Rules and regulations for the construction and
classification of submersibles and diving
systems
London: Lloyd's Register of Shipping

1980, 1 volume [loose leaf]

A1974

Rules for building and classing underwater
vehicles, systems and hyperbaric facilities
Paramus NJ: American Bureau of Shipping
1990, 1 volume

A1975

Safety standard for pressure vessels for human
occupancy
*Fairfield NJ: American Society of Mechanical
Engineers*
1990 [ASME PVHO-1-1990]

Unmanned Submersibles

A1976

Autonomous underwater vehicle technology,
symposium; Washington DC, 5–6 June 1990
*Piscataway NJ: Institute of Electrical &
Electronic Engineers*
1990, 317 pages

A1977

Guidance note on the safe and efficient
operation of remotely operated vehicles
*London: Association of Offshore Diving
Contractors*
1989, 35 pages + appendices [AODC 051]

A1978

Guide to low-cost remotely operated vehicles,
1989–90
Given D (editor)
Spring Valley CA: Windate Enterprises
1989, v + 34 pages

A1979

Intervention 88 / ROV '88, conference; Bergen
17–20 April 1988
Atteraas L (editor)
Bergen: Norwegian Petroleum Society
1988, 663 pages

A1980

Intervention '89 / ROV '89, conference; San
Diego 14–16 March 1989
Washington DC: Marine Technology Society
1989, x + 282 pages

A1981

Intervention '90 / ROV '90, conference;
Vancouver, 25–27 June 1990
Washington DC: Marine Technology Society
1990

A1982

Intervention '91 / ROV '91, conference;
Hollywood, FL, 21–23 May 1991
*Washington DC: Marine Technology Society
ROV Committee*
1991, 451 pages

A1983

An introduction to ROV operations
Last G and Williams P
Ledbury: OPL Ltd
1991, 300 pages

A1984

Jet-propelled remote-operated underwater
vehicles guided by tilting nozzles
Gangadharan S N and Krein H L
Marine Technology
1989, Volume 26 (2) pages 131-144

A1985

Lower-cost hydrocarbon production: an
achievable objective utilising ROV tooling
systems
White J
London: Marine Management (Holdings)
1989, 11 pages

A1986

Operational guidelines for remotely operated
vehicles
Washington DC: Marine Technology Society
1984, xxi + 197 + 29 pages

A1987

ORIA 1985: Offshore robotique intelligence
artificielle; Marseilles, 17–18 June 1985
*Chambre de Commerce et d'Industrie de
Marseilles et IIRIAM avec le patronage de
l'ASTEO et de Petrole Information*
1985, 20 papers

A1988

Remotely operated vehicles
Busby R F (for NOAA)
US Department of Commerce
1979 [PB 80-131881]

A1989

Remotely-operated vehicles: technology,
applications and markets: a state-of-the-art
review
Cunningham P
Bedford: BHRA, Elsevier
1989, IX + 91 pages

A1990
ROV '83: remotely operated vehicles: a
technology whose time has come; San Diego,
14–17 March 1983
*San Diego: Marine Technology Society, San
Diego Section*
1983, ix + 312 pages

A1991
ROV '84: remotely operated vehicles
technology update, a international perspective;
San Diego, 14–18 May 1984
*San Diego: Marine Technology Society, San
Diego Section*
1984, xi + 375 pages

A1992
ROV '85: Remotely operated vehicles, realizing
operational value; San Diego, 2–4 April 1985
*San Diego: Marine Technology Society, San
Diego Section*
1985, xi + 242 pages

A1993
ROV '86: Remotely operated vehicles:
technology requirement - present and future
(MTS, SUT and AODC) ; Aberdeen, 24–26
June 1986
London: Graham & Trotman
1986, 366 pages

A1994
ROV '87: 5th remotely operated vehicle
conference; San Diego, 10–12 March 1987
*San Diego: Marine Technology Society, San
Diego Section*
1987, ix + 337 pages

A1995
ROV Review, 4th edition
Ledbury: OPL Ltd
1991, 300 pages

A1996
Unmanned free-swimming submersible (UFSS)
system description
Johnson H A et al
Washington DC: Naval Research Laboratory
1980, 117 pages [NTIS AD-A092 539]

A1997
Unmanned untethered submersible technology,
3rd international symposium: Durham NH, 6–9
June 1983
Durham NH: University of New Hampshire
1984, 255 pages

A1998
Unmanned untethered submersible techniques,
4th International symposium ; Durham NH,
24–27 June 1985
*Durham NH: University of New Hampshire
Marine Systems Engineering Laboratory*
1985, 372 pages

A1999
Unmanned untethered submersible technology,
5th symposium; Marrineck, 22–24 June 1987
Durham NH: University of New Hampshire
1987, 2 volumes

A2000
Unmanned, untethered submersible
technology, 6th international symposium; Ellicot
City MD, 12–14 June 1989
*Durham N: New Hampshire University Marine
Systems Laboratory*
1989, 584 pages

COMMUNICATIONS SYSTEMS, COMPUTERS & ELECTRONICS

A2001
Automation for safety in offshore petroleum operations: IFIP/IFAC Symposium; Trondheim, 25–27 June 1985
Oslo: Norwegian Society for Automatic Control
1985, 1 volume preprints + 1 volume papers

A2002
Automation in the design and manufacture of large marine systems: 16th MIT Sea Grant College Program lecture and seminar
Chryssostomidis C (editor)
New York: Hemisphere
1990, 298 pages

A2003
CRIOP: A scenario-method for evaluation of offshore control centres
Ingstad O and Bodsberg L
Trondheim: SINTEF
1990, 94 pages [STF75 A89028]

A2004
Computing and data management: information technology offshore
Canterbury: Smith Rea Analysts
1988, iv + 38 pages

A2005
Developments in electronics for offshore fields 1
Bedwell C
Barking: Elsevier Applied Science
1978, x + 230 pages

A2006
Electronic aids to navigation: position fixing, 2nd edition
Tetley L and Calcutt D
Sevenoaks: Edward Arnold
1991, ix + 386 pages

A2007
Electronics for ocean technology: conference; Birmingham 8–10 September 1981
London: Institution of Radio & Electronic Engineers
1981, 348 pages

A2008
Electronics for ocean technology; conference; Edinburgh, 24–26 March 1987
London: Institution of Electronic & Radio Engineers
1987, 223 pages

A2009
Electronics in oil '82: conference; London, 12–14 October 1982
Luton: Benn Electronics Publications
1982, 227 pages

A2010
Global maritime distress and safety system
London: International Maritime Organization
1987, 90 pages [Publication 970.86.20.E]

A2011
Information technology for offshore development: seminar; London 2 May 1991
London: IBC Technical Services
1991

A2012
International code of signals
London: International Maritime Organization
1985, vii + 175 pages

A2013
The merchant shipping (radio installations) regulations 1992
London: HMSO
1992, 36 pages
[Statutory Instrument 1992 No 3]

A2014
Microprocessor applications in underwater technology: seminar; London, 6 November 1980
London: Society for Underwater Technology
1981, 93 pages

A2015
Offshore computers, 1st conference; London, 3–5 November 1981
London: Offshore Conferences & Exhibitions
1982, 3 volumes (26 papers)

A2016
Offshore computers, 2nd conference; London, 16–17 February 1983
London: Offshore Conferences & Exhibitions
1983, 18 papers

A2017
Offshore computers, 3rd conference; Aberdeen
5–7 June 1984
Kingston-upon-Thames: Offshore Conferences
& Exhibitions
1984, 3 volumes

A2018
Offshore computers, 4th conference;
Aberdeen, 8–10 October 1985
Kingston-upon-Thames: Offshore Conferences
1985, 3 volumes

A2019
ORIA '91; 4th offshore, robotics and artificial
intelligence symposium: telerobotics in hostile
environments; Marseille, 11–12 December 1991
Marseille: IIRIAM
1991, 550 pages

A2020
Submarine telecommunication and power
cables
Barnes C C
Stevenage: Peter Peregrinus (for IEE]
1977, xii + 206 pages

A2021
Subsea control and data acquisition:
technology and experience, international
conference; London, 4–5 April 1990
Society for Underwater Technology
London: Kluwer Academic
1990, viii + 203 pages
[Advances in Underwater Technology and
Offshore Engineering Vol 22]

A2022
Subsea control and data acquisition for oil and
gas production systems: SUT conference;
London, 22–23 April 1986
London: Graham & Trotman
1986, 265 pages
[Advances in Underwater Technololgy and
Offshore Engineering Vol 7]

A2023
Underwater signal and data processing
Hassab J C
Boca Raton: CRC Press
1989, 357 pages

INSTRUMENTATION

A2024
Catalogue of marine instrumentation
Weydert M (editor)
Luxembourg: Commission of the European
Communities
1992, vi + 61 pages
[Marine Science & Technology EUR 13808]

A2025
Current measurements offshore: conference;
London, 17 May 1984
London: Society for Underwater Technology
1984, 10 papers

A2026
Evaluation, comparison and calibration of
oceanographic instruments: SUT Ocean data
conference: London, 4–5 June 1985
London: Graham & Trotman
1985, 267 pages
[Advances in Underwater Technology &
Offshore Engineering, Volume 4]

A2027
Instrumentering offshore conference; Bergen,
9–11 December 1987
Oslo: Norwegian Society of Chartered
Engineers
1987

A2028
IOC-UNEP Group of experts on methods,
standards and inter-calibration, 9th session;
Villefrance-sur-Mer, 5–9 December 1988
Paris: UNESCO
1989, 51 pages
[Intergovernmental Oceanographic
Commission Reports of Meetings of Experts
and Equivalent Bodies, 45]

A2029
Marine Instrumentation '90 conference; San
Diego, 27 February–1 March 1990
Spring Valley CA: West Star Productions
1990, 242 pages

A2030
Measuring ocean waves: ocean instrumentation to serve science and engineering:
symposium; Washington DC, 22–24 April 1981
Washington DC: National Research Council,
Marine Board
1982, 300 pages [NTIS AD-A123791]

A2031
Minimum performance standard for underwater
locating devices (acoustic) (self-powered)
*New York NY: American National Standards
Institute*
[ANSI/SAE AS8045]

A2032
Oilfield instrumentation & control
Deighton P & Robinson G (editors)
Canterbury: Smith Rea Energy Associates Ltd
1986, 440 pages

A2033
Shell Brent 'B' Instrumentation Project:
presentation of, and discussion on, projects
results: seminar, London, 14 November 1979
London: Society for Underwater Technology
1980, 7 papers

A2034
Underwater acoustic modelling: principles,
techniques and applications
Etter P C
London: Elsevier Applied Science
1990, xvi + 305 pages

A2035
Underwater acoustic system analysis, 2nd
edition
Burdic W S
Englewood Cliffs: Prentice-Hall
1991, xiii + 466 pages

A2036
Underwater imaging system performance
characterization
Jaffe J S
*Woods Hole: Woods Hole Oceanographic
Institution*
1988, 48 pages
[Technical Report WHO1-88-33]

INSPECTION, MAINTENANCE & REPAIR

A2037
Advances in underwater inspection and
maintenance, conference; Aberdeen, 24–25
May 1989
Society for Underwater Technology
London: Graham & Trotman
1990, xiv + 175 pages
[Advances in Underwater Technology, Ocean
Science and Offshore Engineering Vol 21]

A2038
An introduction to offshore maintenance
Ledbury: OPL LTd
1991, 219 pages

A2039
Asian inspection, repair and maintenance for
the offshore and marine industries: conference;
Singapore, 26–28 February 1985
Singapore: Institute for International Research
1985, 2 volumes (418 + 99 pages)

A2040
Asian conference on inspection, maintenance
& repair for the offshore & marine industries;
Jakarta, 26–27 May 1987
Institute for Industrial Research and NACE
1987

A2041
The inspection, maintenance and repair of
concrete sea structures: state of the art report
*Wexham Springs: Fédération Internationale de
la Precontainte*
1982, 44 pages

A2042
IRM AODC '86: offshore inspection, repair,
maintenace and underwater engineering
industries conference; Aberdeen, 3–6
November 1986
*Kingston-upon-Thames: Offshore Conferences
& Exhibitions*
1986, 4 volumes

A2043
IRM '88, day 3: IRM in the 1990s – subsea
installations, design for IRM, diving operations,
the role of ROVs and robotics; day 3;
Aberdeen 10 November 1988
*Kingston-upon-Thames: Offshore Conference
& Enxhibitions*
1988, 12 papers

A2044
IRM '90 / ROV 90, The practical approach to
offshore safety, conference; Aberdeen 6–9
November 1990
Kingston-upon-Thames: Spearhead
1990, 4 volumes

A2045
Optimising offshore maintenance, conference;
Aberdeen 28–29 January 1992
London: Henry Stewart Conference Studies
1992, 2 volumes

A2046
Review of offshore structure and pipeline
repairs, research report
London: Marine Technology Directorate
due late 1992

A2047
Ships and mobile offshore units: planned
maintenance systems as a basis for survey for
retention of class
Høvik: Det norske Veritas
1983 [Classification Note 10.1]

Inspection and testing

A2048
Alternatives for inspecting outer continental
shelf operations
Marine Board, National Research Council
Washington DC: National Academy Press
1990, ix + 112 pages

A2049
Arctic undersea inspection of pipelines and
structures
Busby R F & Associates
US Department of the Interior Minerals
Management Service
1983, xviii + 176 pages

A2050
Development of inspection techniques for
reinforced concrete: a state of the art survey of
electrical potential and resistivity measurement
for use above water level
Ove Arup & Partners
London: HMSO
1985, viii + 66 pages [OTH 84 205]

A2051
The effectiveness of underwater
non-destructive testing: summary report of a
programme of tests
Techword Services (for Department of Energy)
London: HMSO
1984, 67 pages [OTH 84 203]

A2052
Guidance manual for the inspection and
condition assessment of tanker structures
International Chamber of Shipping, Oil
Companies International Marine Forum
London: Witherby & Co
1986, vii + 200 pages

A2053
A handbook for underwater inspectors
Orca Ltd for the Department of Energy
London: HMSO
1988, xii + 518 pages [OTI 88 539]

A2054
The implications of the use of linear elastic
fracture mechanics in the design and
inspection of welded offshore structures
United Kingdom Atomic Energy Authority
London: HMSO
1987, 28 pages [OTH 87 221]

A2055
IRM '88, day 2: Underwater inspection
techniques, inspection repair and maintenance
philosophy, underwater repair techniques;
Aberdeen, 9 November 1988
Kingston-upon-Thames: Offshore Conferences
& Exhibitions
1988, 11 papers

A2056
Modes of inspection offshore: tried and tested
techniques and their future: IRM '82 Offshore
inspection, maintenance and repair conference;
Edinburgh, 2–4 November 1982
London: Offshore Conferences & Exhibitions
1982, 1 volume + 1 loose paper

A2057
Offshore inspection: change, growth and new
philosophies, IRM '84, 5th Offshore inspection
repair and maintenance conference;
Aberdeen,5–6 November 1984
Kingston upon Thames: Offshore Conferences
& Exhibitions
1984, 11 papers

A2058
Proability-based fatigue inspection planning
London: MTD Ltd
1992, 37 pages [MTD 92/100]

A2059
Recommended practice for ultrasonic
examination of offshore structural fabrication
and guidelines for qualification of ultrasonic
technicians, 2nd edition
Dallas: American Petroleum Institute
1988 [API RP 2X]

A2060
Structural assessment of an early Southern
North Sea Platform
Lloyd's Register of Shipping
London: HMSO
1986, 42 pages [OTH 86 212]

A2061
SUBMATIC: remotely operated subsea
machine tool and inspection system
Ø Bjørke et al
Trondheim: SINTEF
1990, 25 pages [STF10 A90036]

A2062
Undersea inspection of subsea production
systems
Busby Associates
*Reston VA: Minerals Management Service,
Technology Assessment & Research Branch*
1985, x + 176,1,16,25,26 pages

A2063
Underwater inspection
Bayliss M R, Short D and Bax H
Andover: E & F N Spon (ABP)
March 1988, 240 pages

A2064
Underwater inspection and NDT, 7th edition
Haywood M G and Mathers N
Falmouth: Prodive
1986, 480 pages

Maintenance & repair

A2065
Assessment of materials for repair of damaged
concrete underwater
Perry S H and Holmyard J M
London: HMSO
1990, 97 pages [OTH 90-318]

A2066
Cost and safety effective maintenance for the
offshore industry, workshop; Aberdeen 13
November 1990
London IIR Industrial
1990

A2067
Efficient R & M operations offshore – are
operators' needs being met? IRM '82 Offshore
inspection, maintenance & repair conference;
Edinburgh 2–4 November 1982
London: Offshore Conferences & Exhibitions
1982, 1 volume + 2 loose papers

A2068
Grouted repairs to steel offshore structures: a
summary report of a programme of tests
carried out by Wimpey Laboratories for the
Department of Energy
Techword Services
London: HMSO
1985, 56 pages
[Offshore Technology Report OTH 84 202]

A2069
Grouts and grouting for construction and repair
of offshore structures: a summary report
*London Centre for Marine Technology, Imperial
College of Science & Technology*
London: HMSO
1988, 198 pages

A2070
The influence of methods and materials on the
durability of repairs to concrete coastal and
offshore structures
Leeming M B
London: Underwater Engineering Group
1986, 45 pages [UEG publication UR36]

A2071
Maritime and offshore structure maintenance:
2nd international conference; London, 19–20
February 1986
Institution of Civil Engineers
London: Thomas Telford
1986, 361 pages

A2072
Offshore repair and maintenance: efficiency
and cost effectiveness; IRM '84 5th Offshore
inspection repair & maintenance conference;
Aberdeen, 5–6 November 1984
*Kingston-upon-Thames: Offshore Conferences
& Exhibitions Ltd*
1984, 10 papers

A2073
Planned maintenance and repair of offshore oil
and gas facilities: recent advances, best
practices, conference; Aberdeen 26 March 1991
London: Henry Stewart Conference Studies
1991, 7 papers

A2074
Repair of major damage to the prestressed
concrete towers of offshore structures
Wimpey Laboratories Ltd
London: HMSO
1988, 61 pages [OTH 87 250]
[Concrete in the Oceans Technical Report 27]

A2075
Repairs to North Sea offshore structures: a review
London: Underwater Engineering Group
1983, 40 pages [Report UR21]

A2076
Underwater inspection of steel offshore installations: implementation of a new approach (UEG project)
London: Marine Technology Directorate
1989, xii + 237 pages [MTD 89/104]

FATIGUE, CORROSION & FOULING

A2077
Advanced tool for fast assessment of fatigue under offshore random wave stress histories
Kam J C and Dover W D
Proceedings of the Instn ofCivil Engineers 2
1989, Volume 87 pages 539–649
(pages 645–649, discussion)

A2078
Analysis of uncertainties in fatigue life of tubular joints in offshore structures
Carr P
London: HMSO
1988 [OTH 88 190]

A2079
A compilation of fatigue test results for welded joints subjected to high stress/low cycle conditions, stage 1
Rosenberg T D et al
London: HMSO
1991, v + 65 pages [OTI 91 552]

A2080
Corrosion and marine growth on offshore structures: seminar; Aberdeen, September 1982
Lewis J R and Mercer A D (editors)
Chichester: Ellis Horwood (for Society of Chemical Industry)
1984, 156 pages

A2081
Corrosion atlas
During E D D
Amsterdam: Elsevier Science Publishers
1988, 280 loose leaf pages + 242 colour

A2082
Corrosion fatigue crack with growth in BS4360 50D structural steel in seawater under narrow band variable amplitude loading
Thorpe T W and Rance A (for AERE, Harwell)
London: HMSO
1986, x + 73 pages
[Offshore Technology Report OTH 86 232]

A2083
Corrosion fatigue of metals in marine environments
Jaske C e et al
New York: Springer-Verlag
1981, 245 pages

A2084
Corrosion in natural waters, symposium; Atlanta, 8 November 1988
Baloun C H (editor)
Philadelphia PA: American Society for Testing & Materials
1990, 155 pages [STP 1086]

A2085
Corrosion in seawater systems
Mercer A D
Chichester: Ellis Horwood
1990

A2086
Cracking and corrosion
Beeby A W
Wexham Springs: Cement & Concrete Association
1979, 54 pages
[Concrete in the Oceans Technical Report No 1]

A2087
Detection of corrosion in submerged reinforced concete
Carney R F A et al
London: HMSO
1990, 63 pages [OTH 87-239]
[Concrete in the Oceans Technical Report 16]

A2088
Effectiveness of concrete to protect steel reinforcements from corrosion in marine structures
Taywood Engineering Research Laboratories
London: HMSO
1988, 177 pages [OTH 87-247]

A2089
Electro-chemical methods of corrosion monitoring for marine concrete structures
Page C L and Cunningham P J
London: HMSO
1988 [OTH 87-245]
[Concrete in the Oceans Technical Repor 22]

A2090
FAROW: a computer program for dynamic response analysis and fatigue life estimation of offshore structures exposed to ocean waves
Olufsen A et al
Trondheim: SINTEF
1986, 70 pages [Report STF71 A86040]

A2091
Fatigue and corrosion: fatigue behaviour of offshore steel structures
de Back J and Vaessan G H G
NTIS
1981, 551 pages [NTIS PB83-146316]

A2092
Fatigue and crack growth in offshore structures: conference; London, 7–8 April 1986
London: Mechanical Engineering Publications (MEP)
1986, 220 pages

A2093
Fatigue assessments of North Sea fixed platforms: summary report
Lloyd's Register of Shipping
London: HMSO
1986, vii + 28 pages [Offshore Technology report OTH 86 207]

A2094
Fatigue behaviour of offshore structures
Gupta A and Singh R P
Berlin: Springer-Verlag
1986, xii + 309 pages
[Lecture Notes in Engineering, Volume 22]

A2095
Fatigue correlation study, semi-submersible platforms
Potthurst R et al
London: HMSO
1989, v + 64 pages [OTH 88-288]

A2096
Fatigue handbook, offshore steel structures
Almar-Naess A (editor)
Trondheim: TAPIR
1985, 536 pages

A2097
Fatigue in offshore structural steels: implications of the Department of Energy's research programme, conference; London, 24–25 February 1981
Institution of Civil Engineers
London: Thomas Telford
1981, 130 pages

A2098
Fatigue of marine structures, final report
Eide O I and Berge S
Trondheim: SINTEF
1986, 167 pages [Report STF71 A86022]

A2099
Fatigue of offshore structures, conference; London 19–20 September 1988
Dover W D amd Glinka G (for Engineering Integrity Society)
Warley: Engineering Materials Advisory Services
1988, 351 pages

A2100
Fatigue performance of repaired tubular joints
London: HMSO
1990 [OTH 89 307]

A2101
Fundamental mechanisms of corrosion of steel reinforcement in concrete immersed in seawater: results from phase II
Wilkins N J W and Lawrence P F
London: HMSO
1989, 12 pages [OTH 87-238]
[Concrete in the Oceans Technical Report 15]

A2102
General guidelines for corrosion testing of materials for marine applications
Working Party on Marine Corrosion
London: Institute of Metals
1988, 32 pages
[European Federation of Corrosion Reviews & Reports No 3]

A2103
The influence of stress spectrum, seawater and cathodic protection on fatigue crack propagation in structural steels
Thorpe T W and Carney R F A
London: HMSO
1990, 48 pages [OTH 90 326]

A2104
IRM '88, day 1: Topsides, splash zone and remote monitoring systems; Aberdeen, 8 November 1988
Kingston-upon-Thames: Offshore Conferences & Exhibitions
1988, 9 papers

A2105
Marine and offshore corrosion, 2nd edition
Chandler K A
Sevenoaks: Butterworth
1984, 420 pages

A2106
Marine biodeterioration
Costlow J D and Tripper R C (editors)
London: E & F N Spon (ABP)
1984, 512 pages

A2107
Marine corrosion: causes and prevention
LaQue F L
New York, Chichester: John Wiley & Sons
1975, 332 pages

A2108
Microbial problems in the offshore oil industry: conference; Aberdeen, 15–17 April 1986
Hill E C (editor)
Chichester: John Wiley (for Institute of Petroleum)
1987, xvi + 257 pages

A2109
Microbiology influences on marine corrosion
Walch M (editor)
Marine Technology Society Journal
1990, Volume 24(3) pages 3–44

A2110
On cumulative fatigue damage in steel welded joints
Eide O I
Trondheim: University of Trondheim, Marine Technology Centre
1983, x + 448 pages [Report UR-83-30]

A2111
Remaining life of defective tubular joints – an assessment based on surface crack growth in tubular joint fatigue tests
Marine Technology Support Unit (for the Department of Energy)
London: HMSO
1987, 76 pages [OTH 87 259]

A2112
Residual and fatigue strength of grout-filled damaged tubular members
London: HMSO
1990 [OTH 89 314]

A2113
Review of fatigue in concrete marine structures
Price W I J et al
London: HMSO
1989, 106 pages [OTH 87-235]
[Concrete in the Oceans Technical Report 12]

A2114
A review of information of hydrogen induced cracking and sulphide stress corrosion cracking in linepipe steels
Dewsnap R F et al (for the Department of energy)
London: HMSO
1987, 215 pages [Offshore Technology Report OTH 86 256]

A2115
Seawater corrosion handbook
Schumacher M (editor)
Park Ridge NJ: Noyes Data Corp
1979, 494 pages

A2116
A simple model of fatigue crack growth in welded joints
Thorpe T W (for AERE Harwell)
London: HMSO
1986, vii + 34 pages [OTH 86 225]

A2117
Strength & fatigue analysis: the North Sea experience
Bainbridge C A
London: Lloyd's Register of shipping
1987, 14 pages [Technical Paper No 92]

Corrosion control

A2118
Cathodic protection of production platforms in cold seawaters: a collection of papers from NACE files
Houston: National Association of Corrosion Engineers
1983, 192 pages, 19 papers

A2119
Cathodic protection of reinforcement steel in concrete
Berkeley K G C and Pathmanaban S
Sevenoaks: Butterworth
1990, 168 pages

A2120
Cathodic protection of steel in real and simulated seawater environments
Carney R F A et al
London: HMSO
1989, 37 pages
[OTH 89-542]

A2121
Cathodic protection: theory & practice (updated conference papers, Manchester 1982)
Ashworth V and Booker C J L (editors)
Chichester: Ellis Horwood (for Institution of Corrosion Science & Technology
1986, 357 pages

A2122
Corrosion and corrosion control for offshore and marine construction, conference; Xiamen, 6–9 September 1988
Jimei X (editor)
Oxford: Pergamon for Chinese Society of Corrosion & Protection
1989, 677 pages

A2123
Corrosion and water technology for petroleum producers
Jones L W
Tulsa: OGCI
1988

A2124
Corrosion control in offshore environment, seminar; Stavanger, 23 August 1984
Aasvistad T (editor)
Hagfors, Sweden: Uddeholm Stainless with Swedish Trade Council
1984

A2125
Corrosion fatigue crack growth in BS4360 grade 50D structural steel in sea water under narrow band variable amplitude loading
Department of Energy
London: HMSO
1986, 74 pages [OTH 86 232]

A2126
Corrosion offshore, a losing battle? IRM '82 Offshore inspection, maintenance and repair conference; Edinburgh, 2–4 November 1982
London: Offshore Conferences & Exhibitions
1982, 1 volume

A2127
Corrosion protection of steel structures
Chandler K A and Bayliss D A
Barking: Elsevier Applied Science Publishers
1985: xiii + 40 pages

A2128
A corrosion protection system for a North Sea jacket
Lye R E
Materials Performance
1990, volume 29(5) pages 13–18

A2129
Design and operational guidance on cathodic protection of offshore structures, subsea installations and pipelines
London: MTD Ltd
1990, 277 pages [MTD 90/102]

A2130
Marine and offshore corrosion, 2nd edition
Chandler K A
Sevenoaks: Heinemann-Butterworth
1984, 420 pages

A2131
Marine corrosion: general guidelines for corrosion testing of materials for marine applications; working party report
European Federation of Corrosion
London: Institute of Metals
1989

A2132
Marine corrosion on offshore structures: Society of Chemical Industry's Aberdeen section and Materials Preservation Group seminar; Aberdeen, 13–14 September 1979
London: Society of Chemical Industry
1981, 115 pages

A2133
Marine painting manual
Berendsen A M
London: Graham & Trotman
1989, 328 pages

A2134
Original and maintenance painting systems for
North Sea oil and gas platforms
Atherton D
Journal of Oil and Colour Chemists Association
1979, volume 62 (9) pages 351–358

A2135
Protective painting of ships and of structural
steel
Banfield T A
*Manchester: Selection & Industrial Training
Administration*
1978, 108 pages

A2136
Recommended practice: control of corrosion on
offshore pipelines
*Houston: National Association of Corrosion
Engineers*
1988 [RP-06-75]

A2137
Survey of painting practice for protection of
offshore structures
Whitehouse N R
Teddington: Paint Research Association
1983, 71 pages [Paper OT-O-8320]

Fouling organisms

A2138
Algal biofouling
Evans L V and Hoagland K D (editors)
Amsterdam: Elsevier Science Publishers
1986, 310 pages
[Studies in Environmental Science No 28]

A2139
Appraisal of marine fouling on offshore
structures
Oldfield D G
London: Underwater Engineering Group
1980 [OTP 6]

A2140
Appraisal of marine growth on offshore
structures
London: MTD Ltd
1992, 52 pages

A2141
Fouling and corrosion of metals in sea water:
symposium,; Oban, 6–7 April 1982
*Scottish Marine Biological Assn with Y-Ard
Mauchline J (editor)*
Oban: Scottish Marine Biological Association
1982, 304 pages

A2142
Marine fouling of offshore structures:
conference; London, 19–20 May l981
London: Society for Underater Technology
1981, 2 volumes

A2143
Report of the marine fouling working party for
the Director of Petroleum Engineering Division
*London: Underwater Engineering Group (for
Department of Energy)*
1980, 32 pages [OTP 4]

STRUCTURES

A2144
Advances in marine structures: conference;
Dunfermline, 20–23 May 1986
Smith C S and Clarke J D (editors)
London: Elsevier
1986, xii + 713 pages

A2145
Advances in marine structures, 2nd
conference; Dunfermline, 21–24 May 1991
Smith C S and Dow R S (editors)
Barking: Elsevier Applied Science
1991, 792 pages

A2146
Analysis of offshore structural damping with
non-proportional damping
McCormick M E
*Journal of Waterway, Port, Coastal & Ocean
Engineering*
1989, Volume 115(6) pages 775–792

A2147
Applied offshore structural engineering
Hsu T H
Houston: Gulf Publishing
1984, xii + 204 pages

A2148
Background to new fatigue design guidance for steel welded joints in offshore structures
Guidance Notes Revision Drafting Panel, UK Department of Energy
London: HMSO
1984, xi + 76 pages

A2149
Ballasting: systems and control for offshore oil and gas development: conference; London, 10 November 1981
London: Institution of Mechanical Engineers
1982

A2150
BOSS '82: 3rd international conference on the behaviour of offshore structures; MIT, 2–5 August 1982
Cryssostomidis C and Connor J J (editors)
Washington, London: Hemisphere Publishing
1983, 2 volumes

A2151
BOSS '85: behaviour of offshore structures; Delft, 1–5 July 1985
Battjes J A (editor)
Amsterdam: Elsevier Science Publishers
1985, xiii + 1012 pages
[Developments in Marine Technology 2]

A2152
BOSS '88, behaviour of offshore structures; Trondheim, 21–24 June 1988
Moan T et al (editors)
Trondheim: Tapir
1988, 3 volumes

A2153
Buckling of offshore structures: a state-of-the-art review of buckling of offshore structures
Ellinas C P, Supple W J and Walker A C (J P Kenny & Partners)
London: Granada Technical Books (for UK Department of Energy)
1984, 512 pages

A2154
Buckling of shells in offshore structures
Harding J E, Dowling P J and Agelidis N (editors) (Structures Engineering Services)
London: Granada Technical Books
Houston: Gulf Publishing
1982, 581 pages

A2155
Bulletin on design of flat plate structures, 1st edition
Washington DC: American Petroleum Institute
1987 [API Bull 2V]

A2156
Bulletin on stability design of cylindrical shells, 1st edition
Washington DC: American Petroleum Institute
1987 [API Bull 2U]

A2157
Cadcam Maritime '87: cadcam applications for ships & offshore; London, 13–14 October 1987
Chipping Norton: Hawkedon Conferences
1987

A2158
Cohesive buckling programme report
London: J P Kenny & Partners (distributed by BLLD)
1990, 2 volumes [OTO 90-005]

A2159
Construction of offshore structures
Gerwick B C
New York, Chichester: John Wiley & Sons
1986,xvi + 552 pages

A2160
Design against collision for offshore structures
Mavrikos Y and de Oliveira J G
Cambridge, Mass: MIT
1983, xii + 164 pages [Report MITSG 82-7]

A2161
Design and construction of concrete sea structures, 4th edition
FIP
London: Thomas Telford
1985, 29 pages

A2162
Design of offshore structures for Canadian frontiers,symposium; London, Ontario, 1987
Edmonton: Centre for Frontier Engineering
1987, 108 pages
[C-FER Special Publication No 2]

A2163
Dynamic analysis of offshore structures: recent developments
Kirk C L (editor)
Southampton: CML Publications
1982, 123 pages
[Progess in Engineering Sciences volume 1]

A2164
Dynamics of offshore structures
Patel M H
Sevenoaks: Butterworth
1989, vi + 402 pages

A2165
Dynamics of offshore structures
Wilson J F (editor)
Chichester: John Wiley & Sons
1984, 546 pages

A2166
Estimation of extreme wind force for offshore
structure design
Ochi M K and Murer Y
Gainesville FL: Florida University Coastal &
Engineers Department
1989, 44 pages [UFL/COEL TR/081]

A2167
Floating structures and offshore operations:
workshop, Wageningen, 19–20 November 1987
van Oortmensen G (editor)
Amsterdam: Elsevier Science Publishers
1987, x + 266 pages
[Developments in Marine Technology 4]

A2168
Guidelines for design and analysis of steel
structures in the petroleum activity
Stavanger: Norwegian Petroleum Directorate
1990 [in Norwegian and English]

A2169
Integrity of offshore structures, 2nd inter-
national symposium; Glasgow, 1–3 July 1981
Faulkner D (editor)
London: Elsevier Applied Science
1981, viii + 662 pages

A2170
Integrity of offshore structures, 3rd international
symposium; Glasgow, 28–29 September 1987
D Faulkner et al (editors)
Barking: Elsevier Applied Science
1988, vii + 630 pages

A2171
Integrity of offshore structures, 4th conference;
Glasgow 2–3 July 1990
D Faulkner et al (editors)
Barking: Elsevier Applied Science
1991, x + 630 pages

A2172
The interaction between major engineering
structures and the marine environment;
Nyborg, 27–29 May 1991
Honggerberg, Switzerland: IABSE
1991 (volume 63), 31 papers

A2173
Introduction to offshore structures: design,
fabrication, installation
Graff W J
Houston: Gulf Publishing
1981

A2174
Life expectancy assessment of marine
structures
Ayyub B M and White G J
Marine Structures
1990, Volume 3(4) pages 301–317

A2175
Management of the design and construction of
the North Sea structures
Dept of Civil & Structural Engineering
Manchester: UMIST

A2176
The marine environment and structural design
Gaythwaite J
New York, Wokingham: Van Nostrand Reinhold
1981, xvi + 313 pages

A2177
Marine structural reliability, symposium;
Arlington, 1987
Jersey City NJ: Society of Naval Architects &
Marine Engineers
1987, 245 pages

A2178
ISMS '91: , international symposium marine
structures; Shanghai, 12–15 September 1991
Shengkun Z (editor)
Shanghai: ISMS Secretariat
1991, 405 pages

A2179
Model testing of offshore structures at the
Institute for Marine Dynamics
Murray J J
St John's NF: National Research Council
Canada
1990, 17 pages [Report LR 1990-2]

A2180
Nonlinear methods in offshore engineers
Chakrabarti s K
Barking: Elsevier Applied Science
1990, 525 pages
[Developments in Marine Technology 5]

A2181
Offshore Structures: 2nd international
symposium; Rio de Janeiro, October 1979
Carneiro F L L B and Brebbia C A (editors)
London: Pentech Press (John Wiley)
1980, 925 pages

A2182
Offshore Engineering: 3rd international
symposium; Rio de Janeiro, September 1981
Carneiro F L L B (editor)
London: Pentech Press (John Wiley)
1982, 592 pages

A2183
Offshore engineering: 4th international
symposium; Rio de Janeiro, September 1983
Carneiro F L L B (editor)
London: Pentech Press (John Wiley)
1984, 857 pages

A2184
Offshore engineering: 5th international
symposium; Rio de Janiero, Sepstember 1985
Carneiro F L L B (editor)
London: Pentech Press (John Wiley)
1986, xii + 832 pages

A2185
Offshore engineering: 6th international
symposium; Rio de Janeiro, August 1987
Carneiro F L L B (editor)
London: Pentech Press (John Wiley)
1988, 821 pages [Offshore Engineering 6]

A2186
Offshore engineering: 7th international
symposium; Rio de Janeiro, August 1989
Carneiro F L L B (editor)
London: Pentech Press (John Wiley)
1990, 812 pages

A2187
Offshore platforms and pipelines
Mazurkiewicz B K
Clausthal-Zellerfeld, Switzerland: Trans Tech
Publications
1987,390 pages

A2188
Offshore structural engineering
Dawson T H
Englewood Cliffs NJ: Prentice-Hall
1983, xiv + 346 pages

A2189
Offshore structures
Reddy D V and Ariockiasamy M
Malabar FL: Krieger
1991 2 volumes

A2190
Offshore structures: design/fabrication interface
London: UEG
1988 [UR 37]

A2191
Platform and module construction
Canterbury: Smith Rea Energy Analysts
1991 [Offshore Business No 35]

A2192
Procedure for nonlinear dynamic response
analysis of offshore structures
Spidsøe N et al
Trondheim: SINTEF
1989, 39 pages [STF A89047]

A2193
Programme for marine structures: summary
report
Eide O I and Spidsøe N
Trondheim: SINTEF
1990, 43 PAGES [STF71 A89033]

A2194
Recommended practice Volume D: structures –
design against accidental loads
Høvik: Det norske Veritas
1981, 21 pages [RP D204]

A2195
Recommended practice Volume D: structures –
impact loads from boats
Høvik: Det norske Veritas
1981 [RP D205]

A2196
Rules and regulations for the classification of
steel ships and offshore units: material:
amendments and additions No 1
Paris: Bureau Veritas
63 pages

A2197
Rules for building and classing offshore installations, part 1: structures
Paramus NJ: American Bureau of Shipping
1983, 1 volume

A2198
Spon's fabrication norms for offshore structures: a handbook for the oil, gas and petrochemical industries
London: Chapman & Hall
1992, 640 pages

A2199
Strength of tubular members in offshore structures
Warwick D H and Faulkner D (Dept of Naval Architecture & Ocean Engineering)
Glasgow: University of Glasgow
1988, 162 pages [Report NAOE 88-36]

A2200
Study on buckling safety factors for offshore structures
London: J P Kenny & Partners (distributed by BLLD)
1991, various pagings [OTO 90-030]

A2201
Tubular members in offshore structures
Chen W G and Han D J
Boston, London: Pitman
1985, xi + 172 pages

A2202
Ultimate load analysis of marine structures
Soreide T H
Trondheim: TAPIR
1981, 296 pages

A2203
Wave and wind directionality: applications to the design of structures: conference; Paris, 28 September–1 October 1981
Paris: Éditions Technip
1982, 600 pages

Fixed platforms

A2204
Annual offshore platform installations worldwide, 1975 to present
Houston: Offshore Data Services
[routinely updated]

A2205
Code of practice for fixed offshore structures
London: British Standards Institution
1982 [BS6235: 1982]

A2206
Complex tubular joints: assesssment of stress concentration factors for fatigue analysis
Lloyd's Register of Shipping (for Department of Energy)
London: HMSO
1985, 145 pages [OTH 84 200]

A2207
Construction specification for fixed offshore structures in the North Sea
London: Engineering Equipment & Materials Users Association
1989, iv + 124 pages [Publication 158]

A2208
Cost effective topside design and construction, conference; London, 27–28 October 1987
London: IBC Technical Services
1987

A2209
The design of fixed offshore structures
Billington C J and Osborne-Moss D M
London: Granada Technical Publishing
1983, 280 pages

A2210
Design of tubular joints for offshore structures
London: UEG, Offshore Research
1985, 3 volumes [UR33]

A2211
Draft recommended practice for planning, designing and construction of fixed offshore platforms: load and resistance factor design
Washington DC: American Petroleum Institute
1989 [Recommended Practice 2A-LRFD]

A2212
Dynamics of fixed marine structures, 3rd edition
Barltrop N D P and Adams A J (Atkins Oil & Gas Engineering)
Guildford: Butterworth-Heinemann for MTD Ltd
1991, xx + 764 pages

A2213
Guide for building and classing fixed offshore structures
American Bureau of Shipping
1978, xx + 50 + 5 pages

A2214
Guyed tower design and analysis
Dallas: Energy Publications
1984

A2215
Ice interaction with offshore structures
Cammaert A B and Muggeridge D B
New York: Van Nostrand Reinhold
1988, xvi + 432 pages

A2216
Ice mechanics: risks to offshore structures
Sanderson T J O
London: Graham & Trotman
1987, 272 pages

A2217
The influence of node flexibility on jacket
structures: a pilot study
London: Underwater Engineering Group
1984, 116 pages [Report UR22]

A2218
The integrity of platform superstructures:
analysis in accordance with API RP2A
Bunce J W
London: Harper Collins
1985, iv + 108 pages

A2219
A management approach to weight engineering
London: Underwater Engineering Group
1984, 38 pages [Report UR24]

A2220
The marine installation and offshore completion
of the protective barrier around the Ekofisk 2/4
stoage tank
Broughton P and Davies R L
London: Marine Management (Holdings) Ltd
1990, 17 pages

A2221
Noise and vibration control procedures for
offshore installations: guidance for project
managers
London: Underwater Engineering Group
1981, 31 pages [Report UR19]

A2222
NPD's guidlines for safety evaluation of
platform conceptual design: seminar; Hankø,
25–27 May 1981
Oslo: Norske Siviligeniørers Forening
1981, loose-leaf

A2223
North Sea, West Sole field platform WE:
post-removal condition report and lifetime
performance review
London: BP Trading
1979

A2224
Offshore installations (design, construction and
survey) regulations, 1982 and amendment
*St John's: Gazetted 21 July 1982 and 13
August 1982, 33 pages*
[Newfoundland Regulations 183/82]

A2225
Offshore installations: guidance on design,
construction and certification, 4th edition
UK Department of Energy
London: HMSO
1990, looseleaf in binder

A2226
Offshore installations: guidance on design and
construction: final draft
*St John's: Newfoundland & Labrador
Petroleum Directorate*
1982, 165 + 82 pages

A2227
Offshore platform layout: seminar; Bergen,
28–29 September 1981
Oslo: Norwegian Petroleum Society
1981, 11 papers

A2228
Offshore platforms monitoring programmes,
use of full-scale data: a preliminary investigation
Birkinshaw M and Atkins N
London: HMSO
1990, 56 pages [OTH 89-295]

A2229
OWEC '82: Offshore weight engineering
conference; London, 22–23 September 1982
London: Offshore Conferences & Exhibitions
1982, 13 papers

A2230
OWEC '85: 3rd Offshore weight engineering
conference: Aberdeen, 24–25 April 1985
*Kingston-upon-Thames: Offshore Conferences
& Exhibitions*
1985, 11 papers

A2231
OWEC '87: Offshore weight engineering
conference; London, 7–8 April 1987
*Kingston-upon-Thames: Offshore Conferences
& Exhibitions*
1987, 11 papers

A2232
Planning and design of fixed offshore platforms
McClelland B and Reifel M D (editors)
New York, Wokingham: Van Nostrand Reinhold
1986, x + 1023 pages

A2233
Planning, designing and constructing fixed
offshore structures in ice environments
Washington DC: American Petroleum Institute
1982, 49 pages [API Bull 2N]

A2234
Platform completion conference; Stavanger,
25–26 April 1990
Oslo: Norwegian Petroleum Society
1990

A2235
Platform superstructures: design and
construction
Boswell L F (editor)
London: Granada Technical Publishing
1984, 296 pages

A2236
Protection of offshore installations against
impact; background report
J P Kenny & Partners Ltd
London: HMSO
1988, 266 pages

A2237
Recommended practice for planning, designing
and constructing fixed offshore platforms, 19th
edition
Washington DC: American Petroleum Institute
1991, 159 pages' [API RA 2A]

A2238
Recommended practice for planning, designing
and constructing fixed offshore structures in ice
environments, 1st edition
Washington DC: American Petroleum Institute
1988 [API RP 2N]

A2239
Recommended practice for planning, designing
and constructing heliports for fixed offshore
platforms, 3rd edition
Dallas: American Petroleum Institute
1986 [API RP 2L]

A2240
Regulations for the structural design of fixed
structures on the Norwegian Continental Shelf
Norway: Norwegian Petroleum Directorate
1977, 79 pages [unofficial translation]

A2241
Report on guyed tower platforms: state of the
art
US National Bureau of Standards
1983, 128 pages [NTIS PB83-253005]

A2242
Rules and regulations for fixed offshore
installations 1989
London: Lloyd's Register of Shipping
1989, 8 volumes

A2243
Rules and regulations for the construction and
classification of offshore platforms
Paris: Bureau Veritas
1975, 154 pages

A2244
Rules for the construction and inspection of
offshore installations: volume 1, offshore units
(marine structures), preliminary edition
Hamburg: Germanischer Lloyd
1976, 220 pages

A2245
Rules for classification of offshore installations
Høvik: Det norske Veritas
1 volume, looseleaf, routinely updated

A2246
Sensitivity of fixed offshore platforms to
uncertainties in design parameters
Lloyd's Register of Shipping, Offshore Division
London: HMSO
1988, lv pages [OTH 88-254]

A2247
Sensitivity of the reliability of a fixed offshore
platform: an investigation using level II
techniques
Williams R and Tsivanidis H
London: HMSO
1989, iii + 71 pages [OTH 89-544]

A2248
Study of sensitivity of a jacket design to sea current profile
London: HMSO
1987, 240 [OTH 87-267]

A2249
Study on topside weight reduction
Global Engineering
Glasgow: Offshore Supplies Office
1987, 3 volumes
[An extended summary is published separately]

A2250
Technical notes for fixed offshore installations: volume A, structures
Høvik: Det norske Veritas
1981 with subsequent amendments, looseleaf

A2251
Technical notes for fixed offshore installations: volume B, facilities
Høvik: Det norske Veritas
1981 with subsequent amendments, looseleaf

A2252
Topside design and construction, 3rd international conference; London 6–7 March 1989
London: IBC Technical Services
1989, 1 volume

A2253
Topside Engineering for offshore Europe; Newcastle, 13–14 June 1990
Newcastle: The Newcastle Oil Symposium
1990, 13 papers

A2254
Weight symposium; Haugesund, 3–5 March 1986
Oslo: Norsk Sivilingeniørers Forening
1986, 13 papers

A2255
Weight engineering, international conference; London 17–19 April 1991
Oslo: Norwegian Society of Chartered Engineers
1991

Tension-leg platforms

A2256
Compliant offshore structures
Patel M H and Witz J A
Guildford: Butterworth
1991, 416 pages

A2257
A guide to the analysis of floating structures, research report
Lopndon: Marine Technology Directorate
due late 1992

A2258
Recommended practice for planning, designing and constructing tension leg platforms, 1st edition
Washington DC: American Petroleum Institute
1987, 145 pages [API RP 2T]

A2259
The response of floating platforms to subsea blowouts
Milgram J H and McLaren W G, Department of Ocean Engineering
Cambridge MA: MIT
1982, 114 pages [Report 82.8]

A2260
System reliability analysis of structural systems
Lee J S and Faulkner D, Department of Naval Architecture & Ocean Engineering
Glasgow: University of Glasgow
1988, 193 pages

A2261
Tension leg platform: a state of the art review
Demirbilek Z
New York: American Society of Civil Engineers
1989, 352 pages

A2262
Tension-leg platforms from Hutton to Joliet, or deep water without deep pockets
Faulkner D, Department of Naval Architecture & Ocean Engineering
Glasgow: University of Glasgow
1988, 20 pages [Report NAOE 88-26]

Mobile platforms

A2263
BP SWOPS – an operator's and shipbuilder's perspective
Parker T J
Naval Architect
1988, March, pages 21–37

A2264
Code for the construction and equipment of mobile offshore drilling units, 1989 (1989 MODU code)
London: International Maritime Organization
1990, 101 pages [IMO-811 09.05 E]

A2265
Construction of a semi-submersible accommodation rig in a Japanese shipyard
Macleod A K
London: Institute of Marine Engineers
1987 [meeting paper preprint]

A2266
Design for reliability in deepwater floating drilling operations
Harris L M
Tulsa: Pennwell
1979, 274 pages

A2267
The design of moonpools for subsea operations
Day A H
Marine Technology
1990, Volume 27(3) pages 167–179

A2268
Guidelines for site specific assessment of mobile jackup units
London: Noble Denton Consultancy
1990

A2269
The jack-up drilling platform: design and operation seminar; London 25–27 September 1985
Boswell L F (editor)
London: Granada
1986, 340 pages

A2270
The jackup drilling platform, 2nd international conference; London, 26–27 September 1989
Boswell L F and d'Mello C A (editors)
Barking: Elsevier Applied Science
1990, xviii + 290 pages

A2271
The jack-up drilling platform, 3rd conference; London, 24–25 September 1991
Boswell L F (editor)
London: City University, School of Engineering
1991, 18 papers

A2272
Long term response analysis of dynamically sensitive jack-up platforms
Karunakaren D
Trondheim: SINTEF
1991, 61 pages [STF25 A91028]

A2273
Mobile offshore structures and risk assessment of buoyancy loss, conference; London 14–18 September 1987
London: City University
1987

A2274
Offshore station keeping, lst symposium; Houston, 1–2 February 1990
Houston TX: Society of Naval Architects & Marine Engineers
1990, 1 volume + 1 loose paper

A2275
PRADS '87, Practical design of ships & mobile units, 3rd international conference; Trondheim, 1987
Trondheim: Norwegian Institute of Technology
1987, 2 volumes

A2276
PRADS '89, Practical design of ships & mobile units, 4th international symposium; Varna, Bulgaria, 1989
Varna: Bulgarian Ship Hydrodynamics Centre
1989, 2 volumes

A2277
Preliminary rules for building and classing accommodation barge and hotel barges, 1989
Paramus NJ: American Bureau of Shipping
1989, 1 volume

A2278
Regulations for mobile offshore units 1991 regulations laid down by the Norwegian Maritime Directorate
Oslo: Fabritius Forlag
1990, 721 pages

A2279
Rules and regulations for the classification of
mobile offshore units
Crawley: Lloyd's Register of Shipping
1989, 6 volumes

A2280
Rules for building and classing mobile offshore
drilling units
Paramus NJ: American Bureau of Shipping
1991, 1 volume

A2281
Rules for classification of mobile offshore units
Høvik: Det norske Veritas
1 looseleaf volume, routinely updated

A2282
Semi-submersibles: the new generations:
internatiional symposium; London, 17–18
March 1983
London: Royal Institution of Naval Architects
1983, 15 papers

A2283
Ship Conversion and repair, 2nd international
conference; London, 24 May 1991
Hitchin: Lorne & MacLean Marine
1991

A2284
Stationing and stability of semi-submersibles:
conference; Glasgow, 16–18 June 1986
Glasgow: University of Strathclyde
1986, 13 papers

A2285
A study of the methods of measuring the
metacentric heights of semi-submersibles
*National Maritime Institute (for the Department
of Energy)*
London: HMSO
1985, v + 41 pages [OTH 84 211]

BUOYS

A2286
Aspects of buoy design
Tsinipizoglou S
*Glasgow: University of Glasgow Marine
Technology Programme*
1983, 10 pages [Report NAOE-83-21]

A2287
Buoy engineering
Berteaux H O
New York, Chichester: John Wiley & Sons
1976, xx + 314

A2288
Buoy technology, 1983 symposium ; New
Orleans, 27–29 April 1983
Washington DC: Marine Technology Society
1983, xi + 416

A2289
Experiments with conical and cylindrical
moored buoys
Yilmaz O
*Glasgow: Glasgow University Department of
Naval Architecture & Ocean Engineering*
1990, 32 pages [Report NAOE 90-20]

Mooring and tanker loading

A2290
Design and construction specification for
marine loading arms
Oil Companies International Marine Forum
London: Witherby
1981

A2291
Design and construction specification for
marine loading arms, 2nd edition
Oil Companies International Marine Forum
London: Witherby
1987

A2292
Dynamic response of a tanker moored to an
articulated loading platform
Gernon B J and Lou J Y K
Ocean Engineering
1987, Volume 14/6 pages 489–512

A2293
Effective mooring
Oil Companies International Marine Forum
London: Witherby
1989, 56 pages

A2294
Guidelines for deepwater port single-point
mooring systems
Flory J F et al
Springfield VA: NTIS
1977 [NTIS AD-A050 182]

A2295
Guide on marine terminal fire protection and
emergency evacuation, 1st edition
Oil Companies International Marine Forum
London: Witherby & Co
1987

A2296
International safety guide for oil tankers and
terminals, 3rd edition
*International Chamber of Shipping, Oil
Companies International Marine Forum and
International Association of Ports and Harbours*
London: Witherby & Co
1988, xvi + 204 pages

A2297
Offshore oil terminals
*Herndon VA: US Dept of the Interior Minerals
Management Service*
1990 [OCS Report MMS 90-0014]

A2298
Offshore terminal loading using dynamic
positioning
Gjelstad B and Hvamb O G
Seaways
1990, July, pages 13–16

A2299
Recommendations for equipment employed in
the mooring of ships at single-point moorings,
2nd edition
Oil Companies International Marine Forum
London: Witherby
1988

A2300
Rules for building and classing single-point
moorings
New York: American Bureau of Shipping
1975, 1985, 1 volume + 1 supplement

A2301
Rules for the construction and classification of
offshore loading systems
Høvik: Det norske Veritas
1980, 11 pages

A2302
Single-point mooring maintenance and
operations guide
Oil Companies International Marine Forum
London: Witherby & Co
1985

A2303
Single point moorings of the world, 2nd edition
Halyard Offshore
Ledbury: Oilfield Publications Ltd
1992, 300 pages

A2304
SPM Hose ancilliary equipment guide: a guide
providing common descriptive terminology and
technical recommendations for the designers
and operators of SPM systems, 3rd edition
Oil Companies International Marine Forum
London: Witherby
1987, 15 pages

PIPELINES

A2305
Advances in offshore oil and gas pipeline
technology
de la Mare R F (editor)
*London: Oyez Scientific & Technical Services
(ASR Books)*
1985, ix + 383 pages

A2306
Aspect '90: advances in subsea pipeline
engineering and technology, SUT conference;
Aberdeen 30–31 May 1990
Ellinas C P (editor)
*Dordrecht, London: Kluwer Academic
Publishers*
1990 XII + 374 pages
[Advances in Underwater Technology &
Offshore Engineering, Volume 24]

A2307
Commissioning and decommissioning of
pipelines; Dyce, 9–11 June 1987
Beaconsfield: Pipes & Pipelines International
1987, 6 papers

A2308
Design methods for multiphase hydrocarbon transpsort systems
Lenn C P and Fairhurst P (editors)
Glasgow: Blackie & Sons Ltd
1990, 320 pages

A2309
The development of guidelines for the assessment of the submarine pipeline spans: overall summary report
Raven P W J
London: HMSO
1986, vi + 56 pages [OTH 86 231]

A2310
Flexible pipe technology: conference; Oslo, 23–24 September 1986
Oslo: Norwegian Petroleum Society
1986, 12 papers

A2311
Gas transmission and distribution pipline systems: ASME code for pressure piping, B31: an American National Standard
New York: American Society of Mechanical Engineers
1986 [ASME/ANSI B31.8-1986 with addenda B31.8a-1986 issued 1987]

A2312
Glossary of onshore and offshore pipelines / Omnium technique des transports par pipelines
Paris: Éditions Technip
1979, 320 pages
[English-French and French-English]

A2313
Guidance notes in support of the Offshore installations (emergency pipe-line valve) regulations 1989: SI 1989/1029
Department of Energy
London: HMSO
1991

A2314
Guidelines for flexible pipes
Høvik: Det norske Veritas
1987, 77 pages [JIP/GFP-02]

A2315
Handbook on design and operation of flexible pipes
Berge S et al
Trondheim: SINTEF
1992, 377 pages [STF70 A92006]

A2316
Hydrodynamic forces on seabed pipelines
Verley R L P et al
Journal of Waterway, Port, Coastal & Ocean Engineering, ASCE
1989, Volume 115/2, pages 172–189

A2317
Installation of steel flowlines from diving support vessels
Sriskandarajab T
London: Institute of Marine Engineers
1988, 11 pages

A2318
The integrity of offshore pipeline girth welds ...
Slater G and Davey T G (for the Department of Energy
London: HMSO
1986, 209 pages [OTH 86 223]

A2319
The integrity of offshore pipeline girth welds: the influence of hydrogen on the fracture toughness of cellulosic coated electrode weld deposits
Carne M M P and Slater G
London: HMSO
1988, vi + 24 pages [OTI 88 524]

A2320
Internal and external protection of pipes, proceedings of 7th international conference; London, 21–23 September 1987
Cranfield: BHRA
1987, 214 pages

A2321
IP Model Code Part 6: Pipeline safety code; supplement to 4th edition
Institute of Petroleum
London: Institute of Petroleum
1988, 60 pages

A2322
Liquid transportation systems for hydrocarbons, liquid petroleum gas, anhydrous ammonia and alcohols: ASME code for pressure piping, B312: an American National Standard
New York: American Society of Mechanical Engineers
1987 [ASME/ANSI B31.4-1986]

A2323
Marine and offshore pumping and piping
systems
Crawford J
Sevenoaks: Butterworth
1981, 392 pages

A2324
Maritime and pipeline transportation of oil and
gas: problems and outlook; conference,
Montréal, 23–24 April 1990
Poirier A and Zaccour G (editors)
Paris: Éditions Technip
1991, 320 pages

A2325
Multi-phase flow in pipeline systems: its
transfer, measurement and handling
King N W (editor)
London: HMSO
1990, various pagings

A2326
North Sea pipelines: a survey of technology,
regulation and use conflicts in oil and gas
pipeline operation
Nothdurft W E
*Boston, MA: New England River Basins
Commission*
1980, x + 68 pages [NTIS PB80-165848]

A2327
North Sea pipeline systems
Canterbury: Smith Rea Energy Analysts
1989 [Offshore Business No 29]

A2328
Offshore oil and gas pipeline technology
seminar; London,, 20–21 January 1987
London: IBC Technical Services
1987, 16 papers
*[subsequent conferences: Offshore pipeline
technology]*

A2329
Offshore pipeline conference; The Hague,
25–26 January 1984
*Delft: Netherlands Industrial Council for
Oceanology*
1984, 9 papers

A2330
Offshore pipeline design, analysis and methods
Mousselli A H
Tulsa: PennWell
1981, xiii + 193 pages

A2331
Offshore pipeline design elements
Herbich J B
New York: Marcel Dekker
1981, xiv + 233 pages

A2332
Offshore pipeline engineering; course, London
17–19 October 1989
London: IBC Technical Services
1989, 16 pages

A2333
Offshore pipeline girth welds: MIG database
Slater G
London: HMSO
1988, vi + 25 pages [OTI 80-525]

A2334
Offshore pipeline girth welds: non-destructive
testing
Mudge P J
London: HMSO
1989, vi + 96 pages [OTI 88-530]

A2335
Offshore pipeline girth welds: the factors
influencing mechanised mig weld metal
toughness
Hart P H M and Hutt G A
London: HMSO
1988, vi + 98 pages [OTI 88-528]

A2336
Offshore pipeline girth welds: vertical–up weld
metal database
Slager G
London: HMSO
1988, vi + 12 pages
[OTI 88-529]

A2337
Offshore pipeline technology conference;
Stavanger, 27–28 January 1988
London: IBC Technical Services
1988, 15 papers

A2338
Offshore pipeline technology European
seminar; Amsterdam, 1–2 February 1989
:London: IBC Technical Services
1989, 1 volume

A2339
Offshore pipeline technology European
seminar; Paris, 809 February 1990
London: IBC Technical Services
1990, 17 papers

A2340
Offshore pipeline technology European
seminar; Copenhagen, 14–15 February 1991
London: IBC Technical Services
1991, 11 papers

A2341
OMAE 1990; 9th offshore mechanics and arctic
engineering conference, Volume V Pipelines;
Houston, 18023 February 1990
*New York: American Society of Mechanical
Engineers*
1990, viii + 198 pages

A2342
OMAE 1991; 10th offshore mechanics and
arctic engineering conference, Volume V:
pipeline technology; Stavanger, 23–28 June
1991
*New York: American Society of Mechanical
Engineers*
1991, 211 pages

A2343
Model Code of Safe Practice, Part 6;
Supplement,   ????
Institute of Petroleum
Chichester: John Wiley
1986, 60 pages]

A2344
Pipe protection bibliography
Guy N G
London: Elsevier Applied Science (for BHRA)
1987, 284 pages

A2345
The pipe protection conference; Cannes,
23–25 September 1991
J Tiratsoo (editor)
London: Elsevier Applied Science
1991, viii + 269 pages

A2346
Pipeline and riser loss study 1990
London: HMSO
due 1992 [OTH 91 337]

A2347
Pipeline engineering symposium 1990
Oliver H G (editor)
New York: ASME
1990, 122 pages [PD Volume 31]

A2348
Pipeline technology conference; Oostende,
15–18 October 1990
Denys R (editor)
Antwerp: Royal Flemish Society of Engineers
1990, 2 volumes + 1 loose paper

A2349
Pipelines and the environment; Bournemouth,
8–10 March 1988
Beaconsfield: Scientific Surveys
1983, 8 papers

A2350
Pipelines and the offshore environment:
seminar; London, 15 February 1983
Beaconsfield: Scientific Surveys
1983, 8 papers

A2351
Pipelines and umbilicals
Robinson G (editor)
Canterbury: Smith Rea Energy Analysts
1990, iv + 64 pages

A2352
Pipeline engineering course (IOE); Riccarton,
26 March–6 April 1984
Geneva: Battelle Geneva Research Centre
1984, looseleaf

A2353
Pipeline engineering symposium 1984: 7th
annual energy sources technology conference;
New Orleans, 12–16 February 1984
Seiders E J (editor)
*New York: American Society of Mechanical
Engineers*
1984, v + 110 pages

A2354
Pipelines for marginal fields; Dyce, 26–28
September 1988
Tiratsoo J N H (editor)
Beaconsfield: Scientific Surveys
1989, 291 pages

A2355
Pipeline inspection, maintenance and repair:
seminar; London, 3 March 1983
London: Society for Underwater Technology
1984, 129 pages

A2356
Pressure testing of liquid petroleum pipelines,
3rd edition
Washington DC: American Petroleum Institute
1991 [API RP 1110]

A2357
Provisional rules for the construction and
classification of submarine pipelines
London: Lloyd's Register
1989, 17 pages

A2358
Recommended practice, Volume E: pipelines
and risers–onbottom stability design of
submarine pipelines
Høvik: Det norske Veritas
1988, vii + 41 pages [RP E305]

A2359
Recommended practice for design and
installation of offshore production platform
piping systems, 5th edition
Washington DC: American Petroleum Institute
1991 [API RP 14E]

A2360
Recommended practice for field inspection of
new line pipe
Washington DC: American Petroleum Institute
1990 [RP 5L8]

A2361
Recommended practice for flexible pipe, 1st
edition
Washington DC: American Petroleum Institute
1988 [API RP 17B]

A2362
Recommended practice 1107, pipeline
maintenance welding practices, 3rd edition
Washington DC: American Petroleum Institute
1991 [RP 1107]

A2363
Regulations concerning pipeline systems in the
petroleum activities
Stavanger: Norwegian petroleum Directorate
1990

A2364
Reliability data for subsea pipelines
Hokstad P
Trondheim: SINTEF
1989, 71 pages [Report STF75 A89037]

A2365
Rules for submarine pipeline systems
Høvik: Det norske Veritas
1981, 88 pages
[reprinted with corrections 1982]

A2366
Scour around submerged pipelines in
alternating current
*Dahl T E and Yanbao L, Norwegian
Hydrotechnical Laboratory*
Trondheim: SINTEF
1988, 15 pages [Report STF60 A88015]

A2367
Spans in subsea pipelines
Kenny J P and Partners
Oxford: Blackwell Scientific Publications
1988, 356 pages

A2368
Specification for CRA line pipe, 2nd edition
Washington DC: American Petroleum Institute
1991 [Spec 5LC]

A2369
Specification for line pipe, 39th edition
Washington DC: American Petroleum Institute
1991 [Spec 5L]

A2370
Static and dynamic analysis of offshore
pipelines during installation
Larsen C M
Trondheim: Norges Tekniske Hogskole
1976 [Report SK/M35]

A2371
Statpipe: application of deep water pipelaying
technology: conference: Stavanger, 17
November 1983
Stavanger: ONS Foundation
1983, 10 papers

A2372
Submarine pipelines guidance notes
Department of Energy
London: HMSO
1982, 51 pages [Sections 1 & 2 only]

A2373
Submarine pipelines (inspectors etc)
regulations 1977
UK Department of Energy
London: HMSO
1977, 8 pages [SI 1977 No 835]

A2374
Submarine pipelines safety regulations 1982
London: HMSO
1982, 10 pages [SI 1982 No 1513]

A2375
Subsea and pipeline engineering training
course notes; London, 6–7 February 1992
London: BPP Technical Services Ltd
1992, 7 papers

A2376
A survey of pipelines in the North Sea:
incidents during installation, testing and
operation
Protech International BV
Netherlands Industrial Council for Oceanology
1979

A2377
The UK pipeline industry offshore: success or
failure? symposium: London, 23 November
1982
London: Pipeline Industries Guild
1983

A2378
Wave force coefficients for submarine pipelines
Shanhar N H and Cheong H-F
*Journal of Waterway, Port, Coastal & Ocean
Engineering*
1988, Volume 114(4), pages 472–486

A2379
Wave forces on pipelines
Wallingford: Hydraulics Research Station
1982 [Offshore Energy Technology Board
Report, OT-R-8232]

A2380
Wave forces on pipelines: addendum, effects of
surface roughness
Walllingford: Hydraulics Research Station
1982 [OT-R-8233]

A2381
Welded and seamless wrought steel pipes: an
American National Standard
*New York: American Society of Mechanical
Engineers*
1985, vii +30 pages
[ANSI/ASME B36.10 M-1985]

A2382
Workshop on ice scouring and design of
offshore pipelines; Calgary, 18–19 April 1990
*Ottawa: Canada Oil & Gas Lands
Administration*
1990, 499 pages

RISER PIPES

A2383
Applied mechanics of marine riser systems
Morgan G W and Peret J W
Dallas: Energy Publications
1975

A2384
Bulletin on comparison of marine drilling riser
analyses, 1st edition
Washington DC: American Petroleum Institute
1977 [API Bull 2J]

A2385
Design and operation of marine drilling riser
systems, 2nd edition
Dallas: American Petroleum Institute
1984 [API RP2Q]

A2386
Dynamics of compliant risers
Patriakalis N M et al
Journal of Ship Research
1990, Volume 34(2), pages 123–135

A2387
Flexible risers and umbilicals training course
notes; London 4–6 Febnruary 1992
London: BPP Technical Services Ltd
1992, 9 papers

A2388
Flexible risers, special issue of papers from
meeting: London 9 January 1989
Taylor R Eatock (editor)
Engineering Structures
1989, Volume 11(4), pages 205–292

A2389
Investigation into optimised design of flexible
riser systems
Brown P A et al
London: Marine Management (Holdings) Ltd
1989, 10 pages

A2390
A mathematical model for compliant risers
Patrikalakis N M and Chryssostomidis C
*Cambridge MA: MIT Sea Grant College
Program*
1985, 79 pages [MITSG Report No 85-17]

A2391
Proposed offshore installations (emergency
pipe-line valve) regulations 1989
Department of Energy
London: Department of Energy
1989, lv (various pagings)

A2392
Recommended practice for design, rating and
testing of marine drilling riser couplings
Washington DC: American Petroleum Institute
May 1984, 15 pages [API RP 2R]

A2393
Subsea isolation systems: design and
installation; seminar, London 3 October 1990
Swindon: Knighton Enterprises Ltd
1990, 10 papers

SUBSEA PRODUCTION
SYSTEMS

A2394
Applications of subsea systems
Goodfellow Associates
Tulsa: PennWell
1990, 300 pages

A2395
A comparison of wet and dry subsea
installations: seminar; London, 13 January 1983
London: Society for Underwater Technology
1984, 100 pages

A2396
Design and installation of subsea systems:
Subsea International '85 (SUT); London, 15–16
January 1985
London: Graham & Trotman
1985, 247 pages
[Advances in Underwater Technology 2]

A2397
Developments in sub-sea engineering systems:
international symposium; London, 14–15 May
1985
London: Royal Institution of Naval Architects
1985, 2 volumes

A2398
Factors governing the progress of diverless
subsea production on the United Kingdom
continental shelf
Glasgow: Offshore Supplies Office
1991, various pagings

A2399
Failure mode and effect analysis of a subsea
production system
Deegan F J and Burns D J
Transactions, Institute of Marine Engineers
1990, Volume 102(3) pages 191–199

A2400
Future development of offshore oil and gas
fields by subsea production methods
Mohr H and Booth D
*Houston, Inverkeithing: Oil Business
Information Service*
1984, 130 pages

A2401
Management of subsea systems workshop '86;
Aberdeen, 19–20 March 1986
London: Goodfellow Assoiciates
1986, looseleaf

A2402
Modular subsea production systems,
conference; London, 25–26 November 1986)
London: Graham & Trotman (for SUT)
1987, 179 pages
[Advances in Underwater Technology 10]

A2403
Progress in subsea engineering
Jones G (editor)
Bedford UK: BHRA
1989, 140 pages [10 papers]

A2404
Production systems conference: Session P,
Offshore Northern Seas; Stavanger, 26–29
August 1986
Oslo: Norwegian Petroleum Society
1986, 8 papers

A2405
Recommended practice for design and
operation of subsea production systems, 1st
edition
Washington DC: American Petroleum Institute
1987 [API RP 17A]

A2406
Recommended practice for installation,
maintenance and repair of surface safety
valves offshore, 3rd edition
Washington DC: American Petroleum Institute
1991, 13 pages [API RP 14H]

A2407
Reliability prediction of subsea oil/gas
production systems
Hokstad P
Trondheim: SINTEF
1989, 145 pages [Report STF75 A89038]

A2408
Requirements for design guidelines on subsea
installations: the report of a project definition
study
London: UEG
1984, 32 pages [Publications UR 27]

A2409
Safety and reliability of subsea production
systems
Høvik: Det norske Veritas
1985, 27 pages [Guideline No 1-85]

A2410
Satellite & marginal field, 2nd conference;
London, 4–5 May 1991
Hitchin: Offshore Conference Services
1991

A2411
Satellite developments in the oil industry,
seminar; London 13–14 May 1991
London: IBC Technical Services
1991

A2412
Specification for wellhead surface safety valves
and underwater safety valves for offshore
service, 8th edition
Washington DC: American Petroleum Institute
1991 [API Spec 14D]

A2413
Subsea equipment reliabiliity and availability:
publications and conference articles
Hokstad P et al
Trondheim: SINTEF
1989, 126 pages [Report STF75 A89040]

A2414
Subsea '87: subsea trends for shallow and
deep water, new fields and satellites;
Heathrow, 3–4 December 1987
Chipping Norton: The Hawkedon Partnership
1986, 13 papers

A2415
Subsea '88: subsea systems: today's
technology–tomorrow's technology; London,
1–2 December 1988
Chipping Norton: Subsea Engineering News
1988, 14 pages

A2416
Subsea '89: subsea in the 1990s: cost effective
solutions; London, 5–6 December 1989
Chipping Norton: Themedia Ltd
1989, 13 pages

A2417
Subsea '90: subsea achievements and
challenges; London, 11–12 December 1990
Chipping Norton: Themedia Ltd
1990, 11 papers

A2418
Subsea '91: the subsea approach; London, 4–5
December 1991
Chipping Norton: Themedia Ltd
1991, 18 papers

A2419
Subsea International '89: Second generation
subsea production systems; London, 25–26
April 1989
S U T
London: Graham & Trotman
1989, xvii + 245 pages
[Advances in Underwater Technology, Ocean
Science & Offshore Engineering, Vol 20]

A2420
The subsea challenge: conference;
Amsterdam, 23–24 June 1983
London: Society for Underwater Technology
1984, 3 volumes

A2421
Subsea completion: Norwegian subsea
installations; Stavanger, 4–5 May 1987
*Oslo: Norwegian Society of Chartered
Engineers*

A2422
Subsea emergency shutdtown systems,
seminar; London 1 June 1989
Swindon: Knighton Enterprises Ltd
1989

A2423
Subsea production systems: prevention of corrosion problems, conference; Trondheim, 26–27 January 1988
Trondheim: Trondelag Korrosjonstekniske Forening
1988, 11 papers

A2424
Subsea production yearbook 1992
Faringdon, Oxfordshire: Knighton Enterprises
1992

A2425
Subsea separation and transport; Oslo, 6 October 1987
Oslo: Norwegian Society of Chartered Engineers
Rugby: Institution of Chemical Engineers
1987, 111 pages

A2426
Subsea separation and transport II; Oslo, 1–2 March 1989
Oslo: Norwegian Society of Chartered Engineers
1989, 11 pages

A2427
Subsea separation and transport III; London, 23–24 May 1991
Rugby: Institution of Chemical Engineers
1991, 191 pages + 2 loose papers

A2428
Subsea standardisation: trends for the 1990s, seminar; 5 June 1991
Swindon: Knighton Enterprises Ltd
1991

A2429
Survey of difficult oil and gas fields for subsea production
Glasgow: Offshore Supplies Office
1991, various pagings

A2430
Tentative rules for the certification of subsea production systems
Høvik: Det norske Veritas
1984

ANCHORING & MOORING

A2431
Anchoring of floating structures
Tirant P Le and Meunier J (editors)
Paris: Éditions Technip
1990, xvii + 225 pages
[Design Guides for Offshore Structures Vol 2]
Translated from the French: Ancres et lignes d'ancrage

A2432
Anchoring systems: seminar; Aberdeen, 17 February 1982
London: Society for Underwater Technology
1983, 75 pages

A2433
Anchors and chain cables Act 1967
UK Parliament
London: HMSO
1967, 3 pages [Chapter 64]

A2434
Ancres et lignes d'ancrage
ARGEMA
Paris: Éditions Technip
1987, 304 pages
[Guides Practiques sur les Ouvrages en Mer]

A2435
Design considerations for the use of ropes and cables in the marine environment: opportunity brief No 42; MIT Marine Industry colloquium
Cambridge MA: MIT Sea Grant Program
1986, 20 pages [MITSG 86–12]

A2436
Draft bulletin on the design of windlass wildcats for floating offshore structures, 1st edition
Washington DC: American Petroleum Institute
1988 [API Bull 25]

A2437
Monohull mooring study for the Department of Energy: final report
Houlder Offshore Engineering
Boston Spa: BLLD
1990, 65 pages [OTO-90-003]

A2438
Mooring chain failure on semi-submersible drilling rigs: a statistical review
Ellix D M
Høvik: Det norske Veritas
Harwell: MaTSU (distributed by BLLD)
1990, 5 pages
[Offshore Technology Report OTO 89-014]

A2439
Mooring dynamics for offshore applications: part 1, theory; part 2, applications
Triantafyllou M S et al
Cambridge Mass: MIT Sea Grant College Program
1986, 2 volumes [MITSG 86-1 and 86-2]

A2440
Mooring equipment guidelines, 1st edition
Oil Companies International Marine Forum
London: Witherby
1992, 125 pages

A2441
Moorings guidance: dynamic aspects and safety factors
Garen O et al; Det norske Veritas
Boston Spa: BLDSC
1990, 35 pages
[Report OTO 90-009]

A2442
Mooring Systems: a state of the art review
Skop R A
Journal of Offshore Mechanics & Arctic Engineering
1988, Volume 110(4) pages 365-372

A2443
Offshore moorings: conference; East Kilbride, 10 March 1982
Institution of Civil Engineers
London: Thomas Telford
1982, 164 pages

A2444
The oil rig mooring handbook, 2nd edition
Vendrell J
Glasgow: Brown Son & Ferguson
1985, 38 pages, 7 charts

A2445
Qualification testing of steel anchor designs for floating structures, 1st edition
Dallas: American Petroleum Institute
1980 [API RP 2M]

A2446
Recommended practice for in-service inspection of mooring hardware for floating drilling unit, 1st edition
Washington DC: American Petroleum Institute
1987, 41 pages [API RP 2I]

A2447
Recommended practice for the analysis of spread mooring systems for floating drilling units, 2nd edition
Dallas: American Petroleum Institute
1987 51 pages [API RP 2P]

A2448
Recommended practices for design, analysis and maintenance of mooring for floating production systems, 1st edition
Washington DC: American Petroleum Institute
1991 [API RP 2FP1]

A2449
Research on behaviour of piles as anchors for buoyant structures
Ove Arup & Partners (for the Department of Energy)
London: HMSO
1986, vii+80 pages
[Offshore Technology Report OTH 86 215]

A2450
Review of current methods for analysing the characteristics of mooring systems with particular reference to non-linear response and extreme mooring forces: part 1, analytical methods
Chan H
London: Lloyd's Register of Shipping
1980, 45 pages [Hull Structures Report 80/21]

A2451
Specification for mooring chain, 5th edition
Dallas: American Petroleum Institute
1988 [API Spec 2F]

A2452
Spread mooring systems: Rotary drilling, unit V lesson 2
Petroleum Extension Service, University of Texas at Austin and International Association of Drilling Contractors
1976

A2453
Survey of mooring systems on floating drilling units
Houston: Offshore Data Services
[routinely updated]

A2454
The use of anchors in offshore petroleum operations
Puech A
Paris: Éditions Technip
1984, 160 pages

A2455
Wire rope in offshore applications: a review of wire rope research
Chaplin C R and Potts A E (University of Reading)
London: Marine Technology Directorate
1988, vii + 166pages [MTD 99/100]

A2456
Wire rope offshore: a critical review of wire rope endurance research affecting offshore applications
Reading University for HSE and MTD Ltd
London: HMSO
1991, 335 pages [OTH 91 341]

GEOTECHNOLOGY

A2457
Acoustic modelling of the seafloor
Berge A M et al
Stavanger: Norwegian Petroleum Directorate
1984, 77 pages [NPD Contribution No 17]

A2458
Applied high-resolution geophysical methods: offshore geoengineering hazards
Trabant P K
Hemel Hempstead: Simon & Schuster for IHRDC
1984, 279 pages

A2459
3rd Canadian conference in marine geotechnical engineering; St John's, 11–13 June 1986
St John's: C–CORE
1986, 2 volumes

A2460
CRC Handbook of techniques for aquatic sediments sampling
Murdoch A and Macknight S D
Boca Raton: CRC Press
1991, 208 pages

A2461
Geotechnical aspects of coastal and offshore structures: symposium; Bangkok, 14–18 December 1981
Yudhbir and Balasubramanamian A S (editors)
Rotterdam: A A Balkema
1983, 280 pages

A2462
Marine geotechnics
Poulos H G
London: Unwin Hyman
1988, xviii + 473 pages

A2463
Marine geotechnology and nearshore/offshore structures; symposium, Shanghai 1–4 November 1983
Chaney R C and Fang H Y
Philadelplhia: ASTM
1986, x + 372 pages [STP 923]

A2464
Recommended practice, Volume D: Structures–site investigation for offshore structures
Høvik: Det norske Veritas
1987, iv + 8 pages [RP D301]

A2465
Seabed mechanics: International Union of Theoretical & Applied Mechanics symposium; Newcastle upon Tyne, 5–9 September 1983
Denness B (editor)
London: Graham & Trotman
1984, xxi + 281 pages

A2466
Seabed stability near floating structures
McDougal W E and Sulisz W
Journal of Waterway, Port, Coastal & Ocean Engineering
1989, Volume 115 (6), pages 727–739

A2467
Sediment transport by currents and waves, handbook
Rijn L C van
Delft: Delft Hydraulics
1989, loose-leaf and floppy disk

A2468
Shear waves in marine sediment, conference; La Spezia, Italy, 15–19 October 1990
Hovem J M et al (editors)
Dordrecht: Kluwer
1991, 612 pages

A2469
The strength of grouted pile–sleeve
connections for offshore structures; static tests
relating to sleeve buckling
Wimpey Offshore Engineers & Constructors
London: HMSO
1986, 59 pages [OTH 85 223]

A2470
Strength testing of marine sediments:
laboratory and in-situ measurements
Chaney R C and Demars
Philadelphia: ASTM
1985, 557 pages [STP 883]

A2471
Understanding the arctic sea floor for
engineering purposes
National Research Council, Marine Board
Washington DC: National Academy Pres
1982, 141 pages

A2472
Wave-induced breakout of half-buried marine
pipes
Foda M A et al
Journal of Waterway, Port, Coastal & Ocean
Engineering (ASCE)
1990, Volume 116(2) pages 267–286

Seabed sampling & data acquisition

A2473
Marine geological surveying and sampling
Hailwood E A and Kidd R B (editors)
Dordrecht: Kluwer Academic Publishers
1990, 168 pages

A2474
Geotechnical properties of deep sea
sediments: a critical review of measurement
techniques
Schultheiss P J
Wormley: Institute of Oceanographic Sciences
1982, 73 pages [Report No 134]

A2475
Geotechnical investigation of UK test sites for
the foundations of offshore structures
Thomas S (McLelland Ltd)
London: HMSO
1990, 287 pages [OTH 89-294]

A2476
Marine geotechnics: an introduction to site
investigation practice
McQuillan R et al
London: Graham & Trotman
1981

A2477
Offshore seismic exploration: data acquisition,
processing and interpretation
Verma R K
Houston: Gulf Publishing
1987, 608 pages

A2478
Offshore site investigation: conference;
London, 6–8 March 1979
Ardus D A (editor)
London: Graham & Trotman
1980, iv + 291 pages

A2479
Offshore site investigation: SUT conference;
London, 13–14 March 1985
London: Graham & Trotman
1985, 310 pages
[Advances in Underwater Technology 3]

A2480
The pressuremeter and its marine applications,
symposium; Paris, 19–20 April 1982
Paris: Éditions Technip
1982: 440 pages

A2481
The pressuremeter and its marine applications,
2nd international symposium
Briaud and Audibert (editors)
Philadelphia: ASTM
1986, 508 pages [STP 950]

A2482
Seabed reconnaissance and offshore soil
mechanics for the installation of petroleum
structures
le Tirant P
Paris: Éditions Technip
1979, 508 pages [English translation]

A2483
Study of the superficial sea sediments by the
measurement of the reflexion co-efficient
Sousounis I G
Washington DC: NTIS
1983, 167 pages [DE 84751541/WOT]

Foundations

A2484
Capacite portante des pieux
ARGEMA
Paris: Éditions Technip
1988, 320 pages
[Guides Practiques sur les Ouvrages en Mer 3]

A2485
Capacity of offshore friction piles in clay:
opportunity brief No 44
MIT Marine Industry Collegium
Cambridge Mass: MIT Sea Grant Program
1986, 21 pages [MITSG 86-14]

A2486
Foundation of offshore structures; bibliography
Stuttgart: Information Centre for Regional
Planning & Building Construction
1989, 76 pages [ICONDA Bibliography No 39]

A2487
Geotechnical and hydraulic aspects with regard
to seabed and slope stability
Kroezen M et al
Delft; Delft Hydraulics Laboratory
1982, 13 pages [Publication 272]

A2488
Geotechnical practice in offshore engineering:
conference; Austin,Texas, 27–29 April 1983
Wright S G (editor)
New York: American Association of Civil
Engineers
1983, 629 pages

A2489
Local scour and scouring protection of drilling
platforms in the arctic sea environment
Rytkoenen J
Springfield VA: NTIS
1983, 91 pages [NTIS PB84-157262]

A2490
Numerical methods in offshore piling: 3rd
international conference; Nantes, 21–22 May
1986
Paris: Éditions Technip
1987, 544 pages

A2491
Pieux dans les formations carbonatées
ARGEMA
Paris: Éditions Technip
1988, 192 pages
[Guides Practiques sur les Ouvrages en Mer 4]

A2492
Reliability of offshore foundations: state of the
art
Wu T H et al
Journal of Geotechnical Engineering
1989, Volume 115(2), pages 157–178

A2493
Research on the behaviour of displacement
piles in an overconsolidated clay
Bond A J and Jardine R J
London: HMSO
1990, x + 22 pages [Report OTH 89.296]

A2494
Research on the behaviour of piles as anchors
for buoyant structures; final report
Building Research Establishment (for the
Department of Energy, BP
International and Britoil)
London: HMSO
1986, 1 folder

A2495
A review of current practice in the design and
installation of piles for offshore structures
St John H D
London: Underwater Engineering Group
1980, 152 pages [OTP 5]

A2496
Scour around an offshore platform
Carstens T
Ocean Science & Engineering,
1983, Volume 8 (2), pages 157–172

A2497
Scour around seafloor structures
Keith Philpott Consulting with Acres Consulting
Calgary: Infopall
1986, 225 pages [ESRF Report 017]

A2498
Scour below pipelines in waves
Sumer B Murtlu and Fredsøe J
Journal of Waterway, Port, Coastal & Ocean
Engineering (ASCE)
1990, volume 116(3) pages 307–323

A2499
Scour prevention techniques around offshore
structures: seminar; London, 16 December
1980
London: Society for Underwater Technology
1980 pages

A2500
Scouring
Breusers H N C and Raudkivi A J
Rotterdam: Balkema
1990, 190 pages
[Hydraulic Structures Design Manual No 8]

A2501
Seafloor scour: design guidelines for
ocean-founded structures
Herbich J B et al
New York: Marcel Dekker
1984, xi + 320 pages

A2502
Soil reaction stresses on offshore gravity
platforms
Lacasse S and d'Orazio T B
Journal of Geotechnical Engineering
1988, Volume 114(11) pages 1277-1299

A2503
Statement of research needs in offshore
foundation design
*London: Society for Underwater Technology
Geotechnics Committee*
1982, 13 pages

A2504
The strength of grouted pile-sleeve connections
Wimpey Offshore Engineers & Constructors
London: HMSO
1986, 152 pages [OTH 86 210]

A2505
A study of length, longitudinal stiffening and
size effects on grouted pile-sleeve connections
Wimpey Offshore Engineers & Constructors
London: HMSO
1987, vi + 53 pages [OTH 86 230]

ELECTRICAL EQUIPMENT & POWER SUPPLY

A2506
Analysis of characteristics for underwater
power systems
Wiborg J
Høvik: Det norske Veritas
1981, 45 pages [Report 81-1163]

A2507
Code of practice for the safe use of electricity
underwater, 2nd edition
*London: International Association of Offshore
Diving Contractors*
1985, 105 pages

A2508
Electric drives on ships and oil platforms
Yacamini R et al
London: Institute of Marine Engineers
1992, 14 pages

A2509
Electrical power system for fixed offshore
installations; course, Oslo 5–6 February 1985
Oslo: Norwegian Petroleum Society
1985, 16 papers

A2510
Electrical safety code, (Model Code of Safe
Practice in the Petroleum Industry, Part 1), 6th
edition
Institute of Petroleum
Chichester: John Wiley
1991, vii + 46 pages

A2511
Electrical systems for oil and gas production
facilities
Bishop D N
*Research Triangle Park, NC: Instrument
Society of America*
1988, iv + 82 pages

A2512
Electricity in water: safety of divers: part 1,
fundamental principles
Diesen A
Ytre Laksevag: Norwegian Underwater Institute
1980, 64 pages [NUTEC Report 5-80]

A2513
Marine electrical practice, 6th edition
Watson G O, revised by BP Shipping Ltd
London: Butterworth
1990, vii + 308 pages

A2514
Offshore electrical engineering
Gerrard G T
Guildford: Butterworth
1989, 272 pages

A2515
Power plants for offshore platforms:
conference; 3 November 1981
*Bury St Edmunds: Mechanical Engineering
Publications*
1982, 70 pages

A2516
Recommended practice for design and
installation of electrical systems for offshore
production platforms, 3rd edition
Washington DC: American Petroleum Institute
1991 [API RP 14F]

A2517
Recommendations for the electrical and
electronic equipment of mobile and fixed
offshore installations
London: IEE
1983, x + 149 pages

A2518
Recommended practice for classification of
locations for electrical installations at petroleum
facilities, 1st edition
Washington DC: American Petroleum Institute
1991 [RP500]

A2519
Regulations for the electrical and electronic
equipment of ships with recommended practice
for their implementation, 6th edition
Hitchin: Institution of Electrical Engineers
1990, xx + 216 pages

A2520
The safe use of electricity under water
Logan C W
London: Institute of Marine Engineers
1987, 10 pages [meeting paper preprint]

A2521
Underwater electric cabling: seminar; London,
1 October 1981
London: Society for Underwater Technology
1984, 82 pages

A2522
Underwater electrical equipment: some
guidance on protection against shock
Mole G
London: Underwater Engineering Group
1979, 79 pages [Report UR14]

A2523
Underwater power source study
*Newhouse H L and Payne P R (for US
Department of the Navy, Naval Air Systems
Command)*
Springfield VA: NTIS
1981, 57 pages [Report No NADC-81157-30]
[NTIS AD-A112644]

A2524
Underwater power sources: a review of current
needs and availability
London: Underwater Engineering Group
1985, 47 pages [Publication UR 32]

MATERIALS

A2525
Composites at sea: structurally used
composites and piping on board vessels and
platforms; conference, Rotterdam, 13–14
March 1990
*Harderwijk: Ketel Consulting BV for
European Pultrusion Technology Association*
1990, 11 papers

A2526
Design guide on the use of aluminium in
offshore structures
Wimpey Offshore and Alcan Offshore
Ledbury: Oilfield Publications Ltd
1990, 3 volumes

A2527
Design of marine structures in composite
materials
Smith C S
Barking: Elsevier Applied Science
1989, xiv + 392 pages

A2528
Elastomers in oil recovery, seminar;
Manchester, 9 September 1991
Shawbury: RAPRA Technology Ltd
1991, 9 papers

A2529
OMAE 1990: 9th offshore mechanics and arctic
engineering conference, Volume III, materials
engineering; Houston, 18–23 February 1990
*New York: American Society of Mechanical
Engineers*
1990, 2 volumes

A2530
OMAE '91: 10th offshore mechanics and arctic engineering conference, Volume III, materials engineering; Stavanger, 23–28 June 1991
New York: American Society of Mechanical Engineers
1991, 2 volumes

A2531
Polymers in a marine environment, 2nd international conference; London, 13–16 October 1987
London: Institue of Marine Engineers
1989, 233 pages

A2532
Polymers in a marine environment, 3rd international conference; London, 23–24 October 1991
London: Institute of Marine Engineers
1992

A2533
Polymers in offshore engineering, 3rd international conference; Gleneagles, 14–16 June 1988
London: Rubber and Plastics Institute
1988, 1 volume + 1 loose paper

A2534
Rubber in offshore engineering: conference; London, 13–15 April 1983
Stevenson A (editor)
Bristol: Adam Hilger
1984, xii + 346 pages

A2535
Study on lightweight materials for offshore structures
Fulmer Research Institute and Wimpey Offshore
Glasgow: Offshore Supplies Office
1987, 2 volumes
[an extended summary is published separately]

A2536
Tribology offshore, Tribology Group seminar; 10 May 1984
Bury St Edmunds: Mechanical Engineering Publications
1985, 83 pages

Concrete

A2537
Applications for concrete offshore: report of project for UEG
Pell Frischmann Offshore under the supervision of Bayly D R
London: Underwater Engineering Group
1982, 147 pages [Report UR20]

A2538
Behaviour of concrete caisson and tower members
Regan P E and Hamadi Y D
Wexham Springs: Cement & Concrete Association
1980
[Concrete in the Oceans technical report No 4]

A2539
Classification and identification of typical blemishes visible on the surface of concrete underwater
McAlpine Sea Services
London: HMSO
1985, ix + 135 pages [OTH 84 206]
[Concrete in the Oceans Technical Report No 9]
1988 Supplement [OTH 87 261]

A2540
Concrete in the marine environment, 2nd international conference; St Andrews, Canada, 21–26 August 1988
Detroit: American Concrete Institute
1988, vii + 739 pages [SP-109]

A2541
Concrete in the oceans programme: a coordinating report on the whole programme
Leeming M B (Department of Energy)
London: HMSO
1989, viii + 225 pages [OTH 87-248]

A2542
Concrete in the oceans: management summary of the whole programme
Sharp J V et al (Department of Energy)
London: HMSO
1989, v + 19 pages [OTH 87 249]

A2543
Concrete offshore in the nineties: COIN
London Centre for Marine Technology
London: HMSO
1990 [OTH 90 320]

A2544
Concrete ships and floating structures
convention; Rotterdam, November 1979
Turner F H (editor)
London: Thomas Reed Publications
1980, 23 papers

A2545
Designing for temperature effects in concrete
offshore oil-containing structures
Richmond B et al
London: Underwater Engineering Group
1980, 116 pages [Report UR17]

A2546
The effect of temperature variation on creep of
concrete
Bamforth P B
London: Underwater Engineering Group
1980, 44 pages [Technical Note UTN 19]

A2547
Effectiveness of concrete to protect steel
reinforcement from corrosion in marine
structures
Taywood Engineering Research Laboratories
London: HMSO
1989 [OTH 87 247]
[Concrete in the Oceans Technical Report 24]

A2548
Effects of temperature gradients on walls of oil
storage structures
Clarke J L and Symmons R M
*Wexham Springs: Cement & Concrete
Association*
1979, 54 pages
[Concrete in the Oceans Technical Report 3]

A2549
Effects of temperature gradients on the walls of
concrete oil storage structures
Clarke J L and Williams A
London: HMSO
1987 [OTH 87-234]
[Concrete in the Oceans Technical Report 11]

A2550
Exposure tests on reinforced concrete in
seawater
Stilwell J A (Department of Energy)
London: HMSO
1989 [OTH 87 246]
[Concrete in the Oceans Technical Report 23]

A2551
Factors affecting the design for corrosion
protection in concrete offshore and marine
structures
Fidjestol P et al
London: HMSO
1987 [Report OTH 87 237]
[Concrete in the Oceans Technical Report 14]

A2552
Fatigue and corrosion fatigue performance of
16mm diameter reinforcing steel
Austen I M
London: HMSO
1988 [OTH 87-241]
[Concrete in the Oceans Technical Report 18]

A2553
Fatigue and corrosion fatigue strength of
reinforced concrete in seawater
Paterson W S et al
London: Cement & Concrete Association
1982
[Concrete in the Oceans Technical Report 7]

A2554
Fatigue strength of reinforced concrete in
seawater: results from Phase II
Paterson W S and Dill M J
London: HMSO
1988, 57 pages [OTH 87-248]
[Concrete in the Oceans Technical Report 20]

A2555
Handbook for design of undersea,
pressure-resistant concrete structures
*Haynes H H and Rail R D (for Naval Civil
Engineering Laboratory)*
Washington DC: NTIS
1986, 105 pages [AD-A176 233/5/WOT]

A2556
Marine concrete
Marshall A L
Glasgow; Blackie
1990, xii + 401 pages

A2557
Marine concrete: concrete in the marine
environment conference; London, 22–24
September 1986
London: The Concrete Society
1986, 446 pages

A2558
Marine durability survey of the Tongue Sands tower
Taylor Woodrow Construction
Wexham Springs; Cement & Concrete
Association
1980
[Concrete in the Oceans Technical Report No 5]

A2559
Modes of failure of concrete platforms
Dowrick D J
Wexham Springs: Cement & Concrete
Association
1979, 35 pages
[Concrete in the Oceans Technical Report No 2]

A2560
Offshore structures of concrete, bibliography
Stuttgart: Information Centre for Regional
Planning & Building Construction (IRB)
1989, 107 pages [ICONDA Bibliography No 44]

A2561
Performance of concrete in marine environment: conference; St Andrews New Brunswick, 1980
Malhotra V M (editor)
Detroit: American Concrete Institute
1980, 627 pages [Publication SP-65]

A2562
Research needs for the application of lightweight aggregate concrete in sea water
Harris R J S
London: Underwater Engineering Group
1981, 48 pages [Technical Note UTN 22]

A2563
A review of methods for design against implosion of concrete cylinders in offshore structures
Chrapowicki K A and Boon R D
London: HMSO
1988, 54 pages [OTH 87-236]
[Concrete in the Oceans Technical Report 13]

A2564
Surveys of existing concrete marine structures
Taylor Woodrow Research Laboratories
London: HMSO
1991 [OTH 87-244]
[Concrete in the Oceans Technial Report 21]

A2565
Tenue des ouvrages en béton en mer /
Behaviour of offshore conrete structures: 2nd international colloquium; Paris, 12–14 October 1982
Brest: CNEXO
1983, 557 pages
[25 papers in French, 7 papers in English]

A2566
Underwater concreting and repair
McLeish A
Sevenoaks: Edward Arnold
due 1992, 288 pages

Steel

A2567
Assemblages tubulaires sourdés
ARSEM
Paris: Éditions Technip
1985, 360 pages
[Guide practique sur les ouvrages en mer]

A2568
Assessment of fracture mechanics fatigue predictions of T-butt welded connections with complex stress fields
Sharples J K et al
London: HMSO
1989, 34 pages [OTI 88-536]

A2569
Background to new fatigue design guidance for steel welded joints in offshore structures
UK Department of Energy
London: HMSO
1984

A2570
Background to the new static strength for tubular joints in steel offshore structures
Wimpey Offshore
London: HMSO
1990 [OTH 89-308]

A2571
Bayesian techniques in remaining life calculations for welded joints of jacket structures
Wickham A H S and Frieze P A
Transactions of the Institute of Marine
Engineers
1990, Volume 102(3) pages 183-189

A2572
Cohesive buckling programme report
Kenny J P & Partners
Boston Spa: BLDSC
1990, 2 volumes [OTO 90-005]

A2573
Compendium of fracture toughness data on
weldments
University of Strathclyde, Division of Mechanics
of Materials, Materials Testing Laboratories
London: HMSO
1988, 59 pages [OTI 88-534]

A2574
Grouted and mechanical strengthening and
repair of tubular steel offshore structures
Harwood R G and Shuttleworth E P
London: HMSO
1988, 251 pages + 17 microfiches
[OTH 88-283]

A2575
Offshore structures of steel, bibliography
Stuttgart: Information Centre for Regional
Planning & Building Construction (IRB)
1989, 78 pages

A2576
OTJ '88: recent developments in tubular joints
technology; Egham, 4–5 October 1989
UEG Offshore Research
London: UEG
1988, 14 papers

A2577
Recommended practice for preproduction
qualification for steel plates for offshore
structures, 1st edition
Washington DC: American Petroleum Institute
1987 [API RP 2Z]

A2578
Residual stresses in y-nodes and PW joints
Porter Goff R F D et al
London: HMSO
1989, iv + 31 pages [OTH 89-315]

A2579
Single-sided welding of closure joints in large
tubular fabrication: final summary report
The Welding Institute
London: HMSO
1990 [OTH 90-3335]

A2580
Specification for carbon manganese steel plate
for offshore platform tubular joints, 6th edition
Washington DC: American Petroleum Institute
1990 [API Spec 2H]

A2581
Specification for fabricated ssstructural steel
pipe, 4th edition
Washington DC: American Petroleum Institute
1990 [API Spec 2B]

A2582
Specification for steel plates for offshore
structures, produced by thermo-mechanical
control processing (TMCP), 2nd edition
Washington DC: American Petroleum Institute
1990 [API Spec 2W]

A2583
Specification for steel plates,
quenched-and-tempered, for offshore
structures, 2nd edition
Washington DC: American Petroleum Institute
1990 [API Spec 2Y]

A2584
Static strength of large-scale tubular joints: test
programme and results
Wimpey Offshore
London: HMSO
1989 [OTI 89-543]

A2585
Steel in marine structures: 3rd international
ECSC offshore conference (SIMS '87); Delft,
15–18 June 1987
Noordhoek C and de Back J (editors)
Amsterdam: Elsevier Science Publishers
1987, xx + 954 pages
[Developments in Marine Technology, 3]

A2586
Stress concentration factor data from
large-scale tubular joints
Ma A
London: HMSO
1988, iii + 49 pages [OTH 88-284]

A2587
United Kingdom Offshore Steels Research
Project: Phase 1
Peckover R et al
London: HMSO
1988 [OTH 88-282]

A2588
United Kingdom Offshore Steels Research
Project: Phase 1: research contract reports
*Marine Technology Support Unit for the
Department of Energy*
London: HMSO
1988, 39 pages [OTI 89-540]

A2589
United Kingdom Offshore Steels Research
Project, Phase II: project task reports
Pergamon Technical Services (editors)
London: HMSO
1990 [OTH 89-310]

A2590
United Kingdom Offshore Steels Research
Project, Phase II: summary of project task
reports
Techword Services
London: HMSO
1991 vi + 160 **pages** [OTH 89-266]

A2591
United Kingdom Offshore Steel Research
project, phase II: final summary report
editorial panel of UKOSRP-II Programme
Steering Committee
London: HMSO
1987, vi + 133 pages [OTH 87-265]

A2592
Welded tubular joints
*Association de Recherche sur les Structures
Metalliques Marines*
Marshall N (translator from French)
Paris: Éditions Technip
1987, xxi + 323 pages

Conferences, Seminars and Symposia
expected to give rise to published papers

*Entries in italics indicate contacts for further information,
not necessarily the actual organisers of the meetings*

1992

Jan 27–29 Houston TX
Petro-Safe 92: 3rd annual
internationa exhibition &
conference
Petro-Safe

Jan 28–29 Aberdeen
Optimising topside
maintenance,conference
Henry Stewart Conference Studies

Jan 28–31 Valletta, Malta
Moex 92: first Mediterranean oil &
gas exhibition & conference
*Mediterranean Oilfield Services
Co Ltd*

Jan 29 Aberdeen
Forum for discussion: the Marine
Technology Directorate and the
Scottish Subsea Technology
Group
Society for Underwater Technology

Jan 30 London
GIS applied to maritime and
coastal engineering
Institution of Civil Engineers

Jan 31 Aberdeen
Microbial enhancement of oil
recovery, seminar
Oil Plus

Feb 4–5 London
Flexible risers and umbilicals,
course
BPP Technical Training

Feb 4–7 San Diego CA
Produced water: international
symposium

Feb 6 Aberdeen
Automation of remote subsea
intervention, conference
Society for Underwater Technology
Feb 6–7 London
Subsea and pipeline engineering,
course
BPP Technical Training

Feb 8–10 Singapore
Asia Pacific oil & gas conference
SPE

Feb 9–12 Dubai, UAE
5th Arab Oil & Gas Show

Feb 11 London
Crisis management for the
maritime industry, conference
Legal Studies and Services

Feb 11–12 Aberdeen
Latest developments in safe
escape, evacuation and
escape from offshore
installations, conference
IIR Ltd

Feb 12 London
Oil industry partnering conference
Spearhead Exhibitions

Feb 12 Aberdeen
European electric submersible
pump conference
SPE

Feb 12–15 Orlando FL
44th annual convention
Pipeline Contractors' Association

Feb 13–14 Amsterdam,
Netherlands
Offshore pipeline technology,
conference
IBC Technical Services

Feb 14–15 Wokefield Park
Understanding oil supply logistics,
course
Petroleum Economist

Feb 17–18 Westin Stamford,
Singapore
Pacific petroleum & energy
finance & investment exchange
Euromoney &Petroleum Economist

Feb 17–20 Houston TX
Pipeline pigging and inspection,
1992 conference
Gulf Publishing

Feb 17–21 Aberdeen
Petroleum exploration and
development economics, course
DCA Consultants

Feb 17–21 New Orleans LA
Drilling conference
IADC/SPE

Feb 18 London
Oil price information, seminar
Information for Energy Group

Feb 18 London
Gas pipelines BS8010 and risk
assessment: a case study, lecture
Pipeline Industries Guild

Feb 18–19 London
Project management techniques,
course
BPP Technical Training

Feb 18–20 Trondheim, Norway
Flexible pipe technology: recent
research and development
Norwegian Institute of Technology

Feb 19 London
High pressure / high temperature
subsea production systems
*Society for Underwater
Technology*

Feb 20 London
Doing business without paper: the impact of Electronic Data Interchange on the upstream and downstrean oil industry
Institute of Petroleum
Feb 20 London
Future new developments and techniques for offshore activities
Institute of Petroleum, E & P Discussion Group

Feb 20–21 London
Formal safety assessments, course and workshop
BPP Technical Training

Feb 24 London
The fluid dynamics of drilling & completion seminar
Institite of Mechanical Engineers

Feb 25–27 London
Oil and gas law: solving the practical problems, course
Euromoney Legal Training

Feb 25–28 Baltimore
US Hydrographic conference
NOAA

Feb 26–27 Aberdeen
Introduction to remotely operated vehicles, course
Society for Underwater Technology

Feb 26–27 Lafayette LA
Formation damage control symposium
SPE

Feb 26–28 Santiago, Spain
7th Deepwater and marginal oilfield Conference *Offshore Conference Services*

Feb 26–28 London
Offshore pipeline engineering course
IBC Technical Services

Feb 27–28 London
Contractual default and risk limitation in offshore supply and construction contracts, conference
Legal Studies and Services

Mar 1–6 New Delhi, India
14th Congress: Civilisation through civil engineering
International Association for Bridge & Structural Engineering

Mar 8–11 Caracas, Venezuela
2nd Latin American petroleum engineering conference
SPE

Mar 9–11 Kristiansand, Norway
Seismic processing and interpretation: the integrated approach, seminar
Norwegian Petroleum Society

Mar 9–11 Geilo, Norway
3rd Oil Field Chemicals, international symposium: scales, corrosion, environment
Norwegian Society of Chartered Engineers

Mar 10–13 Brighton
Oceanology International conference and exhibition
Spearhead Exhibitions Ltd

Mar 12–13 London
Achieving maximum topside safety through design and modification, conference
IIR Ltd

Mar 19 London
Impact of technical developments on safety cases
IBC Technical Services

Mar 19–20 London
Tidal power '92
ICE

Mar 23–24 Aberdeen
Achieving total quality offshore, conference
IIR Ltd

Mar 23–25 Baltimore MD
Cables and connectors

Mar 23–28 Beijing, China
Petroleum equipment and technology international exhibition

Mar 24–26 Coventry
Pipeline systems, international conference
BHR Group

Mar 24–26 Manchester
3rd Decommissioning and demolition conference
ICE

Mar 24–27 Beijing, China
Petroleum engineering conference

Mar 25–27 Monterey CA
Position location and navigation

Mar 27 Aberdeen
Implications of residual chlorine in seawater injection systems, seminar
Oil Plus

Mar 29 London
4th Petroleum geology of NW Europe conference
Conference Associates

Mar 30–Apr 1 Bergen, Norway
UTC '92: Underwater Technology Conference: Subsea production technology
Norwegian Petroleum Society

Mar 30–Apr 1 Bakersfield CA
Western regional conference and exhibition
SPE

Mar 31–Apr 1 Milan, Italy
3rd Pipeline Rehabilitation, European and Middle Eastern Seminar
Pipeline Integrity Management Ltd

Apr 1–3 Aberdeen
Risk analysis for the offshore industry
IBC Technical Services

Apr 1–3 Strasbourg, Framce
Rational use of energy and the environmental benefits
Watt Committee on Energy

Apr 2–3 St John's, Canada
Managing the environmental impact of offshore oil
Fisheries and Oceans Canada, Science Branch

Apr 2–3 Vilamoura, Portugal
Natural gas policies & technologies
EEC

Apr 5–7 Washington DC
Meeting
National Ocean Industries Association

Apr 5–10 Washington DC
10th advanced seminar on petroleum, minerals, energy & resources law
Internationa Bar Assn, Section on Energy & Natural Resources Law

Apr 6–7 Aberdeen
Health and safety for the offshore
oil and gas industries, conference
Health & Safety Executive/IChemE

Apr 6–8 Bergen, Norway
Petropisces II: 2nd international
conference on fisheries and
offshore petroleum exploitation
Scanews

Apr 7–8 Aberdeen
Technical and economic
challenges in cold water marine
aquaculture, conference
CEMP

Apr 7–8 London
7th annual energy prospects:
post-Soviet Republics & East
Europe
*PlanEcon Inc and
DRI?McGraw-Hill*

Apr 7–9 London
Sasmex '92: Safety at Sea and
Marine Electronics conference
*International Trade Publications
Ltd*

Apr 7–10 Bombay, India
World dredging conference

Apr 8 Aberdeen
Clad engineering seminar
Nickel Development Institute

Apr 13 Aberdeen
Implications of the new safety
case regulations, conference
Health & Safety Executive

Apr 13–14 Oklahoma
Enhanced oil recovery symposium
and exhibition
SPE/DOE

Apr 13–14 Amarillo TX
Midcontinent gas symposium
SPE

Apr 14–15 London
Opening up the post-Soviet gas
industry
*Centre for Foreign Investment &
Privitisation (Moscow) and Royal
Institute of International Affairs*

Apr 22 London
Underwater navigation meeting
Royal Institute of Navigation

Apr 23–27 Hamburg, Germany
Maritime Europe: 21st annual
conference
Maritime Information Association

Apr 22–24 Tulsa
8th Enhanced oil recovery
symposium & exhibition
SPE/DOE

Apr 27–28 London
Welding and weld performance
offshore, conference
IBC Technical Services

Apr 27–29 Southampton
Coastal Engineering '92
conference
Wessex Institute of Technology

Apr 28–29 London
1st European conference on oil &
gas economics, finance &
management
SPE Europe Ltd

Apr 28–29 London
Optimising design and
construction of integrated flowline
bundles
IIR Ltd Industrial

Apr 28–29 Milan, Italy
3rd European and Middle Eastern
pipeline rehabilitation seminar
*Pipeline Integrity Management
(PIM)*

Apr 28–30 Houston Tx
International forum on slim hole /
coiled tubing / horizontal well
technology
Texaco & BP Exploration

Apr 29–30 London
Subsea control and data
acquisition, conference
Society for Underwater Technology

May 4–7 Houston TX
24th Offshore Technology
Conference
Society of Petroleum Engineers

May 4–8 Liege, Belgium
24th Ocean Hydrodynamics
colloquium
Liege University

May 7 Aberdeen
Oilfield chemicals, seminar
IBC Technical Services

May 7–9 Houston TX
Tension-leg platform design,
course
ASME

May 11–12 Bergen, Norway
13th Bergen conference on oil &
economics: The long term global
supply of petroleum - Norway's
competititve position
Norwegian Petroleum Society

May 11–13 Blackpool
Coastal management '92
ICE

May 14
Sharing facilities: the legal,
financial & commercial aspects
*European Study Conferences and
Offshore*

May 18–21 Hamburg, Germany
Remediation of oil spills, seminar
CONCAWE/DGMK

May 18–21 Caspar, Wyoming
Rocky Mountain regional
conference & exhibition
SPE

May 18–22 Newcastle upon Tyne
PRADS '92: 5th practical design of
ships and mobile units, symposium
Newcastle University

May 19 London
The international impact of marine
engineering offshore
Institute of Marine Engineers

May 19 London
Offshore oil and gas: the impact of
the internal market, conference
Confederation of British Industry

May 19–20 Aberdeen
Offshore inspection and
maintenance conference
IBC Technical Services

May 19–21 Miami FL
TOI '92: business opportunities in
marine tourism, conference
Spearhead Exhibitions

May 19–22 Quebec City, Canada
3rd Wave hindcasting and
forecasting international workshop

May 20–21 London
Offshore safety: protection of life
and the environment
Institute of Marine Engineers

May 20–21 London
Developing cost-cutting strategies
for offshore production, conference
IIR Ltd

May 24–27 Perth
32nd Annual petroleum conference
*Australian Petroleum Exploration
Assn Ltd*

May 24–27 Stavanger, Norway
European petroleum computer
conference
SPE

May 25–29 Budapest, Hungary
Hydrocomp '92: interaction of
computational methods ... in
hydraulics, conference
IAHR etc

Jun 1–2 Oslo, Norway
Financing the petroleum industry
in the 1990s, conference
Norwegian Petroleum Society

Jun 1–3 Trondheim, Norway
Eurogas '92: European applied
research conference on natural
gas
*Norwegian Institute of Technology,
Sintef, Norwegian Petroleum
Society*

Jun 1–5 Delft, Netherlamds
Marine safety and environment/
ship production
Delft University of Technology

Jun 1–5 Paris, France
Meeting
*European Asociation of
Exploration Geophysicists*

Jun 1–5 Kona, Hawaii
PACON '92

Jun 2 London
Marine growth of offshore
structures, seminar
*Society for Underwater
Technology / MTD*

Jun 2 Aberdeen
The training of engineering
personnel for the offshore
industry, seminar
IMechE

Jun 2–3 Newcastle
Use of composite materials
offshore
IBC Technical Services

Jun 2–3 Washington DC
AUV '92: Autonomous underwater
vehicle technology, symposium
IEEE

Jun 2–4 Noordwijkerhout,
Netherlands
3rd Annual European well control
conference
*International Association of Drilling
Contractors*

Jun 3 London
Safety management systems for
the offshore industry, industrial
briefing
IIR Ltd

Jun 3–4 Aberdeen
Subsea Europe '92 conference
Knighton Enterprises

Jun 7–10 Calgary
43rd Anntual technical meeting
Petroleum Society of CIM

Jun 7–11 Calgary, Canada
OMAE '92: 11th Offshore
mechanics and arctic engineering
international conference
ASME

Jun 8–9 Genoa, Italy
Marine and offshore engineering
symposium
AIOM

Jun 8–10 London
Land pipeline engineering
conference
IBC Technical Services

Jun 9–11 Maracaibo, Venezuela
Latin American petroleum show
Spearhead Exhibitions

Jun 9–12 San Diego CA
Intervention/ROV'92
Marine Technology Society

Jun 10–12 Edmonton, Canada
15th AMOP: Arctic and marine
oilspill program, technical seminar
Environment Canada

Jun 14–19 San Francisco CA
ISOPE '92: 2nd Offshore and
polar engineering conference

Jun 15–16 London
Satellite and marginal oilfield
development conference
Offshore Conference Services

Jun 15–17 New Orleans LA
1st Remote sensing for marine
and coastal environments: needs
and solutions for pollution
monitoring, control and abatement:
thematic conference
*Environmental Research Institute
of Michigan*

Jun 15–Jul 10 Rhode Island
2nd Coastal Management summer
institute
University of Rhode Island

Jun 16 London
Offshore safety: the response to
Cullen, conference
Institute of Petroleum

Jun 16–17 Copenhagen,
Denmark
Environmental issues and the
rational use of energy: energy
outlook conference
DRI / McGraw-Hill

Jun 17–18 Newcastle upon Tyne
Anglo-Norwegian Oil Symposium
Spearhead Exhibitions

Jun 17–19 Delft, Netherlands
3rd Mathematics of Oil Recovery,
European conference
Delft University of Technolgy

Jun 22–23 London
8th Offshore oil and gas
construction and supply contracts,
conference
IBC Legal Studies and Services

Jun 22–26 Genoa, Italy
Ocean management in global
change

Jun 23 London
Offshore ropes: review of current
development
*Society for Underwater
Technolopgy*

Jun 23–24 London
Recent large scale fully
instrumented pile tests in clay,
conference
*Institution of Civil Engineers and
BP Engineering*

Jun 23–25 St John's, Canada
Offshore Newfoundland '92: 8th
International onshore and offshore
petroleum, exhibition and
conference
Atlantic Expositions Ltd

Jun 23–27 Bethesda MD
Meeting
Undersea and Hyperbaric Medical Society

Jun 26 Aberdeen
Cost effective provision of TSR (temporary safe refuge): industrial briefing
IIR Ltd

Jun 30–Jul 2 London
UDT '92: Undersea Defence Technology conference and exhibition
NUSC

Jul 5–8 Ana, Australia
Spillcon '92
Australian Maritime Authority

Jul 6–7 London
Biotechnology in the petroleum industry, conference
Henry Stewart Conference Studies

Jul 6–7 London
North Sea oil and gas conference: new investment challenges
Financial Times

Jul 7–9 Genoa, Italy
3rd Congress on marine and offshore
AIOM

Jul 7–10 London
BOSS '92: International conference on Behaviour of Offshore Structures
BPP Technical Services

Jul 10 Cardiff, Wales
Marine science: the management challenge
Challenger Society

Jul 19–22 Houston
Petroleum computer conference
SPE

Jul 22–24 Houston TX
5th Artificial intelligence in petroleum exploration and production, annual conference

Aug 2–7 Colorado
Multiphase flow, pumping and separation technology, forum
Society of Petroleum Engineers

Aug 2-10 Santa Domingo
7th Pan American coastal and ocean congress

Aug 3–6 Copenhagen, Denmark
JONSMOD '92
University of Plymouth

Aug 11–13 Cambridge MA
3rd Ice technology international conference
Wessex Institute of Technology

Aug 16–21 Colorado
Completion of horizontal wells, forum
Society of Petroleum Engineers

Aug 25–28 Stavanger, Norway
ONS '90: Offshore Northern Seas
ONS Foundation

Sep 14–16 Paris, France
Eurocas III: 3rd European core analysis symposium
Society of Core Engineers and Institut Francais du Petrole

Sep 14–17 Luxembourg
Underwater acoustics
Commission of the European Communities

Sep 14–17 Varna, Bulgaria
Black Sea '92: 2nd marine technology conference and exhibition

Sep 16–17 Manchester
Assessment and control of risks to the environment and to people
Safety and Reliability Society

Sep 20–25 Madrid, Spain
Congress
World Energy Council

Sep 21–22 Røros, Norway
3rd Multiphase transportation conference: present application and future trends (abstract deadline 1.2.92)
Norwegian Petroleum Society

Sep 21–26 Taranto, Italy
8th Marine fouling and corrosion international congress

Sep 22–24 London
Offshore site investigation and foundation behaviour, conference
Society for Underwater Technology

Sep 22–25 Bermuda
26th annual convention
International Pipe-line and Offshore Contractors' Association

Sep 23–24 London
Structural design against accidental loads - as part of the offshore saety case
ERA Technology

Sep 23–25 Southampton
Hydrocyclones, conference
BHR Group

Sep 28–Oct 1 Amsterdam
Pipeline pigging and integrity monitoring conference
Scientific Surveys

Sep 30–Oct 1 Aberdeen
Practicalities and realities of human factors in offshore safety
Business Seminars International

Sep 30–Oct 1 Stavanger, Norway
4th Conference on reservoir management
Norwegian Petroleum Society

Oct 1 Dyce, Aberdeen
Subsea standardisation seminar
Subsea Engineering News

Oct 1–2 London
Reservoir technology conference
IBC Technical Services

Oct 4–7 Washington DC
Annual meeting
Society of Petroleum Engineers

Oct 6–7 Rotterdam, Netherlands
2nd Composites at sea conference and trade show
European Pultrusion Technology Association

Oct 6–8 London
Interspill '92 conference
Xponent Ltd

Oct 7–8 Aberdeen
Advances in solving oilfield scaling problems
IBC Technical Services Ltd

Oct 8–9 Oslo, Norway
7th Floating production systems: blueprints for the 90s
Offshore Conference Services

Oct 12–16 St Petersburg, Russia
International symposium & exhibition on unconventional hydrocarbon accumulations, problems of E&P
AAPG

Oct 13–16 Aberdeen
IOCE '92: International offshore
contracting and subsea
engineering (formerly IRM/ROV
'92)
Spearhead Exhibitions

Oct 14–15 Bergen, Norway
Offshore operation and
maintenance, conference
Norwegian Petroleum Society

Oct 14–16 Athens, Greece
Natural gas policies and
technologies, part 2: technologies:
European (THERMIE) conference
LDK Consultants

Oct 14–16 Rio de Janeiro, Brazil
Latin-American drilling congress
Institute Brasiliero de Petroleo

Oct 15–16 Oslo, Norway
8th Offshore information
conference: information and
innovation management for
industrial development
FoP / Heriot-Watt University

Oct 18–22 Taos, New Mexico
Conference on invasion &
permeability
SPWLA

Oct 18–23 Rio de Janeiro, Brazil
3rd Latin American Petroleum
Congress
Brazilian Petroleum Institute

Oct 19–21 Washington DC
Annual meeting: Global ocean
partnership
Marine Technology Society

Oct 20 London
Business information services in
the oil industry, conference
Information for Energy Group

Oct 20–22 Manchester
Major hazards - onshore and
offshore, conference
Institution of Chemical Engineers

Oct 21–23 Lisbon, Portugal
Offshore Iberica, Conference
Offshore Conference Services

Oct 21–23 Yokohama, Japan
Techno-Ocean '92: 4th
international symposium
*Japan International Marine
Science & Technology Federation*

Oct 22 Ottawa, Canada
Ocean industries and Europe '92
workshop
EAITC Canada

Oct 23–24 London
Polymers in the marine
environment, conference
Institute of Marine Engineers

Oct 27–29 Madrid, Spain
CADMO '92: Computer aided
design, manufacture and
operation in the marine and
offshore industries
Wessex Institute of Technology

Oct 27–30 Newport RI
Oceans '92 conference
IEEE

Oct 29–30 London
Oil and gas opportunities in the
former Soviet Union
IBC Technical Services

Oct 29–30 Aberdeen
Emerging North Sea technologies,
conference
Scientific Surveys

Oct 29–30 Oslo, Norway
Petroleum tax, conference
Norwegian Petroleum Society

Oct 29- Edinburgh
11th Offshore search and rescue
conference
Leith International Conferences

Nov 3–4 London
Oil pollution: claims, liability and
environmental concerns,
conference
IBC Legal Studies and Services

Nov 3–4 Berlin, Germany
European autumn cas conference:
coping with transition
Overview Conferences

Nov 3–5 Amsterdam, Netherlands
Holland Offshore '92 and
Petrotech '92 conference
RAI

Nov 3–5 Berlin, Germany
4th EC oil and gas technology in a
wider Europe, symposium
PSTI for the EEC

Nov 8–11 Galveston TX
Archie conference
APE AAPG, SPWLA and SEG

Nov 10–12 Brussels, Belgium
Protecting the North Sea,
conference
Institution of Civil Engineers

Nov 11–12 London
Quality management systems for
the offshore industry
IBC Technical Services

Nov 16–17 London
Offshore safety cases, conference
IBC Technical Services

Nov 16–18 Cannes, France
EUROPEC 92
SPE

Nov 24–25 Aberdeen
6th Offshore drilling technology
conference
IBC Technical Services

Nov 25–27 London
PETEX 92: applying new
technologies conference
PETEX Ltd

Nov 30–Dec 2 Trondheim,
Norway
3rd North Sea oil and gas
reservoirs, conference
Norwegian Institute of Technology

Nov 30 Copenhagen, Denmark
Hydro '92: 8th biennial
international symposium
Hydrographic Society

Dec 1–2 Aberdeen
3rd Minimum facilities for offshore
oil and gas production, conference
IBC Technical Services

Dec 1–4 Singapore
8th Offshore South East Asia
conference and exhibition
Society of Petroleum Engineers

Dec 8–9 London
Subsea '92 international
conference: operations to new
concepts
Themedia Ltd

Dec 8–10 Margaret River,
Australia
Physics of estuaries and coastal
seas
*Centre for Water Resources,
University of Western Australia*

Dec 10–11 London
7th Floating Production Systems
conference
IBC Technical Services

1993

Jan 11–13 Dubai, UAE
1st annual Middle East petroleum
& gas conference
IBC Conferences Ltd

Jan 18–21 New Orleans LS
Intervention/ROV 1993 conference
Marine Technology Society

Jan 20 London
North Sea innovations &
economics
Institution of Civil Engineers

Jan 26–28 Houston
Petrosafe '93

Feb 1–2 Stavanger, Norway
Sequence stratigraphy: advances
and applications for exploration
and production in North West
Europe, geology conference
Norwegian Petroleum Society

Feb 1–2 Haugesund, NOrway
3rd Gas transport symposium
Norwegian Petroleum Society

Feb 8–10 Singapore
Asia Pacific Oil & Gas Conference
SPE

Feb 17–18 London
Installation of major offshore
structures and equipment,
conference
Institute of Marine Engineers

Feb 24–25 Aberdeen
Introduction to underwater
engin-eering for oilfield operations,
course
Society for Underwater Technology

Feb 28 –Mar 3 New Orleans
Symposium on reservoir simulation
SPE

Mar 2–5 New Orleans
International symposium on oil
field chemistry
SPE

Mar 3–5 Aberdeen
4th offshore loss prevention
conference
BHR Group Ltd

Mar 10–12 Brighton
Maritime Defence International '93
Spearhead Exhibitions

Mar 15–17 Fort Pierce FL
Cables and connectors
conference
Marine Technology Society

Mar 24–25 London
Wave kinematics and
environmental forces, conference
Society for Underwater Technology

Mar 29–30 Dallas TX
Symposium on hydrocarbon
economics & evaluation
SPE

Mar 29–Apr 1 Tampa FL
13th Oilspill conference:
prevention, behavior, control,
cleanup
American Petroleum Institute

Apr 3–6 Bahrain
8th Middle East oil show &
conference
SPE

Apr 6–8 Miami FL
SASMEX '93: Safety at sea and
marine electronics exhibition and
conference
International Trade Publications

Apr 12–13 Denver CO
Rocky Mountain regional meeting
and low permeability reservoirs
symposium & exhibition
SPE

Apr 28–29 London
Subsea International '93
conference
Society for Underwater Technology

May 3–6 Houston TX
25th Offshore Technology
Conference
Society of Petroleum Engineers

May 26–28 Torremolinos, Spain
Aquaculture Europe '93

Jun 6–11 Singapore
3rd Offshore and polar
engineering conference
ISOPE-93

Jun 23-24 London
Pipeline management '93: 3rd
symposium and exhibition for the
pipeline industries
*Institution of Water &
Environmental Management*

Jun 20–24 Glasgow
OMAE 93: 12th international
offshore mechanics & arctic
engineering conference
TWI

Jun 27–30 St John's, Canada
4th Canadian Marine Geotechnical
Conference
*C-CORE, Memorial University of
Newfoundland*

Jun 27–Jul 10 Aberdeen
13th Environmental assessment
and management international
seminar
CEMP

Jul 7–8 London
3rd Maritime communications and
control, conference
Institute of Marine Engineers

Jul 7–11 Nova Scotia, Canada
Meeting
*Undersea and Hyperbaric Medical
Society*

Jul 19–23 New Orleans LA
8th Coastal and ocean
management symposium
Coastal Zone '93

Aug 17–20 Hamburg, Germany
POAC '93: 12th Port and Ocean
engineering under Arctic
Conditions, conference

Aug 24–27 Stavanger, Norway
ENS '93: Environemnt Northern
Seas
ENS Secretariat

Sep 7–10 Aberdeen
Offshore Europe

Sep 21–24 Long Beach CA
Annual meeting
Marine Technology Society

Oct 3–6 Houston TX
Annual technical conference &
exhibition
SPE

Nov 2–4 Aberdeen
Subtech '93, conference
*Society for Underwater
Technology / AODC*

Nov 23–26 Melbourne, Australia
Offshore Australia 93: 2nd
Australian international oil & gas
exhibition & conference

late '93 Marseille, France
ORIA '93: telerobotics in hostile
environments conference
IIRIAM

1994

Mar 8–11 Brighton
Oceanology International '94
Spearhead Exhibitions

May 2–5 Houston TX
26th Offshore Technology
Conference
Society of Petroleum Engineers

Sep 25–28 New Orleans
Annual Technical Conference &
Exhibition
SPE

1995

May 1–4 Houston TX
27th Offshore Technology
Conference
Society of Petroleum Engineers

1996

May 6–9 Houston TX
28th Offshore Technology
Conference
Society of Petroleum Engineers

1997

May 5–8 Houston TX
29th Offshore Technology
Conference
Society of Petroleum Engineers

1998

May 4–7 Houston TX
30th Offshore Technology
Conference
Society of Petroleum Engineers

1999

May 3–6 Houston TX
31st Offshore Technology
Conference
Society of Petroleum Engineers

2000

May 1–4 Houston TX
32nd Offshore Technology
Conference
Society of Petroleum Engineers

Subject Index

Photography - A1927 A1943

Piling - A2449 A2484-A2505

Pipelines - A0301-A0303
A2305-A2393

Platforms fixed - A2204-A2255

Pollution - A0238 A0285
A1287-A1388

Polymers - A2531-A2533

Positioning - A1832 A1835
A1861 A2031 A2274

Power supply - A1959
A2506-A2524

Production engineering - A0315
A0857-A0887

Pumps - A0922 A2323

Q

Quality assurance - A0486-A0488
A0499 A0879

R

Regional geology, general -
A0642-A0644 A0651-A0652

Regional geology, specific areas -
A0646-A0650 A0653-A0655

Remotely operated vehicles -
A1976-A2000

Renewable energy resources -
A1581-A1611

Repair - A2037-A2047
A2065-A2076 A2566

Reservoir characterization - A0716
A0719-A0721

Reservoir engineering - A0316
A0759-A0806 A0859

Reservoir simulation - A0773
A0768 A0789 A0800

Riser pipes - A1765 A2346
A2383-A2393

Risk - A0546 A0574 A0577
A0578 A1053-A1107

Robotics - A1845 A2019

Rock mechanics - A0680 A0683

Rubber - A2534

S

Safety - A0123-A0124 A0310
A0978-A1158 A A1723

Salvaging - A1531-A1540 A1930

Scour - A2366 A2382 A2489
A2496-A2501

Seabed sampling - A2473-A2483

Search and rescue - A1108-A1158

Seawater analysis - A1642 A1649
A1676

Seismicity - A1659 A1677-A1679

Seismology - A688 A0692
A0697-A0698 A0700
A0702-A0712 A0714 A0758

Semi-submersibles - A2263-A2285

Shallow gas - A0992 A1039

Ships - A003 A0951-A0977

Simulation - A0800 A0857
A0867-A0868, - A880

Single point moorings -
A2240-A2304

Site investigation - A2473-A2483

Software - A0186 A0192

Spill control - A0318 A1233-A1286

Steel - A2567-A2592

Structures - A0130 A0226
A2144-A2286

Submarine cables - A1423 A2020
A2521

Submersibles (vehicles) - A0148
A0153 A1593-A2000

Subsea production systems -
A2394-A2430

Supply vessels - A0131 A0144
A0951-A0977

Support craft - A0136 A0151
A0278 A0951-A0977 A1109
A1150-A1153

Surveying - A0152 A1823-A1838
A2473-A2483

Survival - A1111 A1114
A1127-A1131 A1136-A1137

T

Tanker loading - A2290-A2304

Taxation - A0288 A0595-A0622

Tension-leg platforms -
A2256-A2262

Testing - A2048-A2064

Tidal energy - A1582 A1605

Tools - A1940,1951

Tourism - A1492

Transportation - A0931-A0977

Tribology - A2536

Tugs - A131 A0140 A1967

U

Underwater construction - A0032
A0322 A1839-A1875

Underwater vehicles - A0148
A0153 A1953-A2000

Underwater wellheads -
A2394-A2430

Unmanned submersibles - A0148
A1976-A2000

V

Vibration - A0994

W

Waste disposal - A1541-A1580

Wave energy conversion -
A1581-A1611

Wave loading - A1776-A1822
A2378-A2380

Weather - A1612-A1695

Weight control - A2219
A2229-A2231 A2254-A2255

Welding - A1852 A1872-A1874
A2362

Well logging - A0251
A0722-A0757

Well testing - A0725
A0804-A0806 A0887

Wind energy conversion -
A1597-A1598 A1601-A1602

Wind loading - A1800 A1822
A2166

Wireline logging - A0199

Work offshore - A0570
A0978-A1052 A1071

Index to Authors

A

Aasvistad T   A2124
Abbotts I L   A0687
Abdel-Aal H K   A0521
Abel R B   A1427
Aczel T   A0699
Adams A J   A2212
Adams N J   A0825
Agelidis N   A2154
Agg A R   A1182
Ahmed T   A0776
Akramkhodzhaev A M   A0695
Al-Hassani S T S   A1415
Albaiges J   A1340
Alexander L M   A1442
Allan J D   A1099
Allegri T H   A1556
Allen J R   A0642
Allen P A   A0642
Allen T O   A0876 A0877
Almar-Naess A   A2096
Alsharhan A S   A0681
Ambraseys N   A1679
Anand R P   A1448
Andersen R S   A1093
Anderson J A   A0501
Anderson J W   A1373
Anderson S   A1446
Aquilera R   A0865
Archer J S   A0899
Ardus D A   A0930 A1473 –74
   A2478
Arlockiasamy M   A2189
Arnold K   A0883 A0884
Arnould M   A0064
Arthur Lubinski   A0816--17
Ashworth V   A2121
Atherton D   A2134

Atkins N   A2228
Atteraas L   A1979
Audibert   A2481
Austen I M   A2552
Axelesson K B E   A1759
Ayyub B M   A2174
Azar J J   A0827
Aziz K   A0789

B

Bühlmann A A   A1883
Backus R H   A1437
Badley M E   A0705
Baibakov N K   A0886
Bainbridge C A   A2117
Baker A C   A1605
Baker J M   A1382 A1344
Bakr M   A0062
Balasubramanamian A S   A2461
Baloun C H   A2084
Barnforth P B   A2546
Banfield T   A2135
Baptist C N T   A1540
Barbagallo M B   A0639
Barltrop N D P   A1790 A2212
Barnes C C   A2020
Barratt M J   A1692
Barrett B   A1041
Barrett B N   A0999
Bartels A H   A1066
Barwis J H   A0684
Bashford A S   A1423
Bateman R M   A0743
Bates J H   A0636 A1232 A1383
Bates R L   A0067
Battjes J A   A2151
Bau J   A1329
Bax H   A2063

Baxter L   A1847
Bayliss D A   A2127
Bayliss M R   A2063
Bayne B L   A1361
Beal R C   A1625
Beazley R   A1389
Bedwell C   A2005
Beeby A W   A2086
Beggs D H   A0878
Behrens C   A1697
Bennet P B   A1908
Bennett F L   A1704
Bentham R W   A1063
Beredjick N   A0592
Berendsen A M   A2133
Berge A M   A2457
Berge H   A1610
Berge S   A2098 A2315
Berger W H   A1486
Berkeley K G C   A2119
Berry R W   A0084
Berteaux H O   A2287
Bevan J   A0032
Beydoun Z R   A0654
Billington C J   A2209
Bird E C F   A0395
Birkinshaw M   A2228
Bischke   A0689
Bishop D N   A2511
Bishop J R   A1819
Bjerkholt O   A0523
Bjørke Ø   A2061
Bjorlykke K   A0685
Blackbourn G A   A0729
Blake G   A1461
Blunden J   A1430 A1588
Boberg T C   A0802

Cronan D   A1515

Cross T A   A0707

Crouse P C   A0509

Cruikshank M   A1517

Cryssostomidis C   A2150

Cullen, The Hon Lord   A1085

Cunningham P   A1989

Cunningham P J   A2089

D

d'Itri F M   A1394

d'Mello C A   A2270

d'Orazio T B   A2502

Dahl T E   A2366

Daintith T   A0594

Dake L P   A0774   A0792

Daling P S   A1297

Dallmeyer D G   A1487

Dalrymple R A   A1817

Darley H C H   A0815

Davey T G   A2318

Davies M E   A1800   A1822

Davies R L   A2220

Davies S R H   A1228

Dawe R A   A0889

Dawson T H   A2188

Day A H   A2267

de Back J   A1091   A2585

De Figueiredo R J P   A0691

de la Mare R F   A2305

de Oliveira J G   A2160

de Ruijter W P M   A1686

Deakin E B   A0006   A0571

Dean R G   A1817

Deegan F J   A2399

Deighton P   A0775   A2032

Delvigne G A L   A1343

Demars K R   A1555   A2470

Demirbilek Z   A2261

Dempster A   A0029

Denness B   A2465

Dent P   A1288

Denys R   A2348

Dera J   A1647

Desbrandes R   A0730

Devereux M P   A0606

DeVorsey L   A1487

Dewan J   A0731

Dewsnap R F   A2114

Dickey D A   A0718

Dickins D   A1255

Dicks B   A1309   A1379

Dickson M   A0002

Diesen A   A1843   A2512

Dikekrs A J   A0717

Dill M J   A2554

Docherty J I C   A1433

Doelling N   A1858

Dommermuth D G   A1685

Donaldson E C   A0764-65   A0781   A0783

Donohue D A T   A0892

Dover W D   A2077   A2099

Dow R S   A2145

Dowling P J   A2154

Dowrick D J   A2559

Draper L   A1689

Dredall I   A1385

Drinnan R W   A1214   A1228

Duckers L J   A1609

Duckworth D F   A1298

Dupuy R-J   A1466   A1467

During E D D   A2081

Dyke P P G   A1639   A1668

Dyson R J   A1902

Dzienkowski J S   A0605

E

Earlougher R C   A0725

Earney F C F   A1513

Economides M J   A0800

Edwards S Sir   A0780

Eide O I   A2098   A2110   A2193

Ellett D J   A1653

Ellinas C P   A2153   A2306

Elliott A J   A1690

Elliott D H   A1908

Ellis D   A0754

Ellix D M   A2438

Ellwood K B   A0016

Engelhardt F R   A1170   A1175   A1357

Enkvist E   A1716

Eranti E   A1716

Ertas A   A0456-57

Etherton J J   A0012

Etkin D S   A1301

Etnyre L M   A0733

Etter P C   A2034

Evans D V   A1589

Evans L V   A2138

F

Fürstner U   A1304   A1339

Faÿ H   A0956

Fagetti E   A0051

Fairhurst C P   A0786

Fairhurst P   A2308

Falco A F de O   A1589

Faltinsen O M   A1806   A1821

Fang H Y   A2463

Farrington J W   A1359

Farrow R S   A1458

Faulkner D   A0954   A2169 –71   A2199   A2260   A2262

Favero C A   A0615

Fee D   A0098

Fee D A   A0564

Feldman R   A0086

Fenwick J   A0120

Ffooks R C   A0958

Fidjestol P   A2551

Fjaer E   A0680

Flaherty L M   A1268

Fleet A J   A0666   A0679

Fleming N C   A1920

Flory J F   A2294

Foda M A   A2472

Forsberg G   A1539

Foster L A   A0022

Fransson L A   A1759

Fraser K   A0840

Fredsøe J   A2498

Freedman B   A1313

Freeman T J   A1549

Freestone D   A1468–69

Freitag M   A1882

Frieze P A   A2571

G

Gabolde G   A0824

Gage G H   A0530

Gale L   A0030

Gallun R A   A0539

Galperin E I   A0714

Gamble J C   A1925

Gangadharan S N   A1984

Garen O   A2441

Garner J F   A1251

Garushev A R   A0886

Gates E T   A1073

Gaythwaite J   A2176

Geraci J R   A1375

German M I   A0888

Gernon B J   A2292

Gerrard G T   A2514

Gerwick B C   A2159

Geyer R A   A1336 A1509 A1962

Gibson R   A0010 A0011

Gidley J L   A0880

Gilfillan E S   A1381

Gillot R H   A1281 A1368

Ginsburg N   A1476

Giuliano F A   A0896

Given D   A1978

Gjelstad B   A2298

Glas P C G   A1346

Glassner M I   A1464

Glennie K W   A0653 A0678

Glinka G   A2099

Gogonenkov G N   A0712

Golan M   A0887

Graff W J   A2173

Gran S   A1709 A1818

Grandal B   A0502

Grant A   A1597

Grasshoff K   A1649

Gray G   A0815

Green C D   A0930

Greene G   A1175

Greene G D   A1842

Gregory J B   A1522

Griffiths C R   A1621

Grogan W C   A1328

Gross M G   A1564

Gross M Grant   A1666

Grossling B F   A0651 A0652
A0895

Grunau H R   A0682

Gudmestad O T   A1688

Gupta A   A2094

Guy J   A1833

Guy N G   A2344

H

Hackman D J   A1951

Hagoort J   A0772

Hailwood E A   A2473

Halbouty M T   A0660 A0664

Halliwell M   A0559

Halltar R F   A0620

Hallwood C P   A0528

Halmo G   A1231

Halsey S D   A1427

Halstead B W   A1881

Ham C R   A0638

Hamadi Y D   A2538

Han D J   A2201

Hancox D   A1538

Hankinson R L   A0075

Hann D   A0588 A0597

Hansen A   A1762

Harbison I   A0931

Hardage B A   A0692 A0758

Harding E T   A1858

Harding J E   A2154

Hardisty J   A1618

Hardman R F P   A0686

Harris A H   A0536

Harris L M   A2266

Harris R J S   A2562

Harrison P   A0573

Hart P H M   A2335

Harvey W W   A1519

Harwood R G   A2574

Hassab J C   A2023

Hauguel A   A1796

Haux G F K   A1855

Hawkins M   A0760

Haynes H H   A2555

Hays E E   A1847

Haywood M G   A2064

Heal G   A0516

Hearn G E   A1811

Heeps H S   A1639

Heezen B C   A0360

Hekstra G P   A1188

Heleran J van   A1788

Herbich J B   A1719 A1791
A2331 A2501

Higgins S   A1081

Hilchie D W   A0722 A0727

Hill A D   A0749

Hill C   A0629

Hill E C   A2108

Hill G   A1436

Hirakawa S   A0788

Hoagland K D   A2138

Hobson G B   A0677 A0897

Hoefeld J   A1431

Hogben N   A1787

Hohn M E   A0663

Hokstad P   A2364 A2407 A2413

Holmyard J M   A2065

Honarpour M   A0796

Hood D W   A1574

Hooft J P   A1777

Hooper G V   A1568

Hopkins J S   A1614

Horland M   A1675

Horn M K   A1425

House D J   A0939 A1130

House J D   A1106

Hovem J M   A2468

Howarth M J   A1617

Howarth W   A1386

Howells R W L   A0999

Hsu T H   A2147

Hudson F D   A1112

Hueck van der Plas, E H   A1546

Hughes H W D   A1037

Hui D   A0457

Humphrey P B   A1516

Hunter T   A0032

Hurford N   A1342

Hurst A   A0736

Hutt G A   A2335

Hvamb O G   A2298

Hyne N J   A0058

I

Ijstra T   A1468   A1469

Illing L V   A0677

Imarisio G   A0447   A0891   A0894

Ingham A E   A1829

Inglis T A   A0818

Ingmanson D E   A1665

Ingstad O   A0935   A1049   A2003

Iredell R   A0898

Isaacson M   A1798

J

Jackson J A   A0067

Jacobsen E   A1916

Jacobsen N G   A1906

Jaffe J S   A2036

Jain K C   A0691

Jakobsen E   A1918

Jamart B M   A1682

James G W   A1934

Jardine R J   A2493

Jaske C e   A2083

Jenkins G   A0517–18   A0563

Jenyon M K   A0702

Jimei X   A2122

Johansen Ø   A1630

Johnson D E   A0755

Johnson H A   A1996

Johnson S P   A1221

Johnston C S   A1392   A1569

Jones G   A2403

Jones L W   A0860   A2123

Jones M E   A0797

Jones M E W   A1970

Jordan R E   A1320

Jorden J R   A0756–57

Joshi S D   A0866

Joulia J P   A0119

Judd A G   A1675

Justus J R   A1591

K

Kabir C S   A0793

Kagan B A   A1626

Kalim Z   A0589

Kalske S   A1958

Kam J C   A2077

Karayannis-Bacon H   A0015

Karstad O   A1044

Karunakaran D   A1681   A1813   A2272

Kasoulides G C   A1417

Kastor R L   A0830

Kaufman A A   A0739

Keller G V   A0739

Kemp A G   A0511   A0532   A0596   A0603   A1402 –03   A1406

Kemp G   A0844

Kenchington R A   A1456

Kennish M J   A1364

Kester D R   A1385   A1543

Ketchum B H   A1385   A1565

Khandekar M L   A1670

Kidd R B   A2473

Kildow J T   A1571

King N W   A2325

King P R   A0780

King R   A1789

Kinne O   A1196

Kirk C L   A2163

Kleppe C   A0668

Klingstedt J P   A0608–09

Knight H G   A1594

Kobranova V N   A0745

Koederitz L F   A0779

Konuk I   A0455   A0909

Koops W   A1253

Korevaar C G   A1655

Kornberg H   A1354

Koudstaal R   A1388

Kozik T J   A0917

Kramer C J M   A1199

Krein H L   A1984

Kristoferson L A   A1600

Krock Hans-Jurgen   A1592

Kroezen M   A2487

Kumar S   A0863

Kunzendorf H   A1510

Kuznetsov V V   A0745

L

Labo J   A0746

Lacasse S   A2502

Laevastu T   A1644

Lake L W   A0720–21   A0763

Lalwani C S   A1502

Lameijer J N F   A0960

Landry T A   A0366

Landry T W   A1460

Lang K R   A0892

Lange R   A1171

Langenkamp R D   A0069   A0072   A0556

Langeraar W   A1837

LaQue F L   A2107

Larsen C M   A2370

Last G   A1983

Laughton M A   A1599

Lavargne M   A0710

Lavi A   A1595

Lawrence P F   A2101

le Tirant P   A2482

Leanza U   A1462

Lee J S   A2260

Lee M J   A1845

Lee S   A0898

Leeming M B   A2070   A2541

Leitch M   A0092

Lenn C P   A2308

Lenz W   A1657

Lerche I   A0643–44

Letouzey J   A0672

Lewis J R   A2080

Light I M   A1123

Lilley P   A0617

Lindburg B   A0905

Lindeberg Erik G B   A1323

Link P K   A0656

Littmann W   A0790

Lockwood M   A1435–36   A1831

Lodge A E   A0453   A0505   A0525

Logan C W   A2520

Lou J Y K   A2292

Lou Y K   A1714

Lovegrove M   A1080

Lowe A V   A1447

Index to Publishing Organisations

A

AAPG
A0202 A0235 A0660 A0664
A0386

Aberdeen City Libraries,
Commerical & Technical Dept
A0002

Aberdeen Petroleum Publishing
A0150 A0203 A0232

ABN-AMRO Bank A0353

Academic Press
A0643 –44 A0710 A0721
A0737 A1313 A1375 A1564

Acres Consulting A2497

Adam Hilger A2534

Advisory Committee on Pollution
of the Sea A1287 A1332
A1360

AERE Harwell A2116

AIOM A0401

Air Accidents Investigation Branch
A1087

Akademie Verlag A1634

Alcan Offshore A2526

Allen, W H A1001

American Association of Civil
Engineers A2488

American Bureau of Shipping
A0969 A1850 A1974 A2197
A2213 A2277 A2280 A2300

American Chemical Society
A0872 A1358

American Concrete Institute
A2540 A2561

American Gas Association
A0014 A0888

American Geological Institute
A0067

American Geophysical Society
A0360

American Institute of Physics
A1671

American National Standards
Institute
A1027 A1127 A1233 A1852
A1928 A1942 A1948–50
A2031

American Petroleum Institute
A0017 A0048 A0207
A0845–47 A0849–52
A0881–82 A0904 A0924–26
A0946 A0950 A0970
A1028–29 A1104 A1145–46
A1149 A1120 A1159
A1225–26 A1263 A1271–74
A1277 A1310 A1319 A1371
A1381 A2059 A2155–56
A2211 A2233 A2237–39
A2258 A2356 A2359
A2360–62 A2368–69
A2384–85 A2392 A2405–6
A2412 A2436 A2445–48
A2451 A2516 A2518 A2577
A2580–83

American Society for Testing &
Materials
A0695 A0699 A0712 A0715
A1160–61 A1236 A1244
A1252 A1268 A1290–91
A1555 A2084 A2463 A2470
A2481

American Society of Civil
Engineers
A0249 A1592 A1704 A2261

American Society of Mechanical
Engineers
A0279 A0455–57 A0458–60
A0468–69 A0829–31 A0917
A0909 A1021–22 A1584
A1710–12 A1714 A1742–47
A1765 A1975 A2311 A2322
A2341–42 A2347 A2353
A2381 A2529–30

American Welding Society A1874

Andersen - see Arthur Andersen

Ann Arbor Science
A1320 A1193

Annual Review of Fluid Mechanics
A1821

AOTC A1698–99

Applied Ocean Research A1788

Applied Science Publishers
A0789

Aqua Publications A1803

Aquatic Science & Fisheries
Information System A0051

Arab Petroleum Research Center
A0088 A0205

ARAE A1776

Arctic Laboratories A1567

Ardor Publishing A1438

ARGEMA A2434 A2484 A2491

Arguments & Facts International
A0099

Arlington Technical Services
A1519

ARSEM A2567

Arthur Andersen & Co A0535
A0554 A0600–1 A0618

Arthur Young Extractive Industries
Group A0530

Asian Oil & Gas Publications
A0206

ASR Books A0028

Association de Recherche sur les
Structures Metalliques Marines
A2592

Association of British Offshore
Industries A0031

Association of Diving Contractors
A1894

Association of Offshore Diving
Contractors
A1864 A1896 A1909 A1977

Association Technique Maritime et
Aéronautique A1848

Gulf Publishing
A0050 A0075 A0160 A0199
A0272 A0303 A0326 A0349
A0509 A0672 A0694 A0703
A0728 A0776 A0779 A0794
A0811 A0815–17 A0820
A0844 A0863–65 A0883–84
A1073 A1337 A1719 A1791
A2147 A2154 A2173 A2477

Gulf R & D Company/Gulf Canada
A1057

H

Habitat Scotland   A1788

Hall (G K) & Co   A0007

Halyard Offshore   A2303

Harcourt Brace Jovanovich
A0296 A0301 A0832

Harper Collins   A2218

Hawkedon Conferences   A2157

Hawkedon Partnership   A0428
A1769 A2414

Hazardous Cargo Bulletin   A0938

Health & Safety Commission
A1122 A1912

Health & Safety Executive
A0999 A1001 A1003 A1887
A1891 A1912 A2456

Health & Safety Executive, Oil
Industry Advisory Committee
A0981 A0983 A1023 A1025

Heinemann
A0960 A1182 A2130

Helsinki University of Technology
A1869

Hemisphere   A1340 A1580
A1590 A2150

Henry Stewart Conference Studies
A2045 A2073

Her Majesty's Nautical Almanac
Office   A0966

Heriot-Watt University
A0038 A1098

Heyden   A0677

Hilchie Inc   A0722 A0727

Hilger - see Adam Hilger

Hipwell Baume Associates
A1890

HMSO
A0073 A0121–24 A0126
A0233 A0347 A0470 A0566
A0595 A0598–99 A0610–13
A0616 A0625 A0631 A0634
A0700 A0797 A0856
A0913–15 A0919 A0966
A0974 A0998 A1001 A1003
A1008 A1015–16 A1047
A1058–59 A1061 A1066
A1085 A1087–88 A1091
A1095 A1097 A1099 A1100
A1109 A1111 A1113 A1116
A1122 A1134–38 A1154
A1158 A1322 A1354–55
A1400–01 A1526 A1550
A1552 A1569 A1572 A1575
A1597 A1603 A1615 A1617
A1621 A1628 A1631–32
A1648 A1651 A1653 A1659
A1669 A1677 A1689–90
A1692 A1694 A1778
A1780–82 A1790 A1805
A1807–08 A1819 A1853
A1884 A1890–92 A1899
A1902 A1904 A1935 A1938
A2013 A2050 A2053–54
A2060 A2065 A2068–69
A2074 A2078–79 A2082
A2087–89 A2093 A2095
A2100–01 A2103 A2111–14
A2116 A2120 A2125 A2148
A2206 A2225 A2228 A2236
A2246–48 A2285 A2309
A2313 A2318–19 A2325
A2333–36 A2346 A2372–74
A2433 A2449 A2456 A2469
A2475 A2493–94 A2504–05
A2539 A2541–43 A2547
A2549–52 A2554 A2563–64
A2568 A2570 A2573–74
A2578–79 A2584 A2586–91

Hoare Govett & Smith Rea Energy
Associates   A0276

Hodder & Stoughton
A1430 A1588

Hollobone (T A) & Co Ltd
A0004–05

Hollobone Hibbert
A0259 A0573 A0623–24

Houlder Offshore Engineering
A2437

House of Commons Energy
Committee
A1015–16 A1400–01

House of Lords Select Committee
A1552

Hungarian Hydrocarbon Institute
A0778

Hydraulics Research   A1691

Hydraulics Research Station
A2379–80

Hydrographer of the Navy
A0952 A0965

Hydrographic Society
A1824 A1826–28 A1830

Hydrographic Society,
Netherlands Branch   A1529

I

IABSE   A2172

IAGA   A0331

IBC Conferences   A0490

IBC Financial Focus   A0541
A0553

IBC Legal Studies & Services
A1069 A1351 A1352

IBC Technical Services
A0170 A0415 A0423–24
A0444–45 A0473–75 A0499
A0504 A0841 A0486 A0980
A0986 A0988–989 A1014
A1076 A1079 A1094 A1110
A1179 A1207 A1397–98
A2011 A2208 A2252 A2328
A2332 A2337–40 A2411

ICC Information Group   A0095

ICES Advisory Committee on
Marine Pollution   A1370

ICI   A1560

ICS   A0934

IEA/OECD   A0239

IEEE
A0246 A1736–40 A1976

IHRDC   A0853 A2458

IIR Industrial   A0488 A2066

IIR Ltd   A0546

IIRIAM   A1987 A2019

IKU   A0753 A1323

IMO/FAO/UNESCO .. Joint Group
of Experts   see GESAMP

Imperial College
A1514 A1679 A2069

Indian Institute of Technology
A1721–22

Industrieanlagen
Betriebsgesellschaft A1812

Infopall A2497

Information Centre for Regional
Planning & Building
Construction
A1783 A2486 A2560 A2575

Information for Energy Group
A0012 A0031

Inland Revenue A0612

Institute for Fiscal Studies
A0606–07

Institute for Industrial Research
A2040

Institute for International Research
A2039

Institute of Energy A0089

Institute of Geological Sciences
A0385

Institute of Marine Engineers
A0253 A0905–06 A0922
A1011–12 A1186 A1224
A1614 A1815 A1954 A2265
A2317 A2399 A2508 A2520
A2531–32 A2571

Institute of Metals A2102 A2131

Institute of Oceanographic
Sciences Deacon Laboratory
A0332 A1581 A1624 A1694
A2474

Institute of Offshore Engineering
A0024–27 A1062 A1075
A1328 A1344 A1382 A1167
A1391 A1836

Institute of Petroleum
A0031 A0038 A0102 A0299
A0307 A0453 A0505 A0525
A0677 A0822 A0897 A0978
A1020 A1151 A1309 A1248
A1259 A1280 A2108 A2321
A2343 A2510

Institute of Petroleum Accounting,
University of North Texas
A0006 A0571

Institute of Sanitary Engineering,
Polytechnic of Milan A1578

Institute of Social & Economics
Research A1106

Institution of Chemical Engineers
A0479 A0910 A1024 A1741
A2425 A2427

Institution of Civil Engineers
A0524 A1582 A2071 A2077
A2097 A2443

Institution of Corrosion Science &
Technology A2121

Institution of Electrical Engineers
A1605 A2517 A2519

Institution of Engineers of Ireland
A1353

Institution of Mechanical
Engineers
A1036 A1844 A2149

Institution of Radio & Electronic
Engineers
A1140 A2007–08

Instrument Society of America
A2511

Interdepartmental Committee on
Large Experimental Facilities
A0138

International Association for
Hydraulic Research
A1640 A1796 A1809

International Association for Sea
Survival Training A1156

International Association of Drilling
Contractors A1010 A2452

International Association of
Offshore Diving Contractors
A1969 A2507

International Association of Ports
and Harbours A2296

International Association on Water
Pollution Research & Control
A0112

International Atomic Energy
Agency
A1541–42 A1566

International Bar Association
A0584

International Chamber of Shipping
A0934 A0975 A1537 A2052
A2296

International Council for the
Exploration of the Sea
A0333 A1369–70

International Federation of
Institutes for Advanced Study
A1388

International Institute of Welding
A1873

International Labour Office
A1032 A1052

International Maritime
Organisation
A0240 A0331 A0927 A0961
A1128 A1155 A1192 A1195
A1226 A1230 A1260
A1264–66 A1302 A1330
A1366 A1409 A1557 A1880
A2010 A2012 A2264

International Petroleum Industry
Environmental Conservation
Association A1237

International Seabed Authority
A1448

International Society of Offshore &
Polar Engineers
A1717–18 A1756

International Tanker Owners
Pollution Federation A0334
A1311 A1283

International Trade Publ Ltd
A0310

International Transport Workers'
Federation A1013 A0979

International Tribunal for the Law
of the Sea A1449

IOC Committee on Ocean
Processes and Climate
A1636

IRO-Journal A1706

ISMS Secretariat A2178

J

James Capel A1080

Japan Marine Science &
Technology Center
A0242 A1968

Japan Petroleum Energy
Consultants A0243

John Wiley & Sons
A0039 A0563 A0677 A0702
A0802 A0806 A0822 A0897
A0978 A1096 A1108 A1196
A1248 A1259 A1280 A1298
A1309 A1385 A1486 A1545
A1553 A1596 A1662 A1667
A1683–84 A1764 A1814
A1730 A1777 A1779 A1801
A1882 A2107–08 A2159
A2165 A2287 A2343 A2510

Johns Hopkins University Press
A1625

Merrill - see Charles E Merrill

Meteorological Office   A1598

Metocean Consultancy   A1628
A1633 A1645

Meyers R J & Associates   A1276

Middlesex Polytechnic   A0999

Milo (Consultant Engineers) Ltd
A1944

Minerals Management Service -
see US'Dept of the Interior

Ministry for Mines & Energy
A0490

Ministry of Agriculture, Fisheries &
Food   A1572

Ministry of Local Government &
Labour   A1150

Ministry of Transport & Public
Works
A1187 A1190–91 A1198

MIT A0335 A1685 'A2160
A2259

MIT Marine Industry Collegium
A2485

MIT Press A1437 A1797 A1839

MIT Sea Grant College Program
A1571 A1858 A1959 A2390
A2435 A2439 A2485

Monterey Bay Aquarium Research
Institute   A1845

Moscow Board of Scientific &
Technical Society of Oil & Gas
Industry   A0109

MTD Ltd
A0281 A0093 A0108 A0116
A0118 A1037 A1654 A1866
A2058 A2129 A2140 A2456

Myers - see Robert J Myers

N

Nantes Chamber of Commerce &
Industry   A0402

National Academy Press
A1034 A1035 A1194 A1250
A1275 A1303 A1308 A1350
A1481 A1533 A1577 A1593
A2048 A2471

National Association of Corrosion
Engineers A2040 A2118
A2136

National Engineering Laboratory
A1583

National Maritime Institute [now
British Martitime Technology]
A1061 A1787 A1800 A1822
A2285

National Research Council
A1035 A1194

National Research Council
Canada   A2179

National Research Council
Canada Institute for Marine
Dynamics A0130

National Research Council, Marine
Board
A1533 A1577 A1593 A1705
A2030 A2471

National Sea Grant College
Program, National Weather
Service   A1641

National Technical Information
Service (NTIS)
A0043 A1213 A1407 A1525
A2091 A2294 A2483 A2489
A2523 A2555

National Transportation Safety
Board A1060 A1072

Nature Conservancy Council
A1376 A1384

Nautical Institute   A0964

Naval Architect   A2263

Naval Civil Engineering Laboratory
A0337 A0851 A2555

Naval Ocean Systems Centre
A0338

Naval Research Laboratory
A1996

Naval Sea Systems Command
A1932–33

Naval Surface Warfare Center
A1840 A1862

Naval Underwater Systems Centre
A0339

NESDA A0172

Netherlands Industrial Council for
Oceanology
A0173 A2329 A2376

Netherlands Institute for the Law
of the Sea
A1443–45

Netherlands North Sea Directorate
A1253

New England River Basins
Commission A2326

Newcastle Oil Symposium A2253

Newfoundland & Labrador Dept of
Development   A0181

Newfoundland & Labrador
Petroleum Directorate A2226

Newfoundland Ocean Industries
Association A0434 A0451

Newfoundland Offshore Petroleum
Board A0392

Nijhoff - see Martinus Nijhoff

NOAA A1215 A1903 A1988

Noble Denton
A0928 A1631 A2268

Nogepa A0993

NOIA A0171

Nopec a/s A0374 A0381

Norconsultants A0209 A0266

Nordic Council A1345

Norges Tekniske Hogskole
A2370

Noroil Publishing A0228 A0231
A0377

Norsk Olje Revy A0267

Norsk Sivilingeniørers Forening
A2222 A2254

North Sea Books A0132

North Sea Forum A1200

North Sea Monitor A0263

North Sea Observer A0264

Northern Development Company
A0175

Norwegian Fire Research
Laboratory A1101

Norwegian Hydrodynamics
Laboratory
A0341–42 A1793 A2366

Norwegian Institute of Technology
A0526 A0668 A2275

Norwegian Ministry of Justice
A1074

Norwegian Petroleum Directorate
A0046 A0073 A0582 A0614
A0635 A0947 A0962 A1086
A1153 A1673 A1898 A1913
A1939 A2168 A2240 A2363
A2457

Norwegian Petroleum Society
A0113 A0471 A0500
A0507–08 A0662 A0665
A0937 A0941 A1185 A1219
A1979 A2227 A2234 A2310
A2404 A2509

Norwegian Society of Chartered
Engineers
A0871 A0972 A2027 A2255
A2421 A2426

Norwegian State Pollution Control
Authority (STF) A1324

Norwegian Underwater Institute
A1916 A2512

Norwegian Underwater
Technology Centre (NUTEC)
A0343 A0501 A1895

Norwegian University Press
A1171

Noyes Data A1276 A1568
A2115

NTNF A1045

O

Ocean Engineering A1816

Ocean Engineering Information
Centre A0041

Ocean Oil Weekly Report A0327

Ocean Science & Engineering
A2496

Oceana
A0590 A1449 A1462 A1465

OCS A0412–13 A0422

OECD A1249 A1306

OECD Nuclear Energy Agency
A1544

Office for Official Publications of
the European Communities
A0013

Office of Technology Assessment
Task Force A1580

Office of the Oceanographer of the
Navy A1971

Offshore A0178 A0388

Offshore Conference & Exhibitions
A0461 A0907 A0916 A0936
A1007–08 A1410–11 A1531
A1865 A2015–18 A2042–44
A2055–57 A2067 A2072
A2104 A2126 A2229–31
A2410

Offshore Data Services
A0268–69 A0278 A0282–83
A1077 A2204 A2453

Offshore Directories A0174

Offshore Energy Technology
Board
A1915 A1934 A1945

Offshore Engineering Society
A0524

Offshore Industry Liaison
Committee A0570

Offshore Information Centre
A0570

Offshore Northern Seas
Foundation A0183 A0399
A0472 A1757 A2371

Offshore Patrol A0931

Offshore South East Asia
A0476–47

Offshore Supplies Office
A0019 A0029 A0114 A0128
A0196 A0542 A0552 A1729
A2249 A2398 A2429 A2535

OGCI
A0656 A0693 A0713 A0860
A0876–78 A2123

Oil & Chemical Pollution
A1253–54 A1256 A1300
A1321 A1342–43 A1372

Oil & Gas Journal A1415

Oil & Gas Law & Taxation Review
A1389

Oil & Gas News A0289

Oil Business Information Service
A2400

Oil Companies International
Marine Forum
A1275 A1537 A2290–93
A2295–96 A2299 A2302
A2304 A2440

Oil Industry Accounting Committee
A0529

Oil Industry Advisory Committee
A0566

Oil Industry International E & P
Forum A1399

Oil Information Services Norway
AS A0179

Oil News Service A0291

Oilfield Publications
A0057 A0131 A0133–36
A0139–42 A0151–52
A0154–55 A0158 A0165
A0193 A0361 A0363 A0367
A0370–73 A0375 A0378–79
A0384 A0387 A0389
A0393–94 A0398 A0807
A0809–10 A0939 A0967
A1658 A1983 A1995 A2038
A2303 A2526

Ole Camae A0164

OPEC A0293

Open University Course Team
A1424 A1430 A1470 A1619
A1660–61 A1676 A1695

Orca Ltd A2053

Ordnance Survey A0385

Oslo and Paris Commissions
A1365 A1162 A1184 A1188
A1204 A1216–18 A1223
A1235

Oslo Commission A1573

OTC
A1749–54

OTTER Group A1629

OUP A0015 A0547 A0626

Ove Arup & Partners
A1563 A2050 A2449

Oxford Institute for Energy Studies
A0547 A0615 A0626

Oyez Scientific & Technical
Service A0942 A2305

P

PA Technology A1899

Paint Research Association A2137

Pallister Resource Management
A0330 A1018 A1057

Paragon House A1471

Paris Commission
A1163–64 A1324

Patten - see Judith Patten

Seaconsult Ltd   A1615

Seakem Oceanography Ltd
   A1211

Seaways   A2298

SEG   A0691   A0708   A0746

Selection & Industrial Training
   Administration   A2135

Selvig Publishing   A0180

SFT/Statfjord Unit Joint Research
   Project   A1209

Shallow Gas Group, Heriot-Watt
   University   A0992

Sharnel   A0280

Shaw & Sons   A1386

Shell International Petroleum Co
   A0127   A0138   A1245

Simon & Schuster   A0522   A2458

Singapore Association of
   Shipbuilders & Repairers
   A0191

Singapore Exhibition Services
   A0478

SINTEF
   A0348   A0929   A0935   A1030
   A1049   A1053   A1065   A1093
   A1101   A1117   A1165   A1229
   A1297   A1299   A1356   A1630
   A1681   A1813   A2003   A2061
   A2090   A2098   A2191   A2193
   A2272   A2315   A2364   A2366
   A2407   A2413

Smith Rea Energy Analysts
   A0346   A0576   A0701   A0775
   A0799   A0808   A0823   A0842
   A0861   A0890   A0944   A1019
   A1501   A1835   A1854   A2004
   A2191   A2327   A2351

Smith Rea Energy Associates
   A1405   A2032

Society for Underwater
   Technology
   A0323   A0416   A0968   A1039
   A1434   A1585   A1549   A1622
   A1652   A1727   A1771   A1786
   A1825   A1841   A1846   A1856
   A1863   A1875   A1888   A1953
   A1955   A1964–66   A2014
   A2025   A2033   A2037   A2142
   A2355   A2395   A2419–20
   A2432   A2499   A2521   A2021

Society for Underwater
   Technology Geotechnics
   Committee   A2503

Society of Chemical Industry
   A0826   A2080   A2132 6

Society of Core Analysts   A0723
   A0724

Society of Naval Architects &
   Marine Engineers
   A2177   A2274

Society of Petroleum Engineers
   A0107   A0248   A0313–17
   A0403–04   A0406   A0417
   A0419   A0420–21   A0425
   A0427   A0431–33   A0435–39
   A0441–43   A0450   A0466
   A0480–85   A0489   A0491–94
   A0496–98   A0510   A0513–15
   A0569   A0719   A0725   A0744
   A0748–49   A0756–57   A0762
   A0766   A0769–70   A0784–85
   A0801   A0804   A0813   A0819
   A0834–38   A0857   A0859
   A0862   A0867   A0869
   A0873–75   A0880   A0893
   A0900   A0985   A1212

Society of Professional Well-Log
   Analysts   A0750–52

Solar Energy Society   A1609

SPE Education & Accreditation
   Committee   A0903

SPE/DOE   A0767

Spearhead
   A0446   A0464–66   A0990
   A1050   A1492   A1715   A1732
   A1735   A1864   A1875

SPIB   A1205

Spon E & F N
   A1130   A1838   A1861   A2063
   A2106

Springer-Verlag
   A0237   A0674   A0684–85
   A0745   A1006   A1304   A1339
   A1362   A1426   A1441   A1589
   A1657   A1668   A1670   A1789
   A1883   A1905   A2083   A2094

Stanford Maritime Press   A1540

Stanley Paul   A1917   A1921

State of Hawaii Dept of Planning &
   Economic Development
   A1516

State Pollution Control Authority
   A1169

Statfjord Unit Owners   A1210

Steiner P   A0768

Stevens & Sons   A1536

Strathclyde Regional Council
   A0177   A0194

Studia   A0462

Submex
   A0032   A1884   A1911   A1929

Subsea Conferences
   A1767–68

Subsea Engineering News
   A2415

Supervisor of Diving, US Navy
   A1931

Sveriges Mekanfrbund (for
   Swedocean)   A0195

Swedish National Institute of
   Radiation Protection   A1548

Swedish Trade Council   A2124

Swedish Trade Fair Foundation
   A1772   A1871

Swedocean   A0195

Sweet & Maxwell   A0594

T

TAPIR
   A0418   A1610   A1762   A2096
   A2152   A2202

Taylor & Francis
   A0255   A1456   A1458–59

Taylor Woodrow Construction
   A2558

Taylor Woodrow Research
   Laboratories   A2564

Taywood Engineering Research
   Laboratories
   A1601   A2088   A2547

Technica   A1100   A1113   A1938

Technical Indexes Ltd   A0187

Technical Research Centre of
   Finland (VTT)   A1761   A1958

Technical Resources Inc   A1282

Techno-Ocean '88 Secretariat
   A1770

Technology Development Centre
   (TEKES)   A1716

Technology Forum   A0996–97

Techword Services
   A1126   A1650   A2051   A2068
   A2590

Telford - see Thomas Telford

Tetra Tech   A1317

Themedia Ltd
   A0429–30   A2416–18

Thomas Reed
A1538 A1532 A2544

Thomas Telford
A0277 A0503 A0524
A1395–96 A1582 A2071
A2097 A2161 A2443

Thomson Communications
(Scandinavia) A0188

Times Books A0390

TNO A1318 A1362

TNO-IWECO A1867

Total Edition Presse A0798

Trans Tech Publications A2187

Trondelag Korrosjonstekniske
Forening A2423

U

Uddeholm Stainless A2124

Uichida Rokakuho A0788

UK Module Constructors'
Association Ltd A0531

UK Offshore Operators
Association
A0977 A0984 A1026 A1075
A1222 A1068 A1124 A1125
A1412 A1348 A1390 A1962

UK Parliament
A0919 A0920 A1550 A2433

UMIST Dept of Civil & Structural
Engineering A2175

UN A0331

UN Dept of International Economic
& Social Affairs A0586

UN Economic Commission for
Europe A0470

Undersea & Hyperbaric Medical
Society A0245 A0321

Undersea Medical Society A1900
A1907

Underwater Association A1901

Underwater Engineering Group
A0994 A1547 A1878
A1909–10 A1915 A1922
A1924 A1937 A1940 A1945
A1957 A1972 A2070 A2075
A2139 A2143 A2190 A2210
A2217 A2219 A2221 A2408
A2495 A2522 A2524 A2537
A2545–46 A2562 A2576

Underwater Systems Design
A0322 A1528

Underwater Technology A1404

Underwater Technology
Foundation A0502

UNEP A0331

UNESCO
A0074 A0331 A1635 A1680
A1920 A2028

United Kingdom Atomic Energy
Authority A2054

United Nations A0331 A1823

United Nations Environment
Programme A1377

United Nations Ocean Economics
& Technology Branch
A1504 A1508

United Nations Office for Ocean
Affairs and the Law of the Sea
A0020–21 A1422 A1450–52

United States Department of
Energy, Energy Information
Administration A0560

United States General Accounting
Office A1208

Universitetsforlaget A1044 A1918

University College of Cape Breton
A0987

University of Aberdeen
A0511 A0532 A0596 A0603
A1402–03 A1406

University of Alaska A1363

University of Auckland A1700

University of Bristol A1586

University of California A1810

University of California Press
A0350

University of Chicago Press
A1476–79

University of Glasgow
A0527 A0954 A1723 A2196
A2199 A2260 A2262 A2286

University of New Hampshire
A1997 A1998 A1999 A2000

University of Newcastle upon Tyne
A0001

University of Oldenburg
A1368 A1281

University of Strathclyde
A0973 A1857 A2284 A2573

University of Texas at Austin,
Petroleum Extension Service
A0054 A0085 A0921 A0940
A1129

University of Trondheim, Marine
Technology Centre A2110

University of Tulsa A0047 A0066

University of Reading A2455

Unwin & Hyman A0676

Unwin Brothers A1635

Unwin Hyman
A0528 A1177 A1464 A1646
A2462

Urban Verlag
A0087 A0222 A0290

US Coastguard, NTIS A1243

US Congress, Office of
Technology Assessment
A1512

US Dept of Commerce A1831
A1988

US Dept of the Interior
A1831

US Dept of the Interior,
Minerals Management Service
A0097 A0344–45 A0380
A0630 A0638–41 A0908
A0948 A1054–56 A1070
A1206 A1213 A1317 A1407
A1416 A1418–19 A1455
A1517 A1522 A2049 A2062
A2297

US Dept of the Navy A2523

US Dept of Transportation
National Maritime Research
Center A0086

US Environmental Protection
Agency A1282

US Geological Survey
A0049 A0352 A1432

US Government Printing Office
A0560 A1432 A1512 A1932
A1933

US National Bureau of Standards
A2241

US National Oceanic &
Atmospheric Administration
(NOAA)
A1215 A1903 A1988

US National Research Council
A1250 A1303 A1764

Journal of
MARINE ENVIRONMENTAL ENGINEERING

New

Editor in Chief
Michael S. Bruno
Director, Davidson Laboratory, Stevens Institute of Technology, Hoboken, New Jersey 07030, USA

Associate Editors
Iver W. Duedall, Professor, Department of Oceanography, Ocean Engineering and Environmental Science, Florida Institute of Technology, Melbourne, Florida 32901, USA

Albert J. Williams III, Chairman, Applied Ocean Physics and Engineering Department, Woods Hole Oceanographic Institution, Woods Hole, Massachusetts 02543, USA

This international journal provides coverage of the many diverse issues that confront engineers and scientists seeking solutions to environmental problems in ocean and estuary waters, and inland seas. The journal is unique in both its interdisciplinary nature, and its focus on applied science and engineering. As a result, it provides a vehicle for the publication of papers not normally found in traditional development-oriented journals.

Topics of interest include, but are not limited to:

✧ marine waste disposal

✧ dredging

✧ ocean outfalls

✧ beach erosion

✧ habitat enhancement/creation

✧ oil spill prevention and remediation

✧ non-point pollution analysis

✧ aquatic toxicology

✧ fisheries management

Of interest in all cases are the applications of computer models, ocean instrumentation and laboratory experiments to the solution of marine environmental problems.

Call for papers

Contributions to this new journal are welcome. Please send manuscripts to any of the three editors above. All papers are peer-reviewed by experts in the appropriate discipline.

Send now for your FREE SAMPLE COPY
First issue due October 1992

Gordon and Breach Science Publishers,
c/o STBS, Marketing Department

PO Box 786 Cooper Station, New York, NY 10276, USA
Tel: (212) 243-4411/4543
FAX: (212) 645-2459

PO Box 90, Reading, Berkshire, RG1 8JL, UK
Tel: (0734) 560080
FAX: (0734) 568211

STBS - Distributor for Gordon and Breach Science Publishers

Gordon and Breach Science Publishers
Switzerland · Great Britain· Australia · Belgium · France · Germany · India · Japan
Malaysia · Netherlands · Russia · Singapore · USA

Section B

ORGANISATIONS
providing
INFORMATION SERVICES

compiled with Judith Whittick,
Barbara Rodden, Shirley Oue & Margaret Reynolds

MARINE STRUCTURES

*Published by Elsevier Applied Science in association with the
Internatioal Ship and Offshore Structures Congress.*

Consulting Editor: D. Faulkner, *University of Glasgow, UK.*
Editors: P.A. Frieze, *Twickenham, Middlesex, UK,* **A. Mansour,**
University of California, Berkeley, USA and **Y. Ueda,** *Osaka University, Japan.*

AIMS AND SCOPE

This journal provides a medium for presentation and discussion of the latest developments in research, design, fabrication and in-service experience relating to marine structures, i.e. all structures of steel, concrete, light alloy or composite construction having an interface with the sea, including ships, fixed and mobile offshore platforms, submarine and submersibles, pipelines, subsea systems for shallow and deep ocean operations and coastal structures such as piers.

The scope of the journal covers:

• definition of the ocean environment and of loads exerted by waves, currents, winds, tides and ice • seabed foundations and structural interaction • evaluation of static and dynamic structural response including collapse behaviour • collision mechanics • fatigue and fracture • materials (and their selection), corrosion and other forms of degradation • formulation and application of design methods and criteria including use of reliability analysis, optimisation techniques and CAD • inspection and structural monitoring, repair and maintenance • fabrication, launching, installation and decommissioning techniques.

ABSTRACTED/INDEXED IN:

Applied Mechanics Reviews, International Structural Engineering Abstracts, International Civil Engineering Abstracts, Oceanic Abstracts, Offshore Engineering Abstracts, Selected Water Resource Abstracts.

SUBSCRIPTION INFORMATION:

Volume 5 (1992), ISSN 0951 8339
Frequency: 6 issues per volume, 1 volume per year.
Worldwide delivery £177.00/US$336.00.
All subscriptions are airspeeded.

ELSEVIER SCIENCE PUBLISHERS
Crown House, Linton Road, Barking, Essex IG11 8JU, UK.
or for customers in North America
Elsevier Science Publishers, Journal Information Center,
655 Avenue of the Americas, New York, NY 10010, USA.

INTRODUCTION

This Section (which evolved from Heriot-Watt University's *Guide to Information Services in Marine Technology*) has been completely revised and expanded, particularly into North America.

Questionnaires were sent to organisations identified as being of relevance to the users of this Guide and only information provided on returned questionnaires appears in these pages. As well as details of their addresses, contacts, interests, facilities and services, respondents were asked to identify themselves using the following categories:

A   public or national library

B   institution, university, professional association, learned society

C   research establishment

D   Government department or agency having direct or indirect influence on the formulation of regulations or standards relevant to the petroleum and marine technology industries

E   independent consultancy

This information appears in each entry and category listings are provided at the end of the Section.

Organisations are in alphabetical order and consequently they have not been given reference numbers: a checklist precedes the main entries. (Note: addresses from this Section are identified by "¶" in Section D : Directory, where many other useful addresses can be found)

Sybil Richardson, for

Judith Whittick, Barbara Rodden and Shirley Oue
C-Core, Memorial University of Newfoundland

Margaret Reynolds
Well Village, Lincolnshire

August 1992

INTERNATIONAL MARITIME BOUNDARIES

edited by:
Jonathan I. Charney, Vanderbilt University School of Law
and
Lewis M. Alexander, University of Rhode Island

A unique and comprehensive study of every
International Maritime Boundary
in the World

This unique publication systematically studies all of the international maritime boundaries in the world. It contains the text of every modern maritime boundary agreement and descriptions of boundaries that have been judicially established; maps illustrating those boundaries; detailed analyses of them; papers by experts examining the status of maritime boundary delimitations in each of ten regions of the world; and papers analyzing key issues in maritime boundary theory and practice from a global perspective.

The book fills a major gap in the information and analyses available to persons interested in the subject. For the operator in the field it comprehensively collects official descriptions of the existing boundary lines; and identifies the boundaries that are not yet established and, in some cases, why. For the diplomat and technical personnel assisting efforts to resolve maritime boundaries, the book provides comprehensive information on the techniques used by others in the past, and analyzes their problems and successes. For the international lawyer, judge, and academic, the book analyzes the current practices, trends and issues that will contribute to the customary and conventional international law and theory on the subject. For the geographer, the book is essential to an understanding of the partitioning of ocean space and the transboundary issues associated with ocean uses.

Contents

Price: (Two Volume Set) £507/US$850   Pages: 2172
Binding: Hardback   ISBN: 0-7923-1187-6
Publication Date: November 1992

Available from your normal book supplier or directly from the Publishers:
Kluwer Academic Publishers, Order Department, Distribution Centre, PO Box 322, 3300 AH Dordrecht, Netherlands.

Tel: +31 (0) 78 524 400   Fax: +31 (0) 78 524 474.

Contributors

Andronico O. Adede Lewis M. Alexander
David Anderson Peter Beazley Derek W. Bowett
Jonathan I. Charney David Colson Blair Hankey
Giampiero Francalanci Erik Franckx Keith Highet
Eduardo Jimenez de Arechaga Barbara Kwiatkowska
Leonard Legault Kaldone Nweihed Bernard H. Oxman
Choon-Ho Park Robert F. Pietrowski, Jr.
J.R. Victor Prescott Tullio Scovazzi Robert W. Smith
Louis Sohn Elizabeth Verville Prosper Weil

Martinus Nijhoff
A member of Wolters Kluwer Academic Publishers
LONDON/DORDRECHT/BOSTON

SECTION B CONTENTS

In this section the main entries about information services are in alphabetical order by organisation.

Alphabetical List of Entries

A

Aberdeen City Libraries
ABS Europe Ltd
Alberta Oil Sands Technology & Research Authority
American Association of Petroleum Geologists
American Gas Association
American Society for Testing & Materials
American Society of Civil Engineers
American Society of Mechanical Engineers
AODC
Arctic Institute of North America
Association of Consulting Engineers of Canada

B

Bedford Institute of Oceanography Library
BHR Group Ltd - CALtec Ltd
BMT Cortec Ltd
British Cement Association
British Geological Survey
British Institute of Management
British Library, Science Reference & Information Ser
Building Research Establishment

C

Canada Institute for Scientific & Technical Information (CISTI)
Canada-Newfoundland Offshore Petroleum Board
Canadian Association of Oilwell Drilling Contractors
Canadian Centre for Marine Communications
Canadian Geotechnical Society
Canadian Library Association
Canadian Maritime Industries Association
Canadian Petroleum Association
Canadian Petroleum Products Institute (CPPI)
Canadian Society of Petroleum Geologists
Canadian Standards Association
Chartered Institution of Building Services Engineers
Colorado School of Mines
CONCAWE
COPPE/UFRJ -Federal University of Rio de Janeiro

D

Danish Maritime Institute
Department of Energy Library (UK)
Department of Fisheries & Oceans (Canada)
Department of Trade & Industry (UK)
Department of Transport (UK)
Det norske Veritas - Veritec
Duke University Marine Laboratory

E

Empresa National Del Petroleo
Energy Information Administration (USA)
Environment & Natural Resources Institute (USA)
ERA Technology Ltd
ESL Information Services Inc
European Association of Exploration Geophysicists

F

Food & Agriculture Organization of the United
    Nations

G

Geological Society of America
Gulf Canada Resources Ltd

H

Health & Safety Executive (UK)
Highlands & Islands Enterprise
HR Wallingford
Hydrographic Office (UK)

I

IFREMER
IKU - Continental Shelf & Petroleum Research
    Institute
Institute of Gas Technology
Institute of Marine Engineers
Institute of Oceanographic Sciences Deacon
    Laboratory
Institute of Petroleum

Institution of Chemical Engineers

Institution of Civil Engineers

Institution of Electrical Engineers

International Association of Drilling Contractors

International Center for Marine Resource Dev'ment

International Oceanographic Foundation

International Society of Offshore & Polar Engineers

L

Lloyd's Register of Shipping

Louisiana Sea Grant College Program

M

Marine Board (USA)

Marine Technology Directorate (MTD Ltd)

Marine Technology Society

MARINTEK

Massachusetts Coastal Zone Management Office

Memorial University of Newfoundland

Mineralogical Society of America

MIT Sea Grant Program

MITS (Michigan Information Transfer Source)

N

National Academy Press

National Energy Board

National Oceanic and Atmospheric Administration

National Physical Laboratory

National Radiological Protection Board

National Sea Grant Depository

Naval Sea Systems Command

NEL - National Engineering Laboratory Exec Agency

Newfoundland & Labrador Dept of Mines & Energy

Newfoundland Ocean Industries Association

Norwegian Geotechnical Institute

Norwegian Underwater Technology Centre - NUTEC

O

Ocean Engineering Information Centre

Ocean Oil Information

Oceans Institute of Canada

Office of Ocean & Coastal Resource Management

Offshore Data Services Inc

Offshore Engineering Information Service

Offshore Supplies Office Department of Energy

Offshore Technology Conference

P

Paint Research Association

Paleontological Society

Pallister Resource Management Ltd

Petroleum Abstracts

Petroleum Communication Foundation

Petroleum Extension Service (PETEX)

Petroleum Industry Training Service

Petroleum Science & Technology Institute

Petroleum Society of CIM

Plymouth Marine Laboratory

Precision Microfilming Services

R

Rapra Technology Ltd

Royal Institution of Naval Architects

Royal Society for the Prevention of Accidents

S

Sea Fish Industry Authority (UK)

SINTEF Group

Society for Mining, Metallurgy & Exploration (SME)

Society for Underwater Technology - SUT

Society of Naval Architects & Marine Engineers

Standards Council of Canada

T

Texas A&M University

Transportation Research Board

TWI

U

University of Alaska Southeast

University of California, Berkeley Extension

University of Rhode Island

University of Southern California

US Army Cold Regions Research & Engineering
    Laboratory (CRREL)

US Department of Energy

US Department of the Interior, Minerals
    Management Service

W

Warren Spring Laboratory

Water Research Centre

Woods Hole Oceanographic Institution

INFORMATION SERVICES

Aberdeen City Libraries

Business & Technical Department
Rosemount Viaduct
Aberdeen AB9 1GU
Scotland
Tel: (0224) 634622
Contact: Mrs L Smith-Burnett
Tlx: 739196, Mail Box ABD 019
Fax: (0224) 641985

Category: A
The Business & Technical Department is a
specialist unit which provides information on
business, technical and scientific subjects.
Subject interests
Subject interests include oil, gas and all
aspects of company and product information.
Library & Information facilities
Items may be borrowed by people who live,
work or study in Aberdeen. The reference
facility is available to all enquirers and there are
photocopying and microfiche copying facilities.
A database is maintained on Grampian-based
firms; the Department is the Patents
Information Service for the North of Scotland; it
also has a comprehensive collection of British,
American and European standards. Online
search services are available.

ABS Europe Ltd

(A Division of the American Bureau of Shipping)
ABS House
1 Frying Pan Alley
London E1 7HR
UK
Tel: (071) 247 3255
*Contact: James W Templeton, Vice President
of Technology and Business*
Development
Tlx: 8885621
Fax: (071) 377 2455

Category: B E
The American Bureau of Shipping, an
international independent classification society,
has been an integral part of the offshore
industry since 1954. ABS classes (determines
structural and mechanical fitness for intended
service) all types of fixed and mobile units
according to the ABS industry-derived
standards. It also certifies such structures and
components according to other industry
standards as well as national and international
requirements. ABS acts as a certifying authority
for the Australian, Dutch, British, US and other
governments.
Special interests
Offshore services include computer
applications for design analysis, hydrodynamic
studies, classification services during
construction, installation and repair, and
periodic surveys of offshore structures such as
mobile offshore drilling units, fixed platforms,
offshore processing plants, SPMs, underwater
vessels and habitats.
Information services
The ABS technhical advisory services provide
quality assurance and inspection services.

Alberta Oil Sands Technology & Research Authority (AOSTRA)

600 Highfield Place
10010–106 Street
Edmonton, Alberta T5J 3L8
Canada
Tel: (403) 427 8382
Contact: Gene Lau, Library & Information
Services Technical Information Officer
Tlx: 037-3519
Fax: (403) 427 3198 / 422 9112

Category: A
This public library is concerned with the
dissemination of technical information
Subject interests
Energy, fuel, power; mining; geology,
hydrology; occupational and environmental
health and safety
Special interests
Oil sands, heavy oil (world wide), enhanced oil
recovery (mostly Canada).
Library & Information facilities
AOSI (Alberta Oil Sands Index) includes
14,000 documents; HERI (Heavy Oil) includes
11,000 documents. They are available as hard
copies or microfiche.There are no interlibrary
loans, but photocopies can be provided. An
enquiry service is available.
Hours: 8:00am–4:30 pm Monday–Friday
Databases: AOSI and HERI INDICES

Alberta Oil Sands Technology & Research Authority (AOSTRA)

500 Highfield Place,
10010–106th Street
Edmonton, Alberta T5J 3L8
Canada
Tel: (403) 427 7623
Contact: Jackie D Alton, Executive Assistant &
Public Relations Officer
Tlx: 037 3519
Fax: (403) 427 3198

Category: C
Subject interests
Energy, fuel, power.
This research establishment is concerned with
the promotion of oil sands development

American Association of Petroleum Geologists

PO Box 979
Tulsa, OK 74101
USA
Tel: (918) 584 2555
Contact: Larry M Nation, Communications
Director
Fax: (918) 584 0469

Category: A B
Subject interests
Energy, fuel, power; geology, meteorology,
hydrology,oceanography, geodesy,
cartography, business information and
statistics.
The Association aims to further the
educational, scientific and professional goals of
petroleum geologists: it has 34,000 members in
104 countries.
Special interests
Geology as related to exploration for
hydrocarbons.

American Gas Association

1515 Wilson Boulevard
Arlington, VA 22209
USA
Tel: (703) 841 8400
Contact: John Erickson, Vice-President,
Operating & Engineering Group
Fax: (703) 841 8406

Category: B
Natural Gas Trade Association
Library & Information facilities
Library: (703) 841 8415; provides open access
Hours: 9:00am5:00pm EST
Enquiry service: (703) 841 8559
Public Information: (703) 841 8660

American Society for Testing & Materials

1916 Race Street
Philadelphia, PA 19103
USA
Tel: (215) 299 5400
Contact: Philip L Lively, Marketing Director
Fax: (215) 977 9679

Category: B
Subject interests
Freshwater and marine biology, agriculture and fisheries; chemistry; energy, fuel, power; mining, metallurgy; mechanical engineering; civil and hydraulic engineering building; occupational and environmental health and safety, fire prevention, fire fighting.
The ASTM is involved in the development and publication of standard specifications and test methods
Publication: Materials of Engineering
An enquiry service is available to members of ASTM

American Society of Civil Engineers

Civil Engineering
345 East 47th Street
New York, NY 10017
USA
Tel: (212) 705 7463
Contact: Virginia Fairweather
Tlx: 422847 ASCE UI
Fax: (212) 705 7712 (publications)
      (212) 980 4681 (mail room)

Category: B
Subject interests
Civil engineering, building.
Library & Information facilities
The ASCE is a member of the Engineering Societies Library, open from 9:00–5:00 EDT. It holds all Society publications.
Enquiries: (212) 705 7610

American Society of Mechanical Engineers

310 East 47th Street
New York, NY 10017
USA
Tel: (212) 705 7722
Contact: Martin Rogers, Director, Marketing Services
Fax: (212) 705 7674

Category: B
With a professional membership, the Society's function is to disseminate mechanical engineering information
Subject interests
Mechanical engineering; occupational and environmental health and safety, fire prevention, fire fighting.
Library & Information facilities
Information from: Central ASME Service Center
22 Law Drive, Fairfield, NJ 07007-2900
Tel: (800) 321 2633 and (800)843 2763

AODC

177a High Street
Beckenham, Kent BR3 1AH
UK
Tel: (081) 663 3859
Contact: Tom Hollobone
Fax: (081) 663 3860

Category: B
AODC is a trade association representing the International Underwater Engineering Contractors.
Subject interests
Mechanical engineering, civil and hydraulic engineering, building; occupational and environmental health and safety, fire prevention, fire fighting.
Special interests
Diving, submersibles, underwater engineering, safety and health.
Whilst this organisation does not offer an enquiry service as such, it does disseminate information via a quarterly newsletter for members; it also makes available guidance notes and codes of practice to non-members as they are published.

Arctic Institute of North America

The University of Calgary
2500 University Drive NW
Calgary, Alberta T2N 1N4
Canada
Tel: (403) 220 7515
Contact: Ruth Ast
Fax: (403) 282 4609
E-mail: ENVOY 100: ASTIS

Category: B
Development of the North through research
and education is the aim of this organization
Library & Information facilities
40,000 titles, available on Interlibrary loan.
Hours: 8:30-4:30 Monday–Friday
Publications: *Arctic, Information North, The
Komatik Series.*

Association of Consulting Engineers of Canada

130 Albert Street, Suite 616
Ottawa
Canada
Tel: (613) 236 0569
*Contact: Pierre A H Franche, Eng, FCIT,
President*
Fax: (613) 236 6193

Category: B
This is an industry association representing
850 Canadian consulting engineering firms with
a major interest in business information and
statistics.
Library & Information facilities
It provides listings of firms by specialization
upon request and publishes industry
magazines, directories, newsletters, annual
reports.
Publications|: *Communiqué* - a quarterly
members' newsletter;
ExportAction - a quarterly exporters' newsletter
The Membership database is computerized
and an enquiry service is available.

Bedford Institute of Oceanography Library

Department of Fisheries & Oceans Canada
PO Box 1006
Dartmouth, Nova Scotia B2Y 4A2
Canada
Tel: (902) 426 3675
Contact: Anna Fiander
Tlx: 019-31552
Fax: (902)-426-7827
E-mail: DFO.LIB.BIO

Category: A C D
Subject interests
Freshwater and marine biology; agriculture and
fisheries; chemistry; geology, meteorology,
hydrology, oceanography, geodesy,
cartography; mechanical engineering.
The Bedford Institute is a research-oriented
Government organization with special interests
in marine science, oceanography and geology,
chemistry, and fisheries biology.
Full information and library services are
available. Please ask.

BHR Group Ltd - CALtec Ltd

Cranfield
Bedford MK43 0AJ
UK
Tel: (0234) 750422
Contact: Iain C Gordon, Commercial Director
Fax: (0234) 750074

Category: C E
CALtec is the Oil & Gas Division of BHR Group
Ltd, an international technology development
and consultancy group. The company provides
the international industry with research, product
development and consultancy services
Subject interests
Energy, fuel, power; mechanical engineering.
Special interests
Downhole fluid mechanics, pipeline and riser
systems, process equipment, subsea
production technology, safety engineering
including: multiphase flow, fluid sealing,
separation, flow metering, multiphase pumping,
high pressure jetting, pipeline pigging, firewater
system analysis, CFD, system design and audit.
Library & Information services
The Library may be consulted for reference by
appointment; loan and information retrieval
services are available. Enquiries for specialist
advice are referred to specialist engineers.

BMT Cortec Ltd

Wallsend Research Station
Wallsend
Tyne & Wear NE28 6UY
UK
Tel: (091) 262 5242
Contact: Gillian Smith, Head of Information
Tlx: 53476
Fax: (091) 263 8754

Category: E
BMT Cortec offers extensive technical
consultancy and software development
services to marine transport industries, which
are totally independent of manufacturing
interests.
Subject interests
Oceanography, geodesy, marine technology.
Special interests
Ship design and construction, propulsion and
machinery. Offshore technology, ocean
engineering, safety, fluid mechanics.

Library & Information facilities
BMT Cortec has an extensive maritime
technical library, established for over 40 years,
including periodicals, conference proceedings,
reports and standards. There is a photocopy
and enquiry/online searching service. Loans
are to members and registered library users
only.
Publications: Each issue of *BMT Abstracts*
contains 250 abstracts, summaries of
worldwide literature on marine technology.
BMT Abstracts Online (BOATS) covers
abstracts from 1982 to the present.

British Cement Association

Warren Springs
Slough
Bucks SL3 6PL
UK
Tel: (0753) 662727
Contact: Mrs E J Leigh or Ms Doreen Walford
Fax: (0753) 660399

Category: E
This is an independent non-profit research and
development organisation financed
substantially by the Portland cement industry in
the UK.
Subject interests
Civil engineering and all aspoects of concrete
construction.
Library & Information facilities
Available for reference by appointment, there is
a limited loans service to non-members. The
Cement & Concrete Technical Information
Service database has been maintained since
1976, with about 300 new entries being added
each month (this accessions list can be
supplied on subscription). Searches of this and
other external databases can be carried out: a
scale of charges is available on request.
Product and service information relevant to
concrete construction is supplied free of
charge. Civil engineering consultancy and
advice on concrete technology are available for
a fee. Contract R&D is available through the
Association.

CATEGORIES

A	Public or national library
B	Institution, university, professional association, learned society
C	Research establishment
D	Government department or agency having direct or indirect influence on the formulation of regulations or standards relevant to the petroleum and marine technology industries
E	Independent consultancy

British Geological Survey

Keyworth
Nottingham NG12 6GG
UK
Tel: (06077) 6111
Contact: G McKenna, Chief Librarian
Tlx: 378173 BGSKEY G
Fax: (06077) 6602

Category: C
The work carried out by BGS is centred on modern geological surveying of the UK landmass, the UK continental shelf and numerous areas overseas, and the maintenance of national archives of geological data and specimens. Field and laboratory teams embrace projects in geology, geophysics, geochemistry, hydrogeology, engineering geology and economic mineralogy. Contracts with Government Departments include continental shelf geology, geophysics and oil/gas assessment; mineral resources assessment; regional geochemical surveys; waste-disposal feasibility studies; deep geological structures of sedimentay basins; geothermal energy sources; natural hazards assessments; and overseas surveys in developing nations.
Subject interests
Chemistry; energy, fuel, power; mining, metallurgy; geology meteorology, hydrology, seismology.
Special interests
The BGS maintains a collection of worldwide geological materials, including its historical collection of rock samples.
Library & Information facilities
The **Headquarters Library at Keyworth** holds 300,000 bound volumes, 200,000 maps, photograph and archive collections. The **Scotland Regional Office Library** is at Murchison House, West Mains Road, Edinburgh EH9 3LS [*Tel: (031) 667 1000; Fax: (031) 668 2683*]. These libraries are open to the public for reference purposes.
The BGS also publishes a variety of materials; it offers online access to databases and indexes to collections, records and archives held by the National Geosciences Data Centre at BGS Keyworth. Advisory services are provided on a wide range of economic and environment topics.

British Institute of Management

3rd Floor
2 Savoy Court
Strand
London WC2R 0EZ
UK
Tel: (071) 497 0580
Fax: (071) 497 0463

Category: B
The Institute aims to promote good management practice throughout business, and to that end runs a Management Information Centre where enquiries are welcomed.
Special interests
All aspects of management and business.
Publications: *Management News*

British Library

Science Reference & Information Service
25 Southampton Buildings
London WC2A 1AW
UK
Tel: (071) 323 7494/7496 – general enquiries
Tlx: 266959 SCI REF G
Fax: (0712) 323 7930

Category: A B C D
The UK's national library for science, technology, business and commerce, patents, trade marks and designs has the most comprehensive reference collection in Western Europe.
Special interests
Business information, industrial property, Japanese information, biotechnology, online database search and photocopy services; scientific, patent and environmental information.
Library & Information facilities
The Reference Library holds 32,000,000 patents, 67,000 serial titles (27,000 of them current), 235,000 books and over 2,000 market research and industry surveys.
Business Information Service:
free service (071) 323 7454; priced service (071) 323 7457, fax (071) 323 7453
Environmental Information Service:
(071) 323 7959, fax (071) 323 7954
Science & Technology Information Service:
(071) 323 7477, fax (071) 323 7954
Biotechnology Information Service:
(071) 323 7293
Photocopying service (Patent Express):
(071) 323 7927

Building Research Establishment

Bucknall Lane
Garston
Watford WD2 7JR
UK
Tel: (0923) 894040
Contact: Head of Advisory Service
*   or Head or Library*
Tlx: 923220
Fax: (0923) 664010

Category: D
BRE is a government agency which researches all aspects of building and construction. The Fire Research Station is part of BRE (see separate entry).
Special interests
Research covers: geotechnics and structures (including wind engineering), materials, energy and environment, construction and application, fire.
Library & Information facilities
The Library is open for reference and simple enquires; there is a charge for more detailed enquiries and bibliographic searches. Apart from library information, there is an Advisory Service which will answer technical enquiries or provide advice and consultancy on all building and construction topics.

Canada Institute for Scientific & Technical Information (CISTI)

Information, Marine Dynamics Branch
PO Box 12093, Stn A
St John's, Newfoundland
A1B 3T5
Canada
Tel: (709) 772 2468
Contact: David Clark, Librarian
Fax: (709) 772 2462
E-mail: Envoy -CISTI.IMDS; Bitnet -DCLARK@;-
INDIGO.NRC.CA;Internet DCLARK@;
MINNIE.IMD.NRC.CA

Category: A
Serving Canadians requiring information in science and technology, this national library has special interests in Naval Architecture, ice science, offshore engineering, and theoretical hydrodynamics.
Library & Information facilities
The Library holds approximately 180 journals and 8,000 books and reports. Books, but not journals, are available for loan within Canada.
Hours: 9:00am–5:00pm
Database: CAN/OLE. Dobis cataloging system
Main Library: CISTI, National Research Council Canada Building M-55, Montreal Road, Ottawa, Ontario, K1A 0S2, Canada *Tel: (613) 993 1600*

Canada Institute for Scientific & Technical Information (CISTI)

National Research Council Canada
Ottawa, Ontario K1A 0S2
Canada
Tel: (613) 993 1600
Contact: Elizabeth Katz, Publicity and
Communications
Fax: (613) 952 9112

Category: C
This national research establishment runs a science and technology information service.
Special interests
Science, engineering, medicine (all aspects).
Library & Information facilities
The library holds an open, comprehensive collection of science and technology books, journals, conferences, reports.
Hours: 8:30am–4:30 EST
A computer-based enquiry service is available, (see brochures). Services are available to the public as well as businesses and other government departments.

Canada-Newfoundland Offshore Petroleum Board

Suite 500 - TD Place
140 Water Street
St John's, Newfoundland A1C 6H6
Canada
Tel: (709) 778-1400
Contact: Eileen Blanchard
Fax: (709) 778-1473

Category: D
This Board is a regulator of offshore exploration.
Subject interests
The Board is responsible for petroleum resource management and for administering the legislation and regulations governing exploration for and production of hydrocarbons offshore Newfoundland and Labrador.
Library & Information facilities
The library holds a collection of material relating to offshore exploration including standards, geology, engineering, conference proceedings and a variety of journals. These are available for viewing by public but may not to be loaned out, though photocopying is available on a payment basis.
Hours: 8:30am–12:30 1:00–4:30 pm (summer)
8:30am–12:30 1:00–5:00 pm (winter)

Canadian Association of Oilwell Drilling Contractors

800 540–5th Avenue SW
Calgary, Alberta T2P 0M2
Canada
Tel: (403) 264 4311
Contact: Kathryn M. Lundy
Fax: (403) 263 3796

Category: C
This trade association represents the drilling and servicing industry in Canada.
Subject interests
Energy, fuel, power; occupational and environmental health and safety, fire prevention, fire fighting; business information and statistics.
Library & Information facilities
Services are located at the office, but there is some restriction on their use.
Hours: 8:00–4:30 Tuesday–Thursday

Canadian Centre for Marine Communications

PO Box 8454
St John's, Newfoundland A1B 3N9
Canada
Tel: (709) 579 4872
Contact: Ken Butt, Business Development Officer
Fax: (709) 579 0495
E-mail: CCMC@GILL.IFMT.NF.CA

Category: C
Subject interests
Physics, electrical and electronic engineering, telecommunications.
Special interests
To develop marine telecommunications equipment, services and technology.
To promote the Canadian telecommunications industry:
- Radio and Satellite Communications
- EMC and EMI
- Integrated Ship Management Systems
Library & Information facilities
The Marine Communications Resource Centre will carry out research for member companies in Communications and related marine industries.
Hours: 9:00am–5:00 pm (Monday–Friday)
There is a quarterly newsletter

Canadian Geotechnical Society

170 Attwell Drive, Suite 602
Rexdale, Ontario M9W 5Z5
Canada
Tel: (416) 674 0366
Fax: (416)-674-9507

Category: B
This learned society is concerned with all geotechnical matters.

Canadian Library Association

200 Elgin Street, Suite 602
Ottawa, Ontario K2P 1L5
Canada
Tel: (613) 232 9625
<MIContact: Director of Publishing
Fax: (613) 563 9895

Category: B
This is a professional association for librarians
concerned with libraries and related materials
for staff. Hours: 9:00am–5:00pm

Canadian Maritime Industries Association

PO Box 1429, Stn B
801-100 Sparks Street
Ottawa, Ontario K1P 5R4
Canada
Tel: (613) 232 7127
Contact: J Y Clarke, President and CEO
Fax: (613) 232 2490

Category: B E
Subject interests
Freshwater and marine biology, agriculture and
fisheries; energy, fuel, power; mechanical
engineering. This national trade association
covers shipbuilding, ship repairing, and marine
industries.
Library & Information facilities
The library holds 3,000 volumes and tapes
which are available for members only or by
appointment.
Hours: (0900-1700, Monday–Friday)
Publications: Annual statistics report, periodical
statistics
There is an enquiry service. The databases can
be accessed by members only.

Canadian Petroleum Association

3800 150 6th Avenue SW
Calgary, Alberta T2P 3Y7
Canada
Tel: (403) 269 6721
*Contact: Leo Bouckhout, Director, Division
Services*
Fax: (403) 261 4622

Frontier Division
800 TD Place, 140 Water Street
St John's, Newfoundland A1C 6H6
Canada
Tel: (709) 726 7270
*Contacts: L Taylor, Manager, East Coast
Region, Frontier Division
J Rypien, Provincial Administrator
Fax: (709) 726 0232*

Category: B
The Association represents the interests of
major explorers, producers and pipeliners. Its
subject interests are technical and regulatory,
supported by an internal working library.
Special interests
Oil and gas exploration, production and
pipeline companies; policy analysis.
Publications: *Eastern Offshore News*

Canadian Petroleum Products Institute (CPPI)

1000-275 Slater Street
Ottawa, Ontario K1N 5R7
Canada
Tel: (613) 236 3709
*Contact: Brendan P Hawley, Director, Public
Affairs
Victor Bennett, Emergency Response
Coordinator
Fax: (613) 236 4280*

Category: B
CPPI serves and represents the refining and
marketing sectors of the petroleum industry. It
is concerned with the petroleum industry's
effect on environmental, health, safety and
business issues affecting Canadian society.
Subject interests
Energy, fuel, power; occupational and
environmental health and safety; fire
prevention, fire fighting; chemistry; civil and
hydraulic engineering; business information
and statistics.
Special interests
The Association of Petroleum Refiners
- emergency response
- environmental issues
Information & Library facilities
The facilities are intended for internal use only.
Publications: Annual Report, Technical
Bibliography

Canadian Society of Petroleum Geologists

505, 206–7th Avenue SW
Calgary, Alberta T2P 0W7
Canada
Tel: (403) 264 5610
Contact: J D Speelman

Category: B
The aims of the Institute are scientific and educational; to promote petroleum geology as a science, particularly as it relates to Canada.
Subject interests
Geology, meteorology, hydrology, oceanography, geodesy, cartography.
Publications: bulletin, quarterly scientific journal, monthly newsletters. Many books are published through CSPG.

Canadian Standards Association

178 Rexdale Boulevard
Rexdale, Ontario M9W 1R3
Canada
Tel: (416) 737 4098
Contact: Bill Porter
Tlx: 06-989344
Fax: (416) 737 2473

Category: D E
This is a standards development and conformity assessment organization with a design code for offshore structures, etc.
Subject interests
Occupational and environmental health and safety, fire prevention, fire fighting; business information and statistics.
A Standards Development and Conformity Assessment Organization providing a design code for offshore structures, etc.
Library & Information facilities
The library is open between 9:00am–00pm
There is an enquiry service and the Association publishes brochures and videotapes.

Chartered Institution of Building Services Engineers

Delta House
222 Balham High Road
London SW12 9BS
UK
Tel: (081) 675 5211
Contact: P Scurry, Technical Secretary
Fax: (081) 675 5449

Category: B
Special interests
Heating, ventilation, air conditioning; lightingp; public health; security.
Information & Library facilities
Enquiries (preferably written) should be limited the the Institution's specialist interests.
Publications and journals are produced for the guidance of designers and those concerned with building engineering services.

Colorado School of Mines

1500 Illinois Street
Golden, Colorado 80401
USA
Tel: (303) 273 3687
Contact: Arthur Lake Library
Fax: (303) 273 3199

Category: B C
Subject interests
Chemistry, energy, fuel, power; mining, metallurgy; geology, meteorology, hydrology, oceanography, geodesy, cartography; physics, electrical and electronic engineering, telecommunications; mechanical engineering; civil and hydraulic engineering, building; occupational and environmental health and safety, fire prevention, fire fighting.
Special interests
Petroleum, mining, geology, geophysics, offshore technology, arctic.
Library & Information facilities
The School does not run an enquiry service but does have computing and database facilities

CONCAWE

Madouplein 1
B 1030 Brussels
Belgium
Tel: +32 2 2203111
Contact: Technical Coordinator, Oil Spill
Clean-up
Tlx: 20308
Fax: +32 2 2194646

Catgory: B
CONCAWE is the oil companies' European
organisation for environmental and health
protection. In the oil spill clean-up field, it
manages the activities of expert task forces to
provide state-the-art advice on developments.
This is published in CONCAWE reports and
field guides covering planning, equipment,
techniques and strategies.
Subject interests
Energy, fuel, power; occupational and
environmental health and safety, fire
prevention, fire fighting; publishing.
Library & Information facilities
CONCAWE reports are available *gratis* to all
who need such advice and field guides are
supplied at cost price.

COPPE/UFRJ - Federal University of Rio de Janeiro

PO Box 68506
21945 - Rio de Janeiro
Brazil
Tel: +55 21 280 9993
Contact; Professor Nelson Ebecken
Tlx: +55 21 280 9993
Fax: +55 21 290 6626

Category: A B C
Subject interests
Civil and hydraulic engineering, building.
The Center of the Federal University of Rio de
Janeiro undertakes post-graduate teaching,
research and consultancy.
Library & Information facilities
DEEPWATER STRUCTURES: Computer system for
numerical modelling, structural reliability,
experimental models and testing techniques.
The library holds 62,000 volumes, access to
which is unrestricted. Hours: 08:00–16:30.
There is an enquiry service.

Danish Maritime Institute

Hjortekaersvej 99
DK-2800 Lyngby
Denmark
Tel: +45 45 879325
Contact: A H Nielsen, Director
Tlx: 37223 SHILAB DK
Fax: +45 45 879333

Category: C
DMI is an independent non-profit-making
organisation established in 1953 by the Danish
Academy of Technical Sciences. It carries out
studies on ship and ocean engineering, wind
engineering, computations, model tests and
field measurements.
Subject interests
Energy, fuel, power; civil and hydraulic
engineering; oceanography, geodesy, marine
technology.
Special interests
Ship design and performance; physical and
computational modelling of offshore operations;
real-time manoeuvring simulation for ships,
harbours and offshore operations, including full
visual display, interface of any control system
and 6-DOF motions; the effect of wind on
ships, offshore constructions, buildings and
large bridges.
Information facilities
DMI offers consultancy services independently
or jointly with other research institutes.

Department of Energy Library

1 Palace Street
London SW1E 5HE
UK
Tel: (071) 238 3042
Contact: Librarian
Tlx: 918777
Fax: (071) 834 3771

Category: D
This Library provides library and information
services to the Department of Energy
Subject interests
Energy, fuel, power; occupational and
environmental health and safety, fire
prevention, fire fighting.
Special interests
Energy policy, economics and technology
including offshore hydrocarbon exploration and
production.
Library & Information facilities
The Library is available to the public for
reference by appointment only, and loans are
made to other libraries only. *Current Energy
Information*, a current awareness bulletin, is
available on subscription.

Department of Fisheries & Oceans

Physical and Chemical Sciences Directorate
12th Floor
200 Kent Street
Ottawa, Ontario K1A 0E6
Canada
Tel: (613) 990 0298
Contact: G L Holland, Director General
Fax: (613) 990 5510
E-mail: OCEANSCIENCE.OTTAWA OMNET.

Category: A D
Subject interests
Chemistry; geology, meteorology, hydrology,
oceanography, geodesy, cartography.
Special interests
Ocean science policy; fisheries; oceanography;
ocean climate; ocean technology.
Library & Information facilities
This is the headquarters of the Marine
Environmental Data Service. Access to the
library facilities, however, is Departmental only.

Department of Trade & Industry (DTI)

Ashdown House
123 Victoria Street
London SW1E 6RB
UK
Tel: (071) 215 5000, General enquiries
Tlx: 8813148 DTHQ G
Fax: (071) 828 3258

Category: D
This is a major department of UK government.
Contacts for specific enquiries:
Enterprise Initiative   *(0800) 500 200*
Single Market Hotline   *(081) 200 1992*
Innovation Enquiry Line   *(0800) 44 2001*
Environmental Enquiry Point   *(0800) 585 794*

Department of Transport

Marine Library
Sunley House
90 High Holborn
London WC1V 6LP
UK
Tel: (071) 406 5911
Contact: T A Stewart, Librarian
Tlx: 264084
Fax: (071) 831 2508

Category: D
Special interests
All aspect of merchant shipping (including the
fishing fleet), marine safety, oil pollution and
related topics.
Library & Information facilities
The Marine Library provides library information
services primarily to serve the Marine
Directorate of the Department of Transport.
The Library is available to the public for
reference by appointment. Limited bibliographic
searches can be carried out.

Det norske Veritas - Veritec

Veritasveien 1
PO Box 300
1322 Høvik
Norway
Tel: +47 2 477250
Contact: Mrs Mette Nore, Librarian
Tlx: 76-192 Verit N
Fax: +47 2 477474

Category: B E
Det norske Veritas, an independent classification society, acts as a certifying authority for the British, Norwegian and other governments. There is a substantial research programme.
Subject interests
Energy, fuel, power; civil and hydraulic engineering, building; occupational and environmental health and safety, fire prevention, fire fighting.
Special interests
Veritec, subsidiary company of DnV, is concerned with the classification and certification of fixed and mobile offshore structures, marine installations and submersibles.
Library & Information facilities
The Library is intended primarily for Veritec staff, answering questions concerning publications and reports from Det norske Veritas. Det norske Veritas will give advice and consultancy work is taken on.

Duke University Marine Laboratory

Pivers Island
Beaufort, North Carolina 28516-9721
USA
Tel: (919) 728 2111
Contact: Dr Joseph S Ramus, Director
L Lorenzsonn-Willis, Staff Specialist
Fax: (919) 728 2514

Category: B C
Duke University Marine Laboratory is an interdisciplinary research and teaching facility in marine biology, oceanography and limnology, serving undergraduates, graduate students, visiting student groups, and the national as well as international scientific communities.
Subject interests
Freshwater and marine biology, agriculture and fisheries; geology, meteorology, hydrology, oceanography, geodesy, cartography.
Special interests

Thermal dynamics, biotechnology, metal ions in biological systems, ecotoxicology, membrane physiology, sedimentology and palaeoclimatology, estuarine dynamics, marine freshwater systems, development of fish models.
Library & Information facilities
The library holds 60,000 volumes and runs a computer-based enquiry service.
Databases: OCLC
Disseminates knowledge concerning ecological systems and the distribution and abundance of marine plants and animals.

Empresa National Del Petroleo

Ahumala 341 - 3 piso
Santiago
Chile
Tel: +56 2 381845
Fax: +56 2 380164
E-mail: 240427-E - 240447ENAPLL

Category: D
Subject interests
Chemistry; energy, fuel, power; geology; meteorology, hydrology, oceanography; geodesy, cartography.
Special interests
Petroleum and offshore technologies.
Information & Library facilities
The library holds 11,000 books.
Databases: EMIS - DIALOG

Energy Information Administration

National Energy Information Center
1000 Independence Avenue SW
Washington, DC 20585
USA
Tel: (202) 586 8800
Contact: T Christian, Information Specialist
* Telecommunications device for deaf:
* (202) 586 1181

Category: D
Special interests
The acquisition and dissemination of energy statistics.
Library & Information facilities
The Centre produces publications for sale. There is no lending of library holdings but a enquiry service is free to governments and libraries, and available at cost to others.
Hours: Monday–Friday 8am–5pm EST

Environment & Natural Resources Institute (ENRI)

University of Alaska Anchorage
707 A Street
Anchorage, Alaska 99501
USA
Tel: (907) 257 2733
Contact: Juli Braund-Allen, MLS, Research Analyst
Fax: (907) 276 6847

Category: B C
Subject interests
Freshwater and marine biology, agriculture and fisheries; geology, meteorology, hydrology, oceanography, geodesy, cartography.
Special interests
Maritime, resource development, and fisheries policies; federal mining regulations; bowhead and beluga whaling; community and regional planning.
Library & Information facilities
The Institute is a source of Alaska natural resource and environmental knowledge. It is concerned with Natural resource management issues in Alaska.It holds a small focused collection–11,500 government publications and microforms, 8,800 bound volumes, and 65 periodical and newsletter titles.
Access is not restricted; a list is available and there is a computer-based enquiry service.
Hours: Monday–Friday, 8:00am–12:00 and 1:00pm–5:00pm
The Institute is a depository for reports of the Arctic Petroleum Operator's Association, Alaska Oil & Gas Association and Alaska Department of Transportation's Research Division.

ERA Technology Ltd

Cleeve Road
Leatherhead
Surrey KT22 7SA
UK
Tel: (0372) 374151 extension 2224
Contact: D Baynes, Head of Information Services
Tlx: 264045
Fax:(0372) 374496
E-mail: 74:SKK820

Category: C E
ERA provides services on a contract basis to clients all over the world. The staff, supported by laboratory and computer faciliaties, are active in such fields as energy conservation, pollution control, the technological aspects of health and safety, and standardisation.
Subject interests
Physics, electrtical, electronic and telecommunications engineering. All aspects of electrotechnology, including underwater applications.
Library & Information facilities
The Information Centre provides literature and information searches, photocopies, loans and reference facilities. Enquiries are accepted from any source, with preferential rates available to ERA Technical Services Scheme members and also to Information Research Services subscribers (contact extension 2301 for details of IRS).
Consultancy and advisory services are available, drawing on specialist staff in ERA's technical divisions where appropriate.
Publications: technical reports, List of ERA Reports (free), Annual Review (free), ERA News (free); seminar proceedings.

ESL Information Services Inc

345 East 47th Street
New York, NY 10017
USA
Tel: (212) 705 7527
Contact: Lynette Bamfield
Fax: (212) 753 9568

Category: E
Subject interests
Energy, fuel, power; mining, metallurgy; geology, meteorology, hydrology, oceanography, geodesy, cartography; physics, electrical and electronic engineering, telecommunications; mechanical engineering; civil and hydraulic engineering, building.
Library & Information facilities
The Library and technical information service covers all aspects.of engineering. Member loans and a public reading room are available.
Hours: 9:00am–5:00pm Monday–Friday
Tel: (212) 705 7611, reference.

European Association of Exploration Geophysicists

PO Box 298
3700 AG Zeist
The Netherlands
Tel: +31 0 3404 56997
Fax: +31 0 3404 62640

Category: B
The objectives of the Association are to promote exploration geophysics and to foster fellowship and cooperation among those working, studying or being otherwise interested in this field. The Association achieves these objectives by publishing scientific journals, organising Annual Meetings, running a Continuing Education Programme by using other legal means.
Publications: *Geophysical Prospecting* and *First Break* are free to Members; the Business Office will supply information on the availability of other publications

Food & Agriculture Organization of the United Nations (FAO)

Fisheries Information Data & Statistics Service
via delle Terme di Caracalla
00100 Rome
Italy
Tel: +39 6 57971
Contact: Richard Pepe, Fishery Information Officer
Tlx: 6120181 FAOI
Fax: +39 6 57973152

Category: a supranational agency
The FAO is an autonomous agency within the UN system set up to: - raise the levels of nutrition and standards of living - improve the production and distribution of all food and agricultural products - improve the condition of rural perople
Special interests
Fisheries and aquaculture; associated science and technology.
Library & Information facilities
There are four branch libraries in addition to the main library. Retrospective bibliographic retrieval is available from in-house data files (FAO Documentation and Library

Monographs), from AGRIS, from ASFIS and other external data bases in the Organizations's field. Many documents referred to in resulting computer printout can be provided in photocopy or microfiche form.
Cost estimates are available upon request.

Geological Society of America

3300 Penrose Place
PO Box 9140
Boulder CO 80301-9140
USA
Tel: (303) 447 2020
Contact: Ann H Crawford, Marketing Dept
Fax: (303) 447 1133

Category: B
The GSA effects the promotion of earth sciences through publications and technical meetings. It publishes extensively: online searching is minimal and limited to GSA publications and authors. The Library can be referenced on the headquarters site.

Gulf Canada Resources Ltd

401–9th Avenue SW
PO Box 130
Calgary, Alberta T2P 2H7
Canada
Tel: (403) 233 4000
Contact: F E Mitten, Manager Northern Operations
Tlx: 038-24551
Fax: (403) 233 5143

Category: C
Gulf is an oil and gas exploration and producing company active in Canada and internationally. The company operates, through a wholly owned subsidiary, a state-of-the-art drilling system for use in the arctic offshore. The system includes two drilling units and four ice-breaking supply vessels. Gulf is also a participant in.the offshore Hibernia oil project.Information is available on the system and on the individual units.
Subject interests
Energy, fuel, power

Health & Safety Executive

Broad Lane
Sheffield S3 7HQ
UK
Tel: (0742) 768141
Contact: Mrs Sheila Pantry, Head of Library &
information Services
Tlx: 54556
Fax: (0742) 720006

Category: D
The responsibility of the Executive is to
implement the Health & Safety at Work Act and
subsequent regulations. It is concerned with all
aspects of this subject and runs an extensive
information service with two public enquiry
points:
HSE Library & Information Services:
Broad Lane, Sheffield S3 7HQ
Tel: (0742) 752439
12 Chepstow Place, London W2 4TF
Tel: (071) 221 0870

Highlands & Islands Enterprise

20 Bridge Street
Inverness IV1 1QR
Scotland
Tel: (0463) 234171
Contact: R J Ardern, Librarian
Fax: (0463) 244469

Category: D
This is a Government Agency responsible for
the economic, social and environmental
development of the highlands and islands of
Scotland.
Special interests
Industrial development, offshore oil, fisheries,
fish farming, etc.
Library & Information facilities
Admission to the Library for reference purposes
is by appointment only. The Highlands &
Islands Business Information Service is
available to local companies.

HR Wallingford

Hydraulic Research Ltd
Wallingford
Oxfordshire OX10 8BA
UK
Tel: (0491) 35381
Contact: Ian McCarey, Marketing & Business
Development Manager
Tlx: 848552 HRSWAL G
Fax: (0491) 32233

Category: C E
HR Wallingford researches and offers
consultancy in civil engineering, hydraulics and
management of the water environment.
Subject interests
Civil and hydraulic engineering, building;
oceanography, geodesy, marine technology.
Special interests
Environmental modelling studies, hydraulic
testing of offshore structures during towing and
installation.
Library & Information facilities
The Library is intended for HR personnel but
may be consulted by appointment.
HR has its own database of information on
hydrological research; this covers books,
reports and journal articles. Associate
Membership status enables items to be
borrowed. An enquiry service is offered; costs
beyond those of a reasonable nature will be
passed on to the enquirer.

Hydrographic Office

Ministry of Defence (Navy) Taunton
Somerset TA1 2DN
UK
Tel: (0823) 337900
Contact: Librarian or Hydrographic Data Centre
Tlx: 46274
Fax: (0823) 284077/323756

Category: A D
The Hydrographic Office is a Defence Support
Agency, responsible for producing charts and
navigational publications. By means of
international exchange its archives are
world-wide in scope.
Subject interests
Geology, meteorology, hydrology, seismology;
oceanography, geodesy, marine technology.
Specialist interests
Principal subject areas are hydrographic
surveying (including geodetic and tidal
observations), marine cartography and
navigation. Secondary areas are physical
oceanography (including sedimentation),
marine geophysics and photogrammetry.
Library & Information facilities
The Library may be consulted for reference by
appointment (extension 806). The loans service
is restricted to other Government libraries,
though photocopies can be provided. Staff
availability permitting, information retrieval
services will be provided but a charge may be
levied.
Enquirers needing professional advice may be
referred to a specialist where time and other
conditions permit.Enquiries to the Curator,
Hydrographic Data Centre, *(extension 698).*
The Marine Science Branch Four
This Branch maintains three oceanographic
libraries covering: atlases, data and reference
material. Information retrieval services can be
provided from the Oceanographic Library or
from the Digital Data Base (a charge may be
levied). Professional advice on oceanographic
equipment or data handling can be referred to
a specialist.

IFREMER

Institut Français de Recherche pour
l'Exploitation de la Mer
Technopolis 40
155 rue Jean-Jacques Rousseau
92138 Issy-les-Moulineaux
France
Tel: +33 1 46482100
Tlx: 610775
Fax: +33 1 46482296
*Contact: Service de la Documentation et des
Publications (SDP),*

Centre de Brest
BP 70,
29280 Plouzane
France
Tel: +33 98224013
Tlx: 940627
Fax: +33 98224545

Category: C D
IFREMER is a governmental agency placed
under the joint authority of the Ministry for
Resaearch & Technology and the Ministry for
the Sea. The Institute has a very extensive
assignment in research and technological
development for the purpose of advancing
knowledge about the ocean and using marine
resources.
Special interests
Sea resources (mainly in fishing and shellfish
farming), protection of the coastal environment
especially by testing water quality.
Fundamental researach in varying disciplines
such as: marine geosciences, biology,
chemistry; ecotoxicology, pathology and
immunology, biotechnology; physical
oceanography; processing of acoustic and
satellite images, mathematical modelling; data
processing facilities used for processing
images from multibeam echosounder, solar ...
Library & Information facilities
The Library has access to documentary
databases (internal and international bases:
ASFA). Bibliographic references, photocopies
and loans to other libraries are available.
Moreover the SDP edits IFREMER'S scientific
and technical publications and participates in
co-publishing works of interest to the general
public (catalogue on request).

IKU - Continental Shelf & Petroleum Research Institute

S P Andersens vei 15b
N 7034 Trondheim
Norway
Tel: +47 7 591100
Contact: Marthe Granviken, Group Leader,
Information Centre
Tlx: 55434 IKU N
Fax: +47 7 591102

Category: B C
IKU is a Norwegian company specialising in
offshore petroleum exploration and production
research.
Subject interests
Geology, meteorology, hydrology, seismology;
marine technology.
Special interests:
Geology, geophysics, reservoir technology;
organic chemistry, environmental chemistry;
formation physics; offshore instrumentation;
production technology.
Library & Information facilities
Library services are intended primarily for IKU
staff, but visitors are always welcome.
Interlibrary services are executed. IKU has its
own database with relevant information. Most
IKU reports are confidential, but open reports
are for sale.

Institute of Gas Technology

3424 South State Street
Chicago
Illinois 60616-3896
USA
Tel: (312) 567 3650
Contact: J J Kruse, Manager, gaslineR
Services and IGTnet
Fax: (312) 567 5209

Category: C E
Subject interests
Chemistry; energy, fuel, power; geology,
meteorology, hydrology, oceanography,
geodesy, cartography; business information
and statistics.

Special interests
R&D gas industry and environment.Natural gas.
Library & Information facilities
With 100,000 volumes, this is the world's
largest library focused on natural gas.
Available Monday–Friday from 8:15am–5:00pm
CST with a computer-based enquiry service.
Publications: *Gas Abstracts, Gasline, Energy
Statistics.*
Databases: gaslineR and IGTnetR.

Institute of Marine Engineers

Memorial Building
76 Mark Lane
London EC3R 7JN
UK
Tel: (071) 481 8493
Contact: T K Norledge CEng FIMarE
Tlx: 886841
Fax: (071) 488 1854

Category: B
A learned societyi for professional engineers
dedicated to developing maritime engineering.
Subject interests
Marine engineering, ships propulsion and
machinery systems, naval architecture, navigation, offshore technology, ocean engineering,
safety, education and management.
Library & Information facilities
The Marine Information Centre, incorporating
the Library, is primarily for members, but others
may consult it by appointment. A loan service is
available to Members. The MIC has its own
database of information on marine engineering
including the Library's catalogue of books and
material published in IMarE's journal, *Marine
Engineer's Review*, and Transactions. Access
to other databases is charged to the client. An
enquiry service is offered.

Institute of Oceanographic Sciences, Deacon Laboratory

Brook Road
Wormley
Godalming
Surrey GU8 5UB
UK
Tel: (0428) 684141
Contact: Pauline Simpson
Tlx: 858833 Oceans G
Fax: (0428) 683066

Category: C
The Institute of Oceanographic Sciences Deacon Laboratory is a component body of the Natural Environment Research Council (NERC) and is the centre for deep sea oceanography in the UK.
Special interests
Its scientists carry out physical, chemical, geological, geophysical and biological research in the deep ocean. Much of the sophisticated equipment designed to work at oceanic depths is designed and built by IOSDL engineers. The Institute has comprehensive testing and calibration facilities for oceanographic equipment.
Library & Information facilities
The UK National Oceanographic Library is now the major source of oceanographic literature in the UK. Through access to the online library database IOSLIB over 80,000 references can be rapidly searched on any criteria. This service is available to outside organisations with suitable computer communication systems. The Marine Information and Advisory Service (MIAS) supplies information, advice and consultancy to outside organisations based on the unique information resources of the National Oceanographic Library and the expertise of NERC marine scientists. A MIAS Subscription Service offers customers a range of regular information packages and consultancy.
Publications include IOSDL Reports and Annual Report, the Marine Science Alert Service TOPIX and staff publications list.

Institute of Petroleum

61 New Cavendish Street
London WIM 8AR
UK
Tel: (071) 636 1004
Contact: Information Department
Tlx: 264380
Fax: (071) 255 1472

Category: B
The IP is a centre of scientific technical and economic data and expertise for the UK oil industry. It ensures that members (both individual and companies) are informed about their industry and satisfied that the standards and codes are based on good practice and science. The IP has 14 active branches throughout the UK.
Subject interests
Energy, fuel, power; gaeology, meteorology, hydrology, seismology; occupational and environmental health and safety, fire prevention, fire fighting; oceanography, marine technology; publishing and electronic media.
Library & Information facilities
The Library is available to members; there is a daily charge to non-members.
This centre of information for members provides technical, commercial, statistical, economic and market research information on all aspects of the petroleum industry. Online searching (150 international databases), desk research, photocopying, advice and assistance are provided: lower charges for members.
Publications: *Petroleum Review*, monthly; annual *IP Standard Methods*; *IP Codes of Safe Practice*.

Institution of Chemical Engineers

The Davis Building
165-171 Railway Terrace
Rugby CV21 3HQ
UK
Tel: (0788) 578214
Tlx: 311780
Fax: (0788) 560833

Category: B
The IChemE exists to promote the science and practice of chemical engineering; to improve the standards and methods; to increase the facilities for education in the science of chemical engineering and its application; and to act as a qualifying body for chemical engineers.
Special interests
Chemical engineering, safety and loss prevention, environmental protection.
Library & Information facilities
Library services available to Members and non-members of the IChemE include book loans and a photocopying service. Manual and online searches are also available. There is an enquiry service with appropriate referrals.

Institution of Civil Engineers

1-7 Great George Street
London SW1P 3AA
UK
Tel: (0671) 222 7722
Contact: Director-General and Secretary
Fax: (071) 222 7500

Category: B
ICE is a learned society and professional association for civil engineers.
Subject interests
Geology, meteorology, hydrology, seismology; civil and hydraulic engineering, building; occupational and environmental health and safety, fire prevention, fire fighting; publishing.
Special interests
Civil engineering and all related topics, including design and construction of offshore structures and pipelines, wave loading, etc.
Library & Information facilities
The Library is primarily for Members; others may use it for reference only by prior arrangement. Catalogue and publications index is available online; there is also a Bulletin Board. ICE maintains the Coastal Engineering Research Database.

Institution of Electrical Engineers

Savoy Place
London WC2R 0BL
UK
Tel: (071) 240 1871
Contact: John Coupland, Chief Information Officer
Tlx: 261176
Fax: (071) 497 3557

Category: B
Subject interests
Energy, fuel, power; physics, electrical, electronic and telecommunications engineering; occupational and environmental health and safety, fire prevention, fire fighting; publishing, including electronic media.
Special interests
Electrical enginering, electronics, computing, information technology, manufacturing engineering, safety.
Library & Information facilities
Library loan facilities are available only to Members; reference only for non-members. The IEE houses the library of the British Computer Society.
A comprehensive range of information is available, including online searching, CD-Rom databases and market data. Most information is provided on a charged-basis only, with preferential rates for Members.

International Association of Drilling Contractors

PO Box 4287
Houston, Texas 77210
USA
Tel: (713) 578 7171
Contact: Cindy B McManus, Administrator, Office Services
Tlx: 516985 INTDRILL
Fax: (713) 578 0589

Category: B
Trade association for drilling contractors.
Library & Information facilities
The Association holds some historical information but access is limited; it does issue its own publications.

International Center for Marine Resource Development

University of Rhode Island
126 Woodward Hall
Kingsto, RI 02881, USA
Tel: (401) 792 2479
Contact: George Aelion
Fax: (401) 789 3342

Category: B C
CMRD offers training, technical assistance and research services to less developed countries.
Subject interests
Freshwater and marine biology, agriculture and fisheries; geology, meteorology, hydrology, oceanography, geodesy, cartography.
Special interests
Marine sciences/food sciences.
Library & Information facilities
The Centre has two major libraries: the ICMRD Library; and Grey literature on international fisheries. It maintains datasbases and offers an enquiry service.

International Oceanographic Foundation

4600 Rickenbacker Causeway
Miami, Florida 33149
USA
Tel: (305) 361 4888
Contact: Sandra Hersh, Ass Ed, Sea Frontiers
Fax: (305) 361 4711

Category B
The IOF is a non-profit foundation dedicated to informing the public about all aspects of the oceans. It publishes a magazine, offers an information service, runs education travel programs and supports research.
Special interests
Marine biology and fisheries; oceanography.
Publications: Sea Secrets Hotline; Sea Frontiers Magazine.

International Society of Offshore & Polar Engineers

PO Box 1107
Golden, Colorado 80402-1107
USA
Tel: (303) 273 3673
Contact: Publication Department
Fax: (303) 420 3760

Category: B C
Subject interests
Chemistry; energy, fuel, power; mining, metallurgy; physics, electrical and electronic engineering, telecommunications; mechanical, civil and hydraulic engineering, building.
Special interests
Offshore mechanics, offshore technology, polar engineering, ice mechanics, arctic and polar structures, pipelines, materials, welding, geotechnics.
Library & Information facilities
This Engineering Society has an enquiry service.

Lloyd's Register of Shipping

71 Fenchurch Street
London EC3M 4BS
UK
Tel: (071) 709 9166
Tlx: 888379 LR Lon G
Fax: (071) 488 4796

Category: B E
Lloyd's Register, founded in 1760, is the world's leading ship classification and inspection organisation. Complete independence, with no shareholders, ensures third-party inspection and quality assurance without prejudice in a commercial contract. LR's Ship, Industrial and Offshore Divisions are supported by its own Engineering Services.
Special interests
All design aspects of offshore installations including independent structural analysis; control, electrical, mechanical and pipeline engineering; metallurgy; noise and vibration; safety systems and formal safety analysis; stability analysis of floaters and strength analysis of ship-type floating production units.
Publications: LR's Register of Ships (first published 1764); Register of Offshore Units, Submersibles and Underwater Systems (or Offshore Register) is published annually.
With the publishing arm of the Corporation of Lloyds (from whom LR is otherwise quite separate) LR has formed Lloyd's Maritime Information Services Ltd to market their shipping information on printout, tape or disc, and the online SEADATA service.

Louisiana Sea Grant College Program

134 Wetland Resources Building
Louisiana State University
Baton Rouge, Louisiana 70803
USA
Tel: (504) 388 6445
Contact: Ronald E Becker, Associate Director
Fax: (504) 388 6331

Category: C
This is administrative, managerial unit for
marine-related research, education, extension
programs.
Subject interests
Freshwater and marine biology, agriculture and
fisheries; geology, meteorology, hydrology,
oceanography, geodesy, cartography.
Special interests
Marine resources, marine technology, marine
environmental studies. Newsletter, technical
reports, information and library facilities are
through the National Sea Grant Repository.

Marine Board

National Research Council
2101 Constitution Avenue
Washington DC 20418
USA
Tel: (202) 334 3119
Contact: Charles A Bookman
Fax: (202) 334 3789

Category: D
Special interests
Marine and maritime industries, ocean and
coastal developments, environmental
protection, marine engineering.
Library & Information facilities
The facilities of the National Academy of
Sciences Library are available by appointment
only.
Publications: Annual report and project reports.

Marine Technology Directorate Ltd (MTD Ltd)

19 Buckingham Street
London WC2N 6EF
UK
Tel:(071) 321 0674
*Contact: D E Lennard, Director & Chief
Executive*
Fax: (071) 930 4323

Category: B C
MTD supports and manages research, and
promotes education and jtraining in marine
technology. It develops the national capability
in ocean-raelated activities through higher
education institutions. Recently MTD took over
the activities of UEG Offshore Research.
The following personnel welcome enquiries:
Education & training: Mr J A T Grant
Information: Mr B G Richardson
Research: Mr I C Brown

Marine Technology Society

9th Floor, Suite 906
1828 L Street NW
Washington DC 20036
USA
Tel: (202) 775 5966
Contact: Martin J Finerty Jr
Fax: (202) 429 9417

Category: B
This is an international non-profit professional
society involved with interdisciplinary and
scientific issues concerning the oceans and
marine engineering. It publishes a quarterly
Journal and a bi-monthly Newsletter, arranges
an annual Conference and Exhibition and
various annual workshops.
Special interests
Freshwater and marine biology, agriculture and
fisheries; geology, meteorology, hydrology,
oceanography, geodesy, cartography; physics,
electrical and electronic engineering,
telecommunications; mechanical engineering;
civil and hydraulic engineering; occupational
and environmental health and safety, fire
prevention, fire fighting.

MARINTEK

Section for Materials Protection
Norwegian Marine Technology Research
Institute
PO Box 173 Asnes
3201 Sandefjord
Norway
Tel: +47 34 73033
Contact:Per Solheim, Head of Section
Fax: +47 34 73516

Category: C E
This is a corrosion *laboratory* for studying
corrosion and corrosion protection. The work is
mostly related to paints/paint properties,
metallic corrosion/electrochemistry, verification,
site inspection and independent consultancy.
Special interests
Corrosion protection, materials selection,
consultancy, verification, inspection.
MARINTEK is part of the SINTEF Group [*see
entry*] and uses the library facilities of the
Norwegian Institute of Technology.

Massachusetts Coastal Zone Management Office

100 Cambridge Street, 20th Floor
Boston, Massachusetts 02202
USA
Tel: (617) 727 9530
*Contact: Anne Smrcina, Public Outreach
Coordinator*
Fax: (617) 727 2754

Category: D
This is a policy and planning office within the
Executive Office of Environmental Affairs. It
evaluates and reviews US government plans to
lease offshore areas for oil and gas and other
mineral resources. Once leasing occurs, CZM
reviews industry plans for exploration,
development and production activities.
Library & Information facilities
Documents, available for public review, are
restricted to use in the office, 8:45am-5:00pm.
A Permit Advisory Service is offered.
Publications: *Coastlines Newsletter*
Databases: MassGIS

Memorial University of Newfoundland

Queen Elizabeth II Library
St John's, Newfoundland A1B 3Y1
Canada
Tel: (709) 737 7425
Contact: Joy Tillotson, Head, Information Serv
Fax: (709) 737 3118
E-mail: Envoy: qeii.lib Information Services

Category: B
This University library covers the full range of
scientific, technical and commercial subjects of
concern to the petroleum and marine
technology industries.
Library & Information facilities
More than one million volumes are available.
There is a fee-based computer search service.
The Library is open to the public from
8:30am-11:30pm (Monday-Thursday).
8:30am-5:45pm (Friday), 10:00am-5:45pm
(Saturday), 1:30pm-9:45pm (Sunday).

Mineralogical Society of America

1130 Seventeenth Street NW
Suite 330
Washington DC 20036
USA
Tel: (202) 775 4344
Contact: Susan L Myers, Executive Secretary
Fax: (202) 775 0018

Category: B
Subject interests
Geology, meteorology, hydrology,
oceanography, geodesy, cartography.
Professional society for research mineralogists
and petrologists as well as the general public.
Special interests
Mineralogy, petrology.
Publications: *American Mineralogist, Reviews
in Mineralogy.*

MIT Sea Grant Program

Massachusetts Institute of Technology
E38-300
Cambridge, Massachusetts 02139
USA
Tel: (617) 253 5944
Contact: Kathy Seaward, Information Specialist
Fax: (617) 258 8125

Category: C
Subject interests
Freshwater and marine biology, agriculture and
fisheries; geology, meteorology, hydrology,
oceanography, geodesy, cartography; civil and
hydraulic engineering, building.
Special interests
Oceanographic research, ocean engineering,
environment, autonomous underwater vehicles,
fishing gear-nets, aquaculture.
Library & Information facilities
Holdings of books, journals, and technical
reports are open to all but are
non-circulatory.There is a computer-based
enquiry service.
Hours: 9-5 (Monday–Friday) Publications:
technical reports, newsletter, fact sheets.
Databases: DIALOG, SGNET, in-house
publications.

MITS (Michigan Information Transfer Source)

University of Michigan
106 Hatcher Graduate Library
Ann Arbor, Michigan 48109-1205
USA
Tel: (313) 763 5060
Contact: Pamela J MacKintosh, Director
Fax: (313) 936 3630

Category: B
MITS is a research and information service of
the University of Michigan Library system which
serves business and industry on a cost-
recovery basis.
Subject interests
Science, technology, engineering, medicine,
business, social sciences, humanities and
multidisciplinary topics.
Library & Information facilities
MITS has access to the University of Michigan
libraries–the fifth largest research library in the
US: 6.4 million volumes, over 100,000 journal
titles.

Open between 8:00am–5:00 pm, MITS does
database searching, document retrieval and
library research for a fee, rate schedule on
request. MITS has access to over 300
databases and the collections of the University
of Michigan

National Academy Press

2101 Constitution Avenue NW
Washington DC 20418
USA
Tel: (202) 334 3313–800 624 6242
Contact: Kurt Collins
Fax: (202) 334 2793

Category: C D
As publisher for the National Academy of
Science, National Academy Engineering and
Institute of Medicine, the Press covers an
exceedingly wide range of relevant subjects.

National Energy Board

Environment Directorate
311–6th Avenue SW
Calgary, Alberta T2P 3H2
Canada
Tel: (403) 299 3675
Contact: Dr G K Sato
Fax: (403) 292 5503

Category: D
NEB is a Federal Regulatory Agency
responsible for oil and gasexploration and
production activities on Canada's Frontier
Lands.
Special interests
Research activities in marine technology
focussing on sea ice, icebergs, icing, waves,
offshore structures, geotechnical, and
hydrocarbon exploration on the Canadian
Continental Shelf. NEB has a repository for
environmental, geotechnical and engineering
feasibility studies conducted by industry in
connection with marine hydrocarbon activities.
Access to the studies may be restricted.
Library & Information facilities
Regulatory documents are available on request
as are environmental, geotechnical, and
feasibility studies published by NEB. Studies
published under Environmental Studies
Research Fund may be obtained at cost.

National Oceanic and Atmospheric Administration

Sanctuaries & Reserves Division
1825 Connecticut Avenue NW
Washington DC 20235
USA
Tel: (202) 606 4125
Contact: Dr Ervan G Garrison
Fax: (202) 606 4329

Category: D
The Division is concerned with marine
protection and research with a special interest
in National Marine Sanctuaries and Estuaries
Reserves. It offers an information service and
maintains a database.

National Physical Laboratory

Queen's Road
Teddington
Middlesex TW11 0LW
UK
Tel: (081) 977 3222
Contact: Information Services (extension 6352)
or Ms B Sanger,
Technical Enquiries Officer (extension 6880)
Tlx: 262344 NPL G
Fax: (081) 943 2155

Category: C D
NPL, an Executive Agency of the Department
of Trade & Industry research station, is the
national standards laboratory for the United
Kingdom. The National Corrosion Service
(extension 7110) is run by the Division of
Materials Metrology.
Special interests
Physical measurement, standards, calibration,
corrosion including corrosion of materials in
seawater, Information Technology and
accreditation.
Library & Information facilities
The libraries are intended primarily for NPL
staff. All enquiries for technical advice are
referred to the specialist best able to provide it.

National Radiological Protection Board

Chilton
Didcot
Oxfordshire OX11 0RQ
UK
Tel: (0235) 831600
Contact: M J Gains, Head of Information
    (Ex tension 2410)
D A Perry, Librarian (Extension 2526)
Tlx: 837124
Fax: (0235) 833891

Category: D
NRPB is a statutory body, established under
the Radiological Protection Act 1970, with a
duty to further the understanding of the
protection of mankind from radiation hazards
and to provide information and advice to
persons with repsonsibilities in the UK with
respect to protection of public or employees
against radiation.
Special interests
All aspects of health effects of ionising and
non-ionising radiation.
Library & Information facilities
In-house database of literature relating to
radiation hazards enables searches to be
conducted for individuals and organisations by
library staff. Visitors by appointment; loans to
other libaries. Technical enquiries are referred
to the specialist best able to provide the
required information.
A range of free broadsheets is available on
aspects of radiation. Other publications include
Documents of the NRPB, containing formal
advice, *Radiological Protection Bulletin*, reports
and memos.
Radiation protection advisory work is
undertaken under contract.

National Sea Grant Depository

Pell Library Building
URI, Narragansett Bay Campus
Narragansett, Rhode Island 02882
USA
Tel: (401) 792 6114
Contact: Joyce Winn, Loan Librarian
Fax: (401) 792 6160

Category: A
This organisation was set up to provide
information on America's oceans, its Great
Lakes and its coastal zone and to maintain the
collection of publications generated by the
National Sea Grant College Program, NOAA.
Special interests
Oceanography, marine education, aquaculture,
fisheries, coastal zone management, marine
recreation and law.
Library & Information facilities
Open Monday–Friday, 8:30am–4:30pm (closed
weekends and holidays), reference and online
search services are offered.
Publications: *Sea Grant Abstracts*
(PO Box 84, Woods Hole, MA 02543 USA)

Naval Sea Systems Command

NAVSEA
Director of Ocean Engineering
Code 00C
Washington DC 20362-5101
USA
Tel: (703) 697 7403
Contact: Paul Hankins Code 00C25
Fax: (703) 692 5822

Category: D
Special interests
Ocean engineering, salvage, pollution and
hazardous material cleanup. Oil cleanup
equipment Worldwide
Navy equipment supports US government
projects and, if requested, may be available for
commercial projects.

NEL - National Engineering Laboratory Executive Agency

East Kilbride
Glasgow G75 0QU
Scotland
Tel: (03552) 20222
Contact: J Leishman or T McMillan, Marketing
Department
Tlx: 777888
Fax: (03552) 36930

Category: C
NEL is a leading technology organisation
providing comprehensive engineering
technology services to industry and
government in key industry sectors such as
energy, process, defence and transport.
Special interests
NEL's work for the offshore industry includes
structural testing and analysis, failure investigations, offshore component stregth evaluation,
oil and gas multiphase flow measurement;
valve testing to API specification, pump and fan
design, noise and vibration consultancy,
thermal integration, compact heat exchangers,
materials engineering and provision of physical
properties data.
Library & Information facilities
The Library is open to enquirers by appointment. Material that is not readily available from
other sources can be loaned to other libraries
or engineering firms. All enquiries requiring
technical advice on engineering problems are
passed to the specialist best able to deal with
them. Initial enquiries for consultancy services
should be made to the Marketing Department.
Literature searches may be carried out if staff is
available. Commercial online information
retrieval services are available at NEL.

Newfoundland & Labrador Department of Mines & Energy

Energy Resource Centre
6th Floor, Atlantic Place
PO Box 8700
St John's, Newfoundland A1B 4J6
Canada
Tel: (709) 729 2416
Contact: Mary Varghese, Librarian
Fax: (709) 729 2508

Category: D
**The Petroleum Resources and Energy
Branch** is responsible for formulating policy
and advising Government on offshore matters.
**The Petroleum and Energy Economics
Branch** is responsible for formulating policy
and legislation relating to petroleum and energy
activities.
Subject interests
Energy, fuel, power; geology, meteorology,
hydrology, oceanography, geodesy,
cartography; occupational and environmental
health and safety, fire prevention, fire fighting,
energy and environment.
Special interests
Energy in general, petroleum and natural gas,
renewable energy, energy management,
energy economics and energy environment.
Library & Information facilities
The in-house database is available through
search-magic (software). The Library has
access to CAN/OLE database, ILL, and holds
5000 books, 200 periodicals, 100 video tapes,
1000+ slides. Reference library has restricted
access to outside users. Hours: 8:30-12:30,
1:30-4:30 (4:00 in summer). This is a science
and technology library.

Newfoundland Ocean Industries Association

Box 44, Atlantic Place
Suite 805
215 Water Street
St John's, Newfoundland A1C 6C9
Canada
Tel: (709) 753 8123
Contact: Ruth Graham, Executive Director
Fax: (709) 753 6010

Category: B D
Special interests
Ocean industries with particular emphasis on
oil and gas.

Library & Information facilities
Facilities at the NOIA office are available to
members of the Association from 9:00am–
5:00pm (provided the Board Room is not in
use). Publications: *NOIA News*. Detailed
information is held on member companies.

Norwegian Geotechnical Institute

(Norges Geotekniske Institutt)
Sognsveien 72
PO Box 40
Taasen
N-0801 Oslo
Norway
Tel: +47 2 230388
Contact: Knell Hauge, Head of Information
Tlx: 19787 ngi n
Fax: +47 2 230448

Category: B
NGI is a private non-profit foundation; it is a
centre for research and consulting in the
geosciences.
NGI's activities include:
–quaternary geology in the North Sea area,
evaluation of geotechnical properties of submarine soils and sediments, including drilling
supervision, *in situ* and laboratory testing and
development of field and laboratory equipment;
–foundation engineering for pile-supported,
anchored and gravity-base platforms, including
foundation analyses and design, dynamic
earthquake considerations and geotechnical
documentation for platform certificatio;
–rock mechanics and hydraulic properties of
petroleum reservoirs, including cap-rock
integrity studies, compaction, subsidence,
fracturing and stability of wells;
–instrumentation of platforms, buoys and
pipelines for performance observations,
including design, delivery and installation of
field instruments.
Library & Information facilities
NGI's library contains about 23,000 volumes,
including serials, books, reprints, etc, mainly of
geotechnical interest; the library is available for
consultation and loan service. Literature
searches can be carried out, subject to staff
availability. Hours: 9am–3.30pm.

Norwegian Underwater Technology Centre - NUTEC

Gravdalsveien 255
PO Box 6
N05034 Ytre Laksevåg
Bergen
Norway
Tel: +47 5 341600
Contact: T Mellingen, Director, or to the Library
Tlx: 42 892 nutec n
Fax: +47 5 344720

Category: C
NUTEC is a limited company owned by Statoil, Norsk Hydro and Saga Petroleum. Its main concern is research and development within diving medicine and physiology, hyperbaric medical treatment, testing within underwater technology areas, and safety and emergency training for offshore employees.
Special interests
Diver's personal equipment, long term effects of saturation diving on all aspects of the human body, occupational health services for the diving industry, risk analysis in underwater activities and diverless underwater systems and vehicles.
Library & Information facilities
NUTEC has a well-equipped library with its specialities within diving and hyperbaric medicine. The library is also responsible for the published NUTEC reports and has access to several external databases. Although the library is mainly for internal use, visitors may consult it by appointment. Enquirers will be answered as far as possible, and NUTEC is interested in consultancy on relevant projects. Literature searches can be performed in internal databases. There is a mailing list for information on NUTEC reports; some information may be restricted.

Ocean Engineering Information Centre

Bartlett Building
Memorial University of Newfoundland
St John's, Newfoundland A1B 3X5
Canada
Tel: (709) 737 8377/78
Contact: Ms Judith A Whittick, Information Researcher
Tlx: 016-4101
Fax: (709) 737 4706
E-mail: ENVOY 100: C.CORE
Category: D
OEIC's principal role is to provide its sponsoring groups and Associates of the Centre for Cold Ocean Resources Engineering (C-CORE) with an information search service in cold ocean engineering. The service is also available to industry and government under specific contract arrangements.
Subject interests
Geology, meteorology, hydrology, oceanography; civil and hydraulic engineering.
Special interests
A special collection of material relating to ocean engineering and particularly offshore resource development in cold waters has been built up to support the information search service. Subjects include sea ice, icebergs, icing, waves, offshore structures, geotechnics, hydrocarbon exploration, and North Sea offshore technology. The collection covers the Beaufort Sea, East Coast Offshore and the Arctic Islands. Special emphasis has been given to identifying and indexing materials not readily available from other sources.
Documents in OEIC are indexed on the SPIRES retrieval system and are accessible online at the Centre.
The OEIC database, along with several other polar databases, is now available on the Arctic and Antarctic Regions CD-ROM (available from National Information Services Corporation). The collection currently contains: technical report literature from university, industry and government sources; papers from conferences, meetings and workshops; directories, bibliographies and atlases; annual reports, topographic maps and nautical charts; meteorological and oceanographic data in chart and tabulated format.

Information services
OEIC's services include: computer searches, interlibrary loans from outside the Province, retrieval of materials from OEIC and/or other Newfoundland libraries, assistance in locating hard-to-find references and obscure documents.

The bi-monthly *OEIC Information Bulletin* identifies the latest technical literature and includes information on upcoming conferences. It is available free of charge to Associates of C-CORE and to groups and individuals with whom the Centre exchanges information. To others it is available on an annual subscription basis.

Ocean Oil Information

3050 Post Oak (Suite 200)
Houston, Texas 77056
USA
Tel: (713) 621 9720
Contact: Michael Crowden
Tlx: 244-66-77 Pubru
Fax: (713) 963 6285

Category: E
Subject interests
Business information and statistics Ocean Oil Information is a service of *Offshore Magazine* and *Ocean Oil Weekly Report* from PennWell Publishing Co.
Special interests
International offshore oil and gas exploration and development activities.
Publications: a computer-based weekly newsletter and a monthly magazine *Ocean Oil*, maintains the Offshore Mobile Rig database and offers customized reports on specific aspects of offshore oil and gas activity.

Oceans Institute of Canada

5th Floor
1236 Henry Street
Halifax, Nova Scotia B3H 3J5
Canada
Tel: (902) 494 6552
Contact: Judith Swan, Executive Director
Fax: (902) 494 1334

Category: B E
The Institute aims to promote integrated oceans management.
Publication: *InfoOCEANS*.

Office of Ocean & Coastal Resource Management

1875 Connecticut Avenue NW, Room 706
Washington DC 20235
USA
Tel: (202) 606 4111
Contact: Sallie P Cauchon, Coastal Information Specialist
Fax: (202) 606 4329

Category: D
The Office exists to carry out federal laws aimed at protecting, restoring, and developing the nation's coastal resources.
Special interests
Coastal zone management: wetlands, water quality, cumulative impacts, marine sanctuaries.
Library & Information facilities
Coastal Zone Information Center is a Repository for work products done by States under the Coastal Zone Management Act of 1972. The Center is open from 8–4 to the public; publications and information on the Coastal Zone Management Program are sent free to requesters.
The Office administers the National Coastal Zone Management Program, the National Estuarine Research Reserve Program and the National Marine Sanctuary Program.

Offshore Data Services Inc

PO Box 19909
Houston, Texas 77224
USA
Tel: (713) 781 2713
Contact: Ms S Stanton, Research Manager
Fax: (713) 781 9594

Category: E
The Research Department of this independent organisation offers consulting and information-gathering services.
Subject interests
Marine technology and the offshore oil and gas industry; business information and statistics.

Offshore Engineering Information Service

Heriot-Watt University Library
Riccarton
Edinburgh EH14 4AS
Scotland
Tel: (031) 449 5111
Contact: Arnold Myers (extension 4070)
Fax: (031) 451 3164
E-mail: (JANET) LIBAJM@UK.AC.HW.CLUST

Category: B
This external service of the University Library
provides consultancy and information
brokerage in the field of petroleum and marine
technology.
Subject interests
Marine technology, oceanography, petroleum
engineering, naval architecture, materials and
corrosion science, pollution control, underwater
construction,seabed resources, navigation,
geotechnics, environmental protection, law and
economics.
Special interests
Offshore structures and pipelines, diving,
offshore health and safety, petroleum
economics, marine instrumentation and
surveying, renewable energy, subsea
production systems.
Information & library facilities
The Library offers an information service based
on its own holdings, its own offshore literature
database, and access to over 1000 databases
through 23 hosts.
Publications: *Offshore Engineering Information
Bulletin* lists relevant recent publications and
forthcoming meetings; a tailor-made current
awareness service is also offered. Write for
details of Subscribing Membership.

Offshore Supplies Office
Department of Energy

Alhambra House
45 Waterloo Street
Glasgow G2 6AS
Scotland
Tel: (041) 221 8777
Contact: General Enquiries (ex 5746)
or Library & Information Services (ex 5747)
Tlx: 779379 OSOGLW G
Fax: (041) 221 1718

Category: D
OSO is part of the UK Government's Dept of
Energy, with offices in Glasgow, Aberdeen and
London. It provides a point of contact between
the UK offshore supplies industry and oil and
gas operators worldwide.
Special interests
Energy, fuel, power; all aspects of the offshore
oil and gas industry.
Library & Information facilities
The OSO library is available for public use by
appointment. A loan service can be arranged
through other libraries only. Postal and
telephone enquiries are always welcome.

Offshore Technology Conference

PO Box 833868
Richardson, Texas 75083-3868
USA
Tel: (214) 669 0072
Contact: Fred Herbst, Public Relations Manager
Fax: (214) 669 0135

Category: E
Offshore Technology Conference (and
exhibition) represents one of the largest and
most comprehensive interdisciplinary ventures
in the engineering and scientific communities.
The 11 professional OTC Sponsoring
Organizations represent every technology
associated with the exploration, protection and
production of ocean resources. OTC stresses
prudent management of offshore resources
and their development, including cost-saving
technology and techniques used in petroleum
exploration and production operations.
Library & Information facilities
SPE,is the contract manager of OTC. Its
internal library, can be consulted by
appointment. Literature searches can be
carried out, subject to staff availability.

Paint Research Association

8 WaldegraveRoad
Teddington
Middlesex TW11 8LD
UK
Tel: (081) 977 4427
Contact: MissS C Haworth, Librarian
(extenstion 223)
Tlx: 918720
Fax: (081) 943 4705

Category: B C
Paint Research Association is an independent
research and development association.
Subject interests
The formulation, manufacture, application,
wear and resistance of paint coatings in all
environments, including the sea. Investigation
and information covers both anti-corrosive and
anti-fouling coatings.
Library & Information facilities
The Library is available for reference by
appointment; the loan service is restricted to
members. Non-members will be charged for
the use they make of the facilities. An
information retrieval service is charged
according to the time expended and the level of
expertise required.
Computerised processing of abstracts enables
the supply of monthly selections of abstracts on
chosen subjects as well as computerised
retrospective searches. The Association issues
a variety of publications: newsletters, journals,
conference proceedings, reports, etc.
Enquiries needing specialist advice will be
referred to the appropriate specialist within the
Association; translations can be commissioned
from translators with subject knowledge.

Paleontological Society

New Mexico Bureau of Mines & Mineral
Resources
Socorro, New Mexico 87801
USA
Tel: (505) 835 5140
Contact: Dr Donald L Wolberg, Secretary
Fax: (505) 835 6333

Category: B
Subject interests
Geology, meteorology, hydrology, oceano-
graphy, geodesy, cartography.
This is a non-profit scientific society with a
special interest in Paleontology.

Pallister Resource Management Ltd

640, 300 5th Avenue SW
Calgary, Alberta T2P 3C4
Canada
Tel: (403) 233 8480
Contact: Mary Nelson, Administrative Assistant
Fax: (403) 233 8486

Category: E
An enquiry service is available from
Monday–Friday, 8:30am–4:30pm

Petroleum Abstracts

The University of Tulsa
600 S College
101 Harwell
Tulsa, OK 74104-3189
USA
Tel: (918) 631 2297
Contact: David Brown, Marketing Manager
Tlx: 49-7543
Fax: (918) 599 9361
E-mail: question@TUred.pa.utulsa.edu

Category: B
Subject interests
Energy, fuel, power; occupational and
environmental health and safety, fire
prevention, firefighting.
Special interests
Petroleum geology, geophysics, petroleum
engineering, pipelining, ecology and safety.
Library & Information facilities
Petroleum Abstracts is a weekly bulletin
offering Petroleum Abstracts Search Service.
Databases: TULSA (on ORBIT), ERTH (on
ORBIT), PETROLEUM EXPLORATION &
PRODUCTION (DIALOG)

Petroleum Communication Foundation

Suite 1250
633 6th Avenue SW
Calgary, Alberta T2P 2Y5
Canada
Tel: (403) 264 6064
Contact: Gila Naderi, Communication
Coordinator
Fax: (403) 237 6286

Category: B
The Foundation's purpose is to increase public
awareness of Canada's petroleum industry.
Subject interests
Energy, fuel, power; business information and
statistics.
Library & Information facilities
Books and statistics, videos and slides on the
industry are available across Canada to
non-commercial customers

Petroleum Extension Service
(PETEX)

Balcones Research Center
The University of Texas at Austin
Austin, Texas 78712
USA
Tel: (512) 471 3154
Contact: William R Baker, Director
Tlx: 767161
Fax: (512) 471 9410

Category: B
PETEX aims to provide field-level training to
petroleum industry employees. It is a non-profit,
self-supporting part of UT Austin covering all
phases of the petroleum industry except
refining and marketing.
Publications: PETEX sells publications,
audiovisuals, and conducts short courses
mainly for oilfield workers and technicians.
Free catalogue available on request.

Petroleum Industry Training Service

13, 2115 27th Avenue NE
Calgary, Alberta T2E 7E4
Canada
Tel: (403) 250 9606
Contact: Mr Wayne Wetmore, Vice-President,
Training & Development
Tlx: 03-822-806
Fax: (403) 291 9408

Category B
This is a non-profit training organization,
governed by a board made up of senior
representatives from the petroleum industry.
Operating throughout Canada, its objectives
are to identify training needs of the industry,
develop and offer training programs, provide
training advice and guidance and help establish
standards which will be accepted and
supported by industry and regulatory agencies.
Library & Information facilities
The library has a collection of oilfield training
videos available for loan, and a collection of
CGPA/CGPSA research papers presented at
quarterly meetings. Papers will be photocopied
and mailed on request.
The Training Line database lists courses
offered by both private and public agencies on
subjects of interest to petroleum industry
personnel in Alberta. This service is provided
free of charge *(403) 250 2002)*.

Petroleum Science & Technology Institute

Dunedin House
25 Ravelston Terrace
Edinburgh EH4 3EX
Scotland
Tel: (031) 451 5231
Tlx: 9312110965
Contact: Graham Stewart
Fax: (031) 451 5232
E-mail: BT Gold 78 PSL002

Category C
Established by the Department of Energy in
1989 to enhance R&D capability in the UK oil
industry. the Institute conducts its own research
programme, provides information services and
acts as an independent coordinator. Its library
and information facilities are developing.

Its principal strategies are:
—own research programme funded by membership, conducted in UK universities
—research under contract in specialist areas, conducted in UK universities
—effective technology and information transfer
—research coordination and facilitating role.
Publications: *PSTI Technical Bulletin, International Conference Diary.*
The PSTI is an independent, non-profit organisation that acts as an honest broker in the establishment of upstream oil-related R&D projects. The results and most services are availbale to members only and are covered by the membership fee. Membership is open to companies and individuals involved in the upstream oil industry at any level.

Petroleum Society of CIM

320 101 6th Avenue SW
Calgary, Alberta T2P 3P4
Canada
Tel: (403) 237 5112
Fax: (403) 262 4792

Category: B
The Petroleum Society of CIM with 2500 members is a diversified group of professionals concerned with the production, transportation and marketing of petroleum and natural gas in Canada. The Society organizes an Annual Technical Meeting, sponsors the National Petroleum Show, publishes the *Journal of Canadian Petroleum Technology* (JCPT) and is committed to the technical advancement of its members through education, communication and the dissemination of technical information. Its Sections hold local meetings and services in Calgary, Edmonton, Grande Prairie, Regina, Ottawa and Halifax.

Plymouth Marine Laboratory & Marine Biological Association

Library and Information Services
Citadel Hill
Plymouth PL1 2PB
UK
Tel: (0752) 222772
Fax: (0752) 226865
E-mail: (Sciencenet/Omnet): MBA.LIBRARY

Category: B C D
This organisation caries out marine environmental, biological and physiological research.
Subject interests
Freshwater and marine biology, agriculture and fisheries; oceanography, geodesy, marine technology.
Library & Information facilities
The Library has been established for over 100 years and contains one of the world's major collections of publications on marine biology, oceanography, fisheries, ecology and pollution. A range of catalogues and databases are available. The Library is the UK focal point and input centre for the international Aquatic Sciences & Fisheries Information System (ASFIS); it operates a Marine Pollution Information Centre; it provides data for the Environmental Chemicals Data & Information Network (ECDIN) of the European Communities; it provides current awareness services, bulletins and bibliographies for both internal and outside users, and is an active participant in UK, European and international marine information networks.

Precision Microfilming Services

3770 Kempt Road
Halifax, Nova Scotia B3K 4X8
Canada
Tel: (902) 455 5451
Contact: Charles Joyce/Clyde Holland
Fax: (902) 454 5570

Category: E
This Microfilm Service Bureau provides filming/reproducing services and equipment
Library & Information facilities
Call Clyde Holland: B.I.O. for Open File Information, Geological Survey of Canada and Open File Reports.

Rapra Technology Ltd

Shawbury
Shrewsbury
Shropshire SY4 4NR
UK
Tel: (0939) 250383
Contact: Brian Evans, Enquiry Bureau
Tlx: 35134
Fax: (0939) 251118

Category: C E
Rapra is the leading independent consultancy
in Europe specialising in rubber, plastics and
composites.
Special interests
All aspects of the manufacture and application
of rubber and plastics components including
their application in a marine environment.
Library & Information facilities
The Library is available for consultation by
appointment; the loans service is restricted to
members, but copies may be provided. The
Rapra Abstracts Database contains over
350,000 references dating back to 1972. All
references since 1980 (over 200,000) are now
available on CD-ROM to provide an economical
and easy-to-use method of accessing
information. Information retrieval services, in
the form of literature searches and the
compilation of bibliographies, are available and
a technical and commercial enquiry service is
operated.
Additionally, enquirers interested in a specific
field or application may subscribe to the
Selective Alert Service whose contexts are
compiled to meet the specific interests of the
subscriber. More detailed specialist advice and
consultancy are also available.

Royal Institution of Naval Architects

10 Upper Belgrave Street
London SW1X 8BQ
UK
Tel: (071) 235 4622
Contact: J Leather, Assist Secretary (Technical)
Tlx: 265844 SINAI G
Fax: (071) 245 6959

Category: B
RINA is an independent chartered qualifying
body and learned society.
Special interests
Naval architecture, marine enginering and
hydrodynamics including the design and
performance of ships, marine vehicles and
offshore structures.
Library & Information facilities
The Library is available for reference, but only
members of RINA or other EC institutions may
borrow from the Library. Enquiry answering and
information retrieval is undertaken on a small
scale free of charge. If longer searches are
undertaken a donation will be requested. RINA
offers no consulting services.
Publications: journals, transactions,
proceedings and marine technology texts.

Royal Society for the Prevention of Accidents

Occupational Safety Division
Cannon House
The Priory, Queensway
Birmingham B4 6BS
UK
Tel: (021) 200 2461
Contact: Adele Brookes, Librarian
Tlx: 336546
Fax: (021) 200 1254

Category: E
RoSPA is an independent, non-profit-making
company offering consultancy, advice, training
and publications.
Special interests
All aspects of occupational health and safety
including safety on offshore installations.
Library & Information facilities
The Library is freely available for reference.
Loan and photocopying facilities are available
for members only. Information retrieval and
bibliography compilation are also undertaken
for members only.

Sea Fish Industry Authority - Seafish

Marine Farming Unit
Ardtoe, Acharacle
Argyll PH36 4
Scotland
Tel: (096) 785 213/666
Contact: M J S Gillespie, Manager
Fax: (096) 785 343

Category: C
Seafish is the levy-funded organisation which provides promotion, technical services and support for all sectors of the UK fishing industry. The Marine Farming Unit is a specialist branch conducting research and development into the cultivation and stock enhancement of various marine fish and shellfish species.
Special interests
Freshwater and marine biology, fisheries. Farming of sea fish; stock enhancement.
Library & Information facilities
The small library is intended mainly for use by staff and students, though visitors may refer to it by appointment. General information on marine farming can be provided on request. Detailed reports with the results of the Unit's work are available for a small charge. Contract work is undertaken for MAFF, HIE, CEC and commercial organisations.

SINTEF Group

Foundation for Scientific & Industrial Research at the Norwegian Institute of Technology
N-7034 Trondheim
Norway
Tel: +47 7 593000
Tlx: 50 0120 sintf n
Fax: +47 7 592180

Category C
The SINTEF Group is the largest independent research organisation in Scandinavia, spanning most technological disciplines, the natural and social sciences, and medicine. The Group stresses the importance of high quality work and punctual delivery: a staff of 2,000 complete 3,000 projects each year for clients worldwide.

Close cooperation with the Norwegian Institute of Technology (NTH), enables the institutions to share laboratories, equipment and specialist staff; the SINTEF/NTH research community has an established international position as a leading technological innovator.
The SINTEF group consists of 5 organisations:
–**SINTEF**: The Foundation for Scientific & Industrial Research at the Norwegian Institute of Technology
–**EFI**: Norwegian Electric Power Research Institute
–**IKU**: Continental Shelf & Petroleum Technology Research Institute
–**MARINTEK**: Norwegian Marine Technology Research Institute
–**SINTEF MOLAB**: Metallurgical/chemical lab in Mo i Rana

Society for Mining, Metallurgy & Exploration (SME) Inc

PO Box 625002
Littleton, CO 80162-5002
USA
Tel: (303) 973 9550
Contact: Sharon E Connelly, Education Coordinator
Fax: (303) 973 3845

Category: B
Subject interests
Mining, metallurgy; geology, meteorology, hydrology, oceanography; geodesy, cartography. Technical organization for professionals engaged in the mineral

industries. Publishing, conducting technical meetings and enhancing the minerals industry field through service to its participants and the public they serve.
Publications: *Mining Engineering, Minerals & Metallurgical Processing,* also books.

Society for Underwater Technology - SUT

76 Mark Lane
London EC3R 7JN
UK
Tel: (071) 481 0750
Contact: Cdr D Wardle, Executive Secretary
Tlx: 886841 IMARE G
Fax: (071) 481 4001

Category: B
Founded in 1966, the Society is an international learned organisation registered in the UK as a charitable concern and a company limited by guarantee. With headquarters in London, membership is worldwide and multi-disciplinary. It is open to individual and corporate members who are engaged or interested in the wide spectrum of activitites that constitute marine science and technology. Members include: oil companies, engineering construction companies, service contractors, equipment manufacturers, consultancies, government departments, universities, reseach institutes and regulatory authorities.
Special interests
Diving and submersibles technology; geology, geophysics and offshore site investigations; oceanography and environmental forces; ocean energy, mineral and living resources; subsea engineering and operations; marine ecology and pollution; underwater instrumentation, acoustics and data acquisition; underwater structures, pipelines and cables; education and training of future offshore engineers and ocean scientists.
Library & Information facilities
The Society is located in the Institute of Marine Engineers and has facilities for members to use both its own reference collection and that of the IMarE Library. The Society publishes a quarterly journal,*Underwater Technology*, a newsletters, conference proceedings and a careers pack. Technical enquiries are referred to experts within the Society's membership.

Society of Naval Architects & Marine Engineers

601 Pavonia Avenue, Suite 400
Jersey City, NJ 07306
USA
Tel: (201) 798 4800
Contact: Robert G Mende, Executive Director
Fax: (201) 798 4975

Category: B
This is a non-profit, professional association, composed of individual members for the purposes of:
–the exchange of information and ideas among its members .. disseminating the results of research, experience and information;
–promoting the professional integrity and status of its members and advancement in their knowledge;
–the furtherance of education in naval architecture, marine and ocean engineering;
–encouraging and sponsoring research and other inquiries important to the advancement of the art.
Library & Information facilities
Available, for members only, from 9:00am–4:45pm

Standards Council of Canada

45 O'Connor Street
Suite 1200
Ottawa, Ontario K1P 6N7
Canada
Tel: (613) 238 3222
Contact: D C Thompson, Manager Information Division
Tlx: 053-4403
Fax: (613) 995 4564

Category D
The purpose of the Council is to coordinate and promote standards, technical regulations and certification systems in Canada.
Library & Information facilities
400,000 standards and regulations are available for unrestricted reference purposes. The Library opens from 9am–5pm (Monday–Friday) and an enquiry service is available. Databases: Canadian Standards; Referenced Standards–Federal;GATT TBT Notifications/ Draft European Standards: Cost $50/hour connect time, from 1 January 1992.
The Council is also the sales agent in Canada for foreign and international standards.

Texas A&M University

Sterling C Evans Library
College Station, Texas 77843-5000
USA
Tel: (409) 845 8111
Contact: Irene B Hoadley, Director
Fax: (409) 845 6238

Category: B
Set up primarily to provide for the information
needs of the University community, this Library
has special interests in nautical geology, fish
culture and fisheries.
Library & Information facilities
1,900,000 volumes are available for reference
(no restrictions) but opening hours are variable:
check schedule by calling (409) 845 8111. A
computer-based enquiry service is provided.

Transportation Research Board

National Research Council
2101 Constitution Avenue NW
Washington DC 20418
USA
Tel: (202) 334 2934
Tel: (202) 334 3203
Contact: Christina S Casgar, Maritime Specialist
Tel: (202) 334 2005

Transportation Research Board Library:
Tel:(202) 334 2989
Fax: (202) 334 2527.

Catgory: D
The Transportation Research Board
coordinates the Council's activities in the
discipline of transportation research,
engineering, and policy studies. Highways,
Public Transit, Aviation, Rail, Pavements,
Safety, Energy Issues, and Data Needs are
topics in which the Transportation Research
Board has recently completed studies.The
policy study on intermodal marine container
technology was released in 1992 as Special
Report 236.
Library & Information facilities
A holding of 28,000 volumes is available for
interlibrary loans and research. Facilities are
open to TRB Affiliates with some services
available to the general public. An enquiry
service is available. Hours: 8:30am–5:00pm
EST Monday–Friday.
Publications: approximately 100 titles are
published each year, including *Transportation
News* and *Computer*. The Board also serves as

distributor of publications of the Strategic
Highway Research Program.
Databases: The TRIS (Transportation
Research Information Service) Database
contains more than 350,000 records on
domestic and international publications,
articles, and information sources. Searches
may be performed by TRB or on-line at a
nominal fee. Discounts may apply to Affiliates
of TRB. New entries into the database are
published as abstracts in modal specific
journals available by subscription.
Call (202) 334 3262 for further information.
Information packet on Affiliation with the
Transportation Research Board and its benefits
is available by telephoning the Affiliates
Coordinator at (202) 334 3216.

TWI

(formerly The Welding Institute)
Abington Hall
Abington
Cambridge CB1 6AL
UK
Tel: (0223) 891162
Contact: K G Richards, Chief Information Officer
Tlx: 81183 Weldex G
Fax: (0223) 892588

Category: B E
TWI is an independent and professional
association which carries out research into and
development of all aspects of welding
technology, including production, economic and
design considerations.
Special interests
Fatigue, brittle fracture, service failures, the
weldability of ferrous and non-ferrous metals
and alloys, and the behaviour of welds in
service environments.
Library & Information facilities
The TWI Library is intended primarily for
members but others may consult it by
apopointment. The loans service is normally
restricted to members and other libraries, but
copies may be made available. Retrospective
retrieval is available to members and
subscribers to the Weldasearch information
services. A selective dissemination of
information services is provided by
Weldasearch System 1. Facilities are available,
to Industrial Members only, for answering
technical enquiries and solving welding
problems.

University of Alaska Southeast

11120 Glacier Highway
Juneau, Alaska 99801
USA
Tel: (907) 789 4472
Contact: B Meaim, Public Information Specialist
Fax: (907) 789 4595

Category: B
The University offers 2-year and 4-year degrees in the liberal arts. It includes the Egan Memorial Library and publishes the Academic Catalog, Viewbook, and Newsletter.

University of California, Berkeley Extension

Continuing Education in Engineering
2223 Fulton Street
Berkeley, California 94720
USA
Tel: (510) 642 4151
Contact: Nanette Pike, Continuing Education Specialist
Tlx: 9103667114 Berk Berk
Fax: (510) 643 8683

Category: B
Subject interests
Geology, meteorology, hydrology, oceanography, geodesy,cartography; physics, electrical and electronic engineering, telecommunications;mechanical engineering; civil and hydraulic engineering, building.
Special interests
Professional education and various civil engineering topics.
Course brochures are available.

University of Rhode Island

Department of Ocean Engineering
202 Lippitt Hall
Kingston, RI 02881
USA
Tel: (401) 792 2273
Contact: Armand J Silva, Chairman
Fax: (401) 782 1066

Category: A B C
Subject interests
Energy, fuel, power; mining, metallurgy; geology, meteorology, hydrology, oceanography, geodesy, cartography; physics, electrical and electronic engineering, telecommunications; mechanical engineering; civil and hydraulic engineering, building.
Special interests
Education and research, ocean instrumentation and data service; underwater acoustics. marine hydrodynamics and fluid mechanics; coastal and nearshore processes; marine geomechanics; coastal and offshore structures; remote sensing; corrosion and corrosion control.

University of Southern California

Doheny Memorial Library
Government Documents Department
University Park–MC 0182
Los Angeles, California 90089-0182
USA
Tel: (213) 740 2339
Contact: Julia Johnson, Librarian
Fax: (213) 749 1221

Category: B
Subject interests
Freshwater and marine biology, agriculture and fisheries; chemistry; energy, fuel, power; geology, meteorology, hydrology, oceanography, geodesy, cartography; physics, electrical and electronic engineering, telecommunications; mechanical engineering; civil and hydraulic engineering, building; occupational and environmental health and safety, fire prevention, fire fighting; business information and statistics.

US Army Cold Regions Research & Engineering Laboratory (CRREL)

72 Lyme Road
Hanover, NH 03755-1290
USA
Tel: (603) 646 4221
Contact: Nancy Liston, Librarian
Fax: (603) 646 4712

Category: C D
Subject interests
Geology, meteorology, hydrology,
oceanography, geodesy, cartography;
mechanical engineering; civil and hydraulic
engineering.
Special interests
Research in all engineering aspects of cold
regions, snow, ice and permafrost studies.
Library & Information facilities
Available to US citizens: foreigners with prior
approval. There is a limited enquiry service.
Hours: Monday–Friday 8:00am–4:00pm.
Publications: CRREL report series.
Databases: Cold on Orbit Search Service;
Arctic and Antarctic CD-ROM.

US Department of Energy

Bartlesville Project Office
PO Box 1398
Bartlesville, OK 74005
USA
Tel: (918) 337 4401
*Contact: Herbert A Tiedemann, Project
Manager for TechnologyTransfer*
Fax: (818) 337 4418

Category: D
Subject interests
Energy, fuel, power. DOE lead office for
petroleum research. Tertiary (enhanced) oil
recovery; crude oil processing.
Library & Information facilities
The Department issues petroleum-related
publications and the Library is open from
8:00am–4:30 pm.
Database: Crude Oil Analysis Data Bank
(COADB). COADB is a free online service
offering chemical and physical analysis data on
over 9000 crude oils.

US Department of the Interior Minerals Management Service

Alaska OCS Region
949 E 36th Avenue, Room 110
Anchorage, Alaska 99508-4302
USA
Tel: (907) 271 6000
Contact: Christine Huffaker
Fax: (907) 271 6805

Category: D
Subject interests
Freshwater and marine biology, agriculture and
fisheries; geology, meteorology, hydrology,
oceanography, geodesy, cartography;
mechanical, civil and hydraulic engineering.
Special interests
Science and economics.
Library & Information facilities
10,000 volumes are available with Interlibrary
Loans. Hours: 7:30–4:30 Monday–Friday
Publishes studies and reports with an enquiry
service on (907) 271 6435.

Warren Spring Laboratory

Gunnels Wood Road
Stevenage
Hertsfordshire SG1 2BX
UK
Tel: (0438) 741122
*Contact: M Webb, Head of Marine Pollution &
Bulk Materials Division*
Tlx: 82250 WSLDOI G
Fax: (0438) 360858
*DTI Environmental Enquiry Point:
    (0800) 585794*

Category: C D
Warrens Spring Laboratory is the Environmental Technology Executive Agency of the
Department of Trade & Industry. It undertakes
research into environmental problems for
Government departments, the European
Community, local authorities and industry.The
Marine Pollution & Bulk Materials Division
provides reasearch and development in the
general field of oi and chemical pollution of the
sea and shore. Areas of expertise include
remote sensing of oil at sea (for example by
Infra-Red Line Scan), dispersant treatment of
oil spills, and beach cleaning. Other facilities
include a ground-level test tank and a ship's
motion simulator, the latter being used in the
Division's work on cargo stability.

Water Research Centre

WRC Environment
PO Box 16
Henley Road
Medmenham, Marlow
Buckinghamshire SL7 2HX
UK
Tel: (0491) 571531
Contact; A C Jordan, Head of Library &
Information
Fax: (0491) 579094

Category: C E
WRc is a private, contract research
organisation.
Subject interests
All aspects of water pollution including human
health and aquatic life; UK and EEC legislation;
analytical services; marine and land disposal of
wastes; environmental management.
Library & Information facilities
Reference, loan and photocoopy services are
provided on a payment basis. Use of library by
appointment.
Dtabases: WRc produces the database
Aqualine. Online searching is available on a
payment basis. Technical enquiry services
available with specialist advice on a
consultancy basis.

The Welding Institute: *see TWI*

Woods Hole Oceanographic Institution

Woods Hole,
Massachusetts 02543
USA
Tel: (508) 548 1400
Contact: Carolyn P Winn, Research Librarian
Tlx: 951679 - Answer Back
OCEANINSTWOOH
E-mail: OMNET - WHOI.LIBRARY

Category: C
Woods Hole is a non-profit research institution
with special interests in geology and geophysics
Subject interests
Freshwater and marine biology, agriculture and
fisheries; chemistry; geology, meteorology,
hydrology, oceanography, geodesy,
cartography; physics, electrical and electronic
engineering, telecommunications.
Library & Information facilities
The library holds 40,000 volumes and is open
from 7:00am–5:00pm Monday–Friday.
Publications: *Oceanus* (journal) and the Annual
Report.

Information Services
by Category

Category A

Public or national library

Aberdeen City Libraries
Alberta Oil Sands Technology & Research Auth
American Association of Petroleum Geologists
Bedford Institute of Oceanography
British Library
Canada Institute for Scientific & Technical
　　Information
COPPE/UFRJ-
　　Federal University of Rio de Janeiro
Department of Fisheries & Oceans
Food & Agriculture Organization of the
　　United Nations (FAO)
Hydrographic Office
National Sea Grant Depository
University of Rhode Island

Category B

*Institution, University, Professional Association,
Learned Society*

ABS Europe Ltd
American Association of Petroleum Geologists
American Gas Association
American Society for Testing & Materials
American Society of Civil Engineers
American Society of Mechanical Engineers
AODC
Arctic Institute of North America
Association of Consulting Engineers of Canada
British Institute of Management
British Library
Canadian Geotechnical Society
Canadian Library Association
Canadian Maritime Industries Association

Canadian Petroleum Association
Canadian Petroleum Products Institute
Canadian Society of Petroleum Geologists
Chartered Institution of Building Services Enginrs
Colorado School of Mines
CONCAWE
COPPE/UFRJ-
　　Federal University of Rio de Janeiro
Det norske Veritas - Veritec
Duke University Marine Laboratory
Environment & Natural Resources Institute
European Association of Exploration Geophysicists
Geological Society of America
IKU - Continental Shelf & Petroleum Research
　　Institute
Institute of Marine Engineers
Institute of Petroleum
Institution of Chemical Engineers: IChemE
Institution of Civil Engineers: ICE
Institution of Electrical Engineers
International Association of Drilling Contractors
International Center for Marine Resource
　　Development
International Oceanographic Foundation
International Society of Offshore & Polar Enginrs
Lloyd's Register of Shipping
Marine Technology Directorate (MTD Ltd)
Marine Technology Society
Memorial University of Newfoundland
Mineralogical Society of America
MITS (Michigan Information Transfer Source)
Newfoundland Ocean Industries Association
Norwegian Geotechnical Institute
Oceans Institute of Canada
Paint Research Association
Paleontological Society
Petroleum Abstracts

Petroleum Communication Foundation
Petroleum Extension Service (PETEX)
Petroleum Industry Training Service
Petroleum Society of CIM
Plymouth Marine Laboratory & Marine Biological
   Association
Royal Institution of Naval Architects
Society for Mining, Metallurgy & Exploration
Society for Underwater Technology
Society of Naval Architects & Marine Engineers
Texas A&M University
TWI
University of Alaska Southeast
University of California, Berkeley Extension
University of Rhode Island
University of Southern California

Category C

Research Establishment

Alberta Oil Sands Technology & Research Auth
Bedford Institute of Oceanography
BHR Group Ltd - CALtec Ltd
British Geological Survey
British Library
Canada Institute for Scientific & Technical
   Information
Canadian Assocn of Oilwell Drilling Contractors
Canadian Centre for Marine Communications
Colorado School of Mines
COPPE/UFRJ-
   Federal University of Rio de Janeiro
Danish Maritime Institute
Duke University Marine Laboratory
Environment & Natural Resources Institute
ERA Technology Ltd
Gulf Canada Resources Ltd
HR Wallingford
IFREMER
IKU - Continental Shelf & Petroleum Research
   Institute
Institute of Gas Technology
Institute of Oceanographic Sciences Deacon
   Laboratory

International Center for Marine Resource
   Development
International Society of Offshore & Polar Enginrs
Louisiana Sea Grant College Program
Marine Technology Directorate (MTD LTD)
MARINTEK
MIT Sea Grant Program
National Academy Press
National Physical Laboratory
NEL - National Engineering Laboratory
Norwegian Underwater Technology Centre
Paint Research Association
Petroleum Science & Technology Institute
Plymouth Marine Laboratory & Marine Biological
   Association
Rapra Technology Ltd
Sea Fish Industry Authority
SINTEF Group
University of Rhode Island
US Army Cold Regions Research &Engineering
   Laboratory
Warren Spring Laboratory
Water Research Centre
Woods Hole Oceanographic Institution

Category D

*Government Department or Agency having
direct or indirect influence on the formulation of
regulations or standards relevant to the
petroleum and marine technology industries*

Bedford Institute of Oceanography
British Library
Building Research Establishment
Canada-Newfoundland Offshore Petroleum Board
Canadian Standards Association
Department of Energy Library
Department of Fisheries & Oceans
Department of Trade & Industry (DTI)
Department of Transport
Empresa National Del Petroleo
Energy Information Administration
Health & Safety Executive
Highlands & Islands Enterprise
Hydrographic Office

IFREMER

Marine Board

Massachusetts Coastal Zone Man'ment Office

National Academy Press

National Energy Board

National Oceanic &Atmospheric Administration

National Physical Laboratory

National Radiological Protection Board

Naval Sea Systems Command

NEL - National Engineering Laboratory
Executive Agency

Newfoundland & Labrador
Dept of Mines & Energy

Newfoundland Ocean Industries Association

Ocean Engineering Information Centre

Office of Ocean & Coastal Resource Man'ment

Offshore Supplies Office Department of Energy

Plymouth Marine Laboratory & Marine Biological
Association

Standards Council of Canada

Transportation Research Board

US Army Cold Regions Research & Engineering
Laboratory

US Department of Energy

US Department of the Interior, Minerals
Management Service

Warren Spring Laboratory

Category E

Independent Consultancy

ABS Europe Ltd

BHR Group Ltd - CALtec Ltd

BMT Cortec Ltd

British Cement Association

Canadian Maritime Industries Association

Canadian Standards Association

Det norske Veritas - Veritec

ERA Technology Ltd

ESL Information Services Inc

HR Wallingford

Institute of Gas Technology

Gulf Canada Resources Ltd

Lloyd's Register of Shipping

MARINTEK

Ocean Oil Information

Oceans Institute of Canada

Offshore Data Services Inc

Offshore Technology Conference

Pallister Resource Management Ltd

Precision Microfilming Services

Rapra Technology Ltd

Royal Society for the Prevention of Accidents

TWI

Water Research Centre

Section C

ONLINE DATABASES
and
& CD-ROMS

compiled by Diana Edmonds & Gail Lee

infoil
DATABASE
sesame

**Gateway to the European
Hydrocarbon Technology,
Offshore Oil & Gas and
Other Petroleum Related
RESEARCH PROJECTS**

THE NEED

Everyone concerned with sponsoring, carrying out or using results from research and development in any field of activity knows how difficult it is to find out about R&D projects before they have been completed and published. The same problem applies to finding names of possible partners or cooperators to research activities. There are well established methods for searching for published reports of completed projects including the use of internationally available bibliographic online databases. Published information, however, on current research and development projects is usually not available until two or three years after a project has been started. Hence there is a need for up-to-date information on current hydrocarbon and offshore related projects.

With the INFOIL-SESAME database you don't have to wait for three years!

INFOIL-SESAME

The factual and textual database produced by the Directorate-General for Energy (DGXVII) of the Commission of the European Communities (CEC), Health and Safety Executive in the UK (HSE) and the Norwegian government bodies: The Royal Norwegian council for Scientific and Industrial Research (NTNF) and the Norwegian Petroleum Directorate (NPD) is the only online source of the highly innovative European research projects within the hydrocarbon technology, offshore oil & gas and petroleum subject areas.

INFOIL-SESAME contains

3700 project entries (since 1980) and is continuously updated with new project entries.

Details in English cover the administrative, technical and operational aspects of each project and in most cases key information such as names of project managers/institutions with addresses, telephone numbers and addresses of associated equipment manufacturers are given.

INFOIL-SESAME, A merged database

This is a merged database containing part of the SESAME database and INFOIL developed by the UK/Department of Energy/HSE, and the Norwegian NTNF and NPD from 1980. Since June 1990 this new database provides information on hydrocarbon technology, offshore oil and gas and other petroleum subject areas.

DISKETTES AND CD-ROM

In addition to the online service there are two special products available from Infoil/Sesame:

- a diskette version for use in personal computers with half yearly updates and

- a CD-ROM version containing Infoil/Sesame along with data from three other directories on energy, oil and gas.

For Further Information Contact The INFOIL-SESAME Host Centres

In the UK:
On behalf of the UK Health and Safety Executive:
Marine Technology Support Unit (MaTSU)
Culham Laboratory, Abingdon
Oxfordshire OX14 3DB
Telephone: Direct Line 0235-464326
Telefax: 0235-464207
Telex: 83189 FUSION G

In Norway:
SDS - Norwegian Government
Computer Centre Ltd
Ulvenveien 89 B
N-0581 Oslo
Norway
Telephone: 47-2956300
Telefax: 47-2648407

In Germany:
STN International
PO Box 2465
D-7500 Karlsruhe
Germany
Telephone: 49-7247/808-555
Telex: 17724710
Telefax: 49-7247/808-666
Teletex: 724710+FIZKA

INTRODUCTION

The range of online databases covering petroleum and marine technology is constantly changing and has seen a significant increase since the last edition of this directory. In addition, the growth of the CD-ROM industry has produced a variety of unique and complementary information sources which enhance coverage of the field.

This section provides an international survey of online databases, databanks and CD-ROMs of particular relevance to those seeking information on the oil and gas industries, and general aspects of marine technology. For ease of use the entries are classified under a number of major headings and sub-divisions, as shown in the CONTENTS list.

The information provided on each database includes a brief description of the subject coverage, file length and updating frequency, together with details of producers, hosts or suppliers (whose addresses are identified by "§" in Section D : Directory). The equivalent hard copy version is also mentioned where appropriate.

The scope, content and indeed suppliers of individual databases change frequently. We suggest that this directory should be used as a guide to online information in the field, but that potential users should contact database hosts and suppliers directly for detailed up-to-date information.

Diana Edmonds and *Gail Lee*

Instant Library Ltd
Loughborough

SECTION C CONTENTS

GEOBASE

GEOBASE is an online bibliographic database offering rapid access to over 400,000 references to the international literature on geography, geology, ecology and related disciplines.

GEOBASE is a truly interdisciplinary service across language and cultural boundaries. Subjects covered include climatology, ecology, economic geology, geomorphology, geophysics, hydrology, and stratigraphy.

GEOBASE is updated monthly with an annual growth of over 40,000 records. Every record includes an informative abstract as well as a bibliographic citation. Items for inclusion are selected from over 2000 scientific and technical journals, as well as from books, reports, monographs and theses.

GEOBASE is currently available on Dialog, Orbit Search Service and ESA/IRS.

FLUIDEX

FLUIDEX is an online bibliographic database and provides a comprehensive source of information on all aspects of fluid engineering and behaviour and applications of fluids.

FLUIDEX contains over 250,000 records dating from 1974 onwards, and over 15,000 records are added annually in monthly updates. Items are selected from over 500 international scientific and technical journals, together with books, reports, conference papers, standards and other publications.

FLUIDEX is unique, having a comprehensive coverage of many aspect of fluids in engineering. Its interdisciplinary nature includes the following topics: Age Embrittlement, Coastal & Ocean Waters, Corrosion, Fluidics, Gas/solids Mixing, Gas/solid Flows, Gas/liquid Flows, Oil & Gas Industry, Pipeline Systems/risers, Pipes & Piping Systems, Ports and Harbours, Pumping Systems, Pumps & compressors, and Rheology.

FLUIDEX is currently available throughout the world via DIALOG and ESA/IRS.

For more information on **FLUIDEX** and **GEOBASE** please write to:
Elsevier/Geo Abstracts
Regency House, 34 Duke Street, Norwich NR3 3AP, England
Phone: (+44) 603 626327,   Fax: (+44) 603 667934

CLASSIFIED DATABASES

GENERAL & TECHNICAL INFORMATION

Bibliographic databases

C001
Alberta Oil Sands Index

Contains citations and abstracts to literature on the Alberta Oil Sands. Subject coverage includes regional history, environment, commercial development and mining, including *in situ* recovery, analysis and products.
Producer: Alberta Oil Sands Technology and Research Authority
Host: University of Alberta Computing Services; CISTI - CAN/OLE
Updated: quarterly
File length: 1900 to date (with some data from 1800s)
Hard copy version: Alberta Oil Sands Index

C002
APILIT

Worldwide coverage of the literature relating to the petroleum and petrochemical industries including properties and handling of crude oil and natural gas, transport and storage of petroleum, environmental, health and safety matters, refining processes and products.
Producer: American Petroleum Institute (Central Abstracting and Indexing Services)
Host: Dialog; Orbit; STN International
Updated: monthly
File length: 1964 to date
Hard copy version: API Abstracts/Literature

C003
APIPAT

Contains references to international patents on petroleum refining processes, pipelines, petroleum transport, petrochemicals, fuels and products. Most citations include abstracts and provide patent number, owner and index terms.
Producer: American Petroleum Institute (Central Abstracting and Indexing Services)
Host: Dialog; Orbit; STN International
Updated: monthly
File length: 1964 to date
Hard copy version: API Alphabetical Subject Index: Patents

C004
Aquatic Sciences & Fisheries Abstracts

A bibliographic database containing references to the science, technology and management of marine and freshwater environments. Includes chemical and geological oceanography, ocean resources, pollution, ocean engineering and offshore activities.
Producer: US National Oceanic and Atmospheric Administration; Cambridge Scientific Abstracts
Host: Dialog; ESA-IRS
Updated: monthly
File length: Dialog 1978 to date; ESA-IRS 1982 to date
Hard copy version: Aquatic Sciences and Fisheries Abstracts

C005

(ASTIS)
Arctic Science & Technology Information System

Covers the world literature on Arctic regions, with emphasis on the Canadian Arctic. Topics include physical, earth, biological and social sciences, engineering and technology.
Producer: Arctic Institute of North America
Host: QL Systems
Updated: every two months
File length: 1978 to date
Hard copy version: ASTIS Current Awareness Bulletin; ASTIS Bibliography; ASTIS Occasional Publications Series

C006

Australian Earth Sciences information System

References to published and unpublished Australian literature covering all aspects of earth sciences including oceanography, geology, petroleum exploration and drilling.
Producer: Australian Mineral Foundation
Host: INFO-ONE International
Updated: quarterly
File length: 1975 to date
Hard copy version: AESIS Quarterly

C007

BMT Abstracts Online (BOATS)

Provides international coverage of material published on all aspects of maritime technology.
Producer: BMT Cortec Ltd
Host: BMT Cortec Ltd
Updated: monthly
File length: 1982 to date
Hard copy version: BMT Abstracts

C008

COLD

A bibliographic database covering all disciplines relating to the Arctic, Antarctica, the Antarctic Ocean and sub-Antarctic islands. Specific topics include navigation on ice, civil engineering in cold regions, and the behaviour and operation of materials and equipment in cold temperatures.

Producer: Cold Region Research and Engineering Laboratory of the US Army Corps of Engineers; National Science Foundation
Host: Orbit
Updated: quarterly
File length: 1951 to date
Hard copy version: Antarctic Bibliography; Bibliography on Cold Regions Science and Technology

C009

COMPENDEX * PLUS

Covers a wide range of engineering disciplines including marine, petroleum and fuel engineering. Also includes literature from engineering-related areas of science and management including petroleum refining, transportation andenvironmental pollution. The database is also available on magnetic tape direct from the producer.
Producer: Engineering Information Inc
Host: BRS; CISTI- CAN/OLE; CEDOCAR; Data-Star; Dialog; ESA-IRS; Orbit; STN International
Updated: monthly
File length: Varies by host, earliest data from 1970
Hard copy version: Engineering Index

C010

DAUGAZ

Contains references to articles on gas research, production, storage, transport, distribution and use. Primarily in French, with some information in English.
Producer: Gaz de France
Host: Questel
Updated: monthly
File length: 1980 to date

C011

ECOMINE

Provides coverage of world literature on mining and mineral economics relating to metallic and non-metallic ores, coal, oil and gas.
Producer: Bureau des Recherches Geologiques et Minieres
Host: Questel
Updated: monthly
File length: 1984 to date
Hard copy version: Ecomine, Revue de Presse; Ecomine Miniere

C012

ENERGIE

A bibliographic database containing citations and abstracts to German language literature on energy research and technology. Includes fossil fuels, energy consumption, geosciences, and energy-related aspects of environmental sciences. Retrieval is possible in both English and German.
Producer: Fiz Karlsruhe
Host: STN International
Updated: twice a month
File length: 1976 to date

C013

ENERGY

Contains references to the world literature on energy research and technology including fossil fuels, energy conversion, storage and consumption, energy policy and management.
Producer: Contracting parties of the IEA Energy Technology Data Exchange Operating Agent; Fiz Karlsruhe; US Department of Energy, Office of Scientific and Technical Information (OSTI)
Host: STN International
Updated: twice a month
File length: 1974 to date

C014

Energy Bibliography

Contains references to energy-related material held at the Texas A & M University Library. Subjects include the production, utilisation and conservation of all types of fuels, energy policy, and economic, political, environmental and statistical aspects of energy.
Producer: Texas A & M University Library
Host: Orbit
Updated: periodically
File length: 1919 to date
Hard copy version: Energy Bibliography and Index

C015

ENERGYLINE

A bibliographic database containing references to the published material on energy matters. Coverage includes petroleum and gas resources and reserves, research and development, environmental impact, and

international political and economic issues. The database is also available on magnetic tape.
Producer: Bowker
Host: Data-Star; Dialog; ESA-IRS; Orbit
Updated: monthly
File length: 1971 to date
Hard copy version: Energy Information Abstracts (1976-); Energy Index (1971-1976)

C016

Energy Science & technology

Formerly called DOE Energy, this database of the US Department of Energy is one of the world's largest sources of references to energy-related material. Sources include books, reports, conference proceedings, journal articles, dissertations and patents.
Producer: US Department of Energy
Host: Dialog; INKADAT; STN International
Updated: every two weeks
File length: 1974 to date

C017

FLUIDEX

Contains references to books, reports, journal articles, British patents and standards on all aspects of fluid engineering, including offshore technology.
Producer: BHRA
Host: Dialog; ESA-IRS
Updated: monthly on Dialog; quarterly on ESA-IRS
File length: 1974 to date
Hard copy version: Fluidex is the online version of the whole range of BHRA printed abstracting services

C018

FRANCIS: Economie de l'Energie

Covers the world literature on energy economics, with specific studies on gas, oil and new energy technologies. The database is also available on floppy disk and magnetic tape.
Producer: Centre National de la Recherche Scientifique (CNRS/INIST); Universite des Sciences Sociales de Grenoble
Host: G.Cam Serveur; Questel
Updated: quarterly
File length: 1979 to date
Hard copy version: Economie de l'energie

C019

GASLINE

Contains 3 files of information on the gas industry: - Gas Abstracts - covers exploration and drilling, processing, storage and distribution, manufacture of fuels and petrochemical feedstocks, legal and regulatory issues. Energy Statistics - time series on the energy industry worldwide including energy production, reserves, imports and exports, and prices for natural gas, gas liquids and oil in over 70 countries. AGA Catalogue - citations to books, monographs and reports published by the American Gas Association.
Producer: Institute of Gas Technology (IGT)
Host: IGT Net
Updated: every two months
File length: varies by file
Hard copy version: Gas Abstracts; Energy Statistics

C020

GEOARCHIVE

Provides references to books, serials, reports, conference proceedings and 15,000 geological maps from the Institute of Geological Sciences libraries. Coverage includes mineral and petroleum production and resources, new minerals and new stratigraphic names, plus broad coverage of geophysics, geology and geochemistry.
Producer: Geosystems
Host: Dialog
Updated: monthly
File length: 1974 to date

C021

GEOBASE

Contains references from the international literature on geosciences including natural resources, mineralogy, sedimentology and other geology related topics.
Producer: Geo Abstracts Ltd
Host: Dialog; Orbit
Updated: monthly
File length: 1980 to date
Hard copy version: Geological Abstracts; Mineralogical Abstracts; Ecological Abstracts, plus some additional material.

C022

GEOLINE

Citations, some with abstracts, to worldwide literature on geosciences including marine and petroleum geology, mineralogy and petrography.
Producer: GeoFiz
Host: Fiz Technik
Updated: monthly
File length: 1970 to date

C023

Geomechanics Abstracts

Covers the published literature on the mechanical performance of geological materials relevant to the extraction of raw materials from the earth and the construction of structures for energy generation. Includes rock mechanics and excavation, engineering geology, mining and tunnelling.
Producer: Rock Mechanics Information Service
Host: Orbit
Updated: every two months
File length: 1977 to date
Hard copy version: Geomechanics Abstracts

C024

GEOREF

Includes over 5 million records from books, journals, conference proceedings, theses and maps on the subject of geosciences. Coverage includes geology and marine geology, geochemistry, petrology, mineralogy and oceanography.
Producer: American Geological Institute
Host: CISTI-CAN/OLE; Dialog; Orbit; STN International
Updated: monthly
File length: 1785 to date for North American material; 1933 to date for international material
Hard copy version: Bibliography and Index of Geology (1969 to date) and several ceased titles for older material

C025

GEOTECHNOLOGY

A database specialising in publications of Asian origin on all aspects of geotechnology including rock mechanics and engineering geology.
Producer: Library and Regional Documentation Centre of the Asian Institute of Technology
Host: ESA-IRS
Updated: annually
File length: 1973 to date

C026

HEATFLO

References to material covering the practice and theory of thermal design, heat transfer and fluid flow, including fouling, corrosion, pipelines and transport properties of liquids.
Producer: Harwell Laboratory, Heat Transfer and Fluid Flow Service
Host: ESA-IRS
Updated: every two months
File length: 1965 to date

C027

HSELINE
(Health & Safety Executive)

A bibliographic database covering occupational health and safety aspects of industry including mines and quarries, the offshore oil industry and the handling of hazardous substances.
Producer: Health and Safety Executive
Host: Data-Star; ESA-IRS; Orbit
Updated: monthly
File length: 1977 to date

C028

HERI (Heavy Oil/Enhanced Recovery Index)

References to articles and reports dealing with heavy oils and processes to recover them. Includes analyses and properties, economics and commercial development, chemical, miscible gas and thermal recovery processes.
Producer: Alberta Oil Sands Technology and Research Authority
Host: University of Alberta Computing Services
Updated: quarterly
File length: 1975 to date
Hard copy version: Heavy Oil/Enhanced Recovery Index

C029

INFORMCHIMMASH: Database on Chemical Engineering

Contains citations to Soviet and Eastern European literature on all aspects of chemical and petroleum engineering.
Producer: Tsintichimneftemash c/o International Centre for Scientific and Technical Information (ICSTI)
Host: International Centre for Scientific and Technical Information (ICSTI)
Updated: annually

C030

International Petroleum Abstracts (IPABASE)

Covers the world literature on the scientific, technical and economic aspects of the petroleum, petrochemical and related industries. Subjects include exploration, drilling, processing, transportation, storage, analysis, safety and environmental issues.
Producer: John Wiley & Sons Ltd for the Institute of Petroleum
Host: Orbit
Updated: quarterly
File length: 1985 to date
Hard copy version: International Petroleum Abstracts

C031

JICST: File on Science, Technology & Medicine in Japan

Contains English language citations and abstracts of scientific and technical literature published in Japan. Coverage includes geosciences, metallurgical, mining and energy engineering.
Producer: Japan Information Center of Science and Technology (JICST)
Host: STN International
Updated: monthly
File length: 1985 to date

C032

MARINELINE

References to published material on marine mineral resources, marine research and technology, including offshore and diving technology, and marine law.
Producer: Federal Institute for Geosciences and Natural Resources
Host: FIZ Technik
Updated: every two months
File length: 1970 to date

C033

MARNA

Citations to the worldwide literature on marine-related topics including shipping, ports, ocean mining, offshore oil and gas, energy generating floating plants and marine pollution.
Producer: Netherlands Maritime Information Centre
Host: TNO Updated: weekly
File length: 1974 to date
Hard copy version: Maritime Information Review

C034

Meteorological & Geoastrophysical Abstracts

Provides current citations for the most important meteorological and geoastrophysical research published in the world literature. Subject coverage includes hydrosphere, hydrology and physical oceanography.
Producer: American Meteorological Society
Host: Dialog
Updated: irregularly
File length: 1972 to date

C035

MINFINDER

An Australian database providing references to reports on the geology and mineral resources of New South Wales, including petroleum exploration license reports.
Producer: New South Wales Department of Minerals and Energy
Host: INFO-ONE International
Updated: periodically
File length: 1875 to date

C036

NEI (Nordic Energy Index)

Provides abstracts and indexes of energy literature published in the Nordic countries, plus descriptions of ongoing energy research projects. Abstracts are generally in English.
Producer: Secretariat for the Nordic Energy Information Libraries (SNEIL)
Host: Datacentralen
Updated: monthly
File length: full file 1982 to date, with some earlier records

C037

NTIS
National Technical Information Service

Covers completed US government-sponsored research from federal government agencies, with an increasing portion of the database also including unpublished material from Western Europe and Japan. Topics include geosciences, natural resources, marine engineering, oceanography and energy sources.
Producer: National Technical Information Service (NTIS)
Host: Dialog; ESA-IRS; Orbit; STN Internationl
Updated: every two weeks
File length: 1964 to date
Hard copy version: Government Reports; Announcements & Index; Abstract Newsletters

C038

Ocean Engineering Information Centre

The Ocean Engineering Information Centre maintains a database on cold ocean engineering, including resource development in Canadian northern waters. The database is only available online at the Centre, but staff will accept external enquiries which can be dealt with quickly.
Producer: Memorial University of Newfoundland, Ocean Engineering Information Centre

C039

Oceanic Abstracts

References to worldwide technical literature on marine-related subjects including oceanography, ships and shipping, geology and geophysics, meteorology, marine pollution, and governmental and legal aspects of marine resources.
Producer: Cambridge Scientific Abstracts
Host: Dialog; ESA-IRS
Updated: every two months
File length: 1964 to date
Hard copy version: Oceanic Abstracts

C040

OIL INDEX

Coverage of Scandinavian publications on the oil industry including petroleum geology, exploration and production, offshore structures and pipelines. Primarily Scandinavian languages with approximately 30% of records in English. The database in also available on floppy disk.
Producer: Norwegian Petroleum Directorate
Host: Fabritius A.S.
Updated: every two months
File length: 1974 to date
Hard copy version: Oil Index

C041

PASCAL: ENERGIE

A multidisciplinary database covering both earth sciences - including mineralogy and mining economics, and engineering - including energy, fuels and fuel processing. Also available on floppy disk and magnetic tape from INIST-CNRS.
Producer: INIST-CNRS
Host: Dialog; ESA-IRS; Questel
Updated: monthly
File length: 1973 to date
Hard copy version: Bulletin Signaletique (1973-1983); Bibliographie Internationale (1984-)

C042

Petroleum Exploration & Production / Tulsa

Worldwide coverage of technical literature and patents related to the exploration, development and production of oil and natural gas. Areas covered include geology, well logging, drilling and completion, recovery methods, reservoir studies etc. This database is called *Tulsa* on Orbit and *Petroleum Exploration & Production* on Dialog, where it is a restricted access file requiring a license from Petroleum Abstracts.
Producer: Petroleum Abstracts at the University of Tulsa
Host: Dialog; Orbit
Updated: monthly on Dialog; weekly on Orbit
File length: 1981 to date on Dialog, 1965 to date on Orbit
Hard copy version: Petroleum Abstracts

043

POWER

Consists of catalogue records for material held in the US Department of Energy Library. Subject coverage includes energy, physical and environmental sciences, technology, economics, renewable energy and resources.
Producer: US Department of Energy
Host: Orbit
Updated: quarterly
File length: 1950s to date

C044

SIGLE (System for Information & Grey Literature in Europe)

Provides references to European non-conventional literature such as research reports, discussion documents, official and industrial publications. Includes geosciences, energy, power and marine engineering.
Producer: European Association for Grey Literature Exploitation (EAGLE)
Host: BLAISE-LINE; STN International
Updated: monthly
File length: 1980 to date

C045
South Australian Mines Reference (SAMREF)

Contains citations to unpublished reports on mineral and petroleum exploration in South Australia.
Producer: South Australian Department of Mines and Energy (SADME)
Host: INFO-ONE International
Updated: monthly
File length: 1983 to date

C046
Soviet Science & Technology

Provides references to journal articles, technical reports, conference papers and patents published in Soviet-bloc countries. Includes coverage of geosciences, energy use and conservation, fuels and petroleum products.
Producer: IFI/Plenum Data Corporation
Host: Dialog
Updated: monthly
File length: 1975 to date

C047
TRIS

A database of transportation research information which includes contributions from the Maritime Research Information Service (MRIS). Topics covered include regulations and legislation, energy, operations, environmental and maintenance technology.
Producer: US Department of Transportation
Host: Dialog
Updated: monthly
File length: 1968 to date

C048
WPIA/WPILA

This database is the merged version of APIPAT and World Patents Index, containing records for more than 5 million patents covering petroleum processes, fuels, pipelines, tankers, storage, pollution control, petrochemicals, C1 chemistry and other technologies.
Producer: Derwent Publications Ltd, with indexing by the American Petroleum Institute (Central Abstracting and Indexing Services)
Host: Orbit
Updated: weekly
File length: 1963 to date

Factual databases

a. Environmental, health and safety information

C049
CHRIS (Chemical Hazard Response Information system)

Contains data on labelling, physical and chemical properties, reactivity, water pollution, and hazard classifications for over 1,000 chemical substances.
Producer: US Coast Guard
Host: Chemical Information System (CIS)
Updated: periodically
File length: current information

C050
Golob's Oil Pollution Bulletin

Provides full text of a newsletter covering worldwide developments in the control and prevention of oil spills and oil-related pollution.
Producer: World Information Systems
Host: NewsNet Inc
Updated: every two weeks
File length: 1991 to date
Hard copy version: Golob's Oil Pollution Bulletin

C051
International Tanker Owners Pollution Federation (ITOPF)

The Information Services section of ITOPF maintains an in-house database on tanker oil spills which permits analysis of spill trends.
Producer: International Tanker Owners Pollution Federation Ltd (ITOPF)
Host: not available on-line, ITOPF staff provide a search service to tanker owners, insurers and government agencies
Updated: daily
File length: 1974 to date

C052
Lloyd's Maritime

Provides 5 files of information on ships, including casualties to drilling rigs, ships of 1,000 or more GRT, tankers, liquid gas carriers and combination tankers/carriers over 6,000 DWT.
Producer: Lloyd's Maritime
Host: Lloyd's Maritime
Updating: varies from daily to monthly according to file
File length: 1976 to date

C053
National Emergency Equipment Locator System

Identifies Canadian locations of emergency and protective equipment, including air and water craft, for use in the event of environmental emergencies such as oil or hazardous material spills. Includes information on handling of previous spills.
Producer: Developed by I P Sharp Associates under the direction of Environment Canada, National Environmental Emergency Centre
Host: I P Sharp Associates
Updated: periodically
File length: current information

C054
OCEANROUTES

Oceanroutes Inc is a worldwide organisation providing data to aid safety and efficiency in marine operations. Services include daily operational weather and wave forecasts to over 100 offshore projects; ocean tow and rig-move strategies and enroute monitoring; offshore instrumentation; offshore and hindcast studies and weather-related simulations for major oil producers and contracters. Databases include weather archives, daily spectral wave and wind data, ocean currents data, and a marine environmental database.
Producer: Oceanroutes Inc
Host: Oceanroutes Inc
Updated: continuously
File length: varies according to data

C055
OHM-TADS (Oil & Hazardous Materials-Technical Assistance Data System)

A collection of chemical, physical, biological, toxicological and commercial data on almost 1,500 materials, with emphasis on their environmental effects and emergency response. Provides technical support for dealing with potential or actual dangers resulting from the discharge of oil or hazardous materials. This database is also available on floppy disk from Fein-Marquart Associates.
Producer: US Environmental Protection Agency
Host: Chemical Information System (CIS)
Updated: periodically
File length: current information

C056
Oil Spill Database

Information on worldwide oil spills since 1967, including details of site, source, date, cause, substance involved, clean-up and damage.
Producer: Center for Short-Lived Phenomena
Host: not available online, staff carry out searches on behalf of clients
Updated: periodically
File length: 1967 to date

C057
Oil Spill Intelligence Report

Provides full text coverage of a weekly newsletter on oil spill clean-up, prevention and control.
Producer: Cutter Information Group
Host: NewsNet Inc
Updated: weekly
File length: 1990 to date
Hard copy version: Oil Spill Intelligence Report

C058

Principal Offshore Oil-spill Accidents & Tanker Casualties Data Bank

Provides accident synthesis data sheets for each incident, comprising tanker name, type, size and flag, date, location, events, weather and consequences.
Producer: Institut Francais du Petrole (IFP)
Host: Institut Francais du Petrole (IFP)
Updated: approximately 30 reports per annum
File length: 1955 to date

C059

Principal Offshore Platform Accidents

Information on approximately 900 accidents, each causing work stoppage of 24 hours or more, involving offshore oil drilling and production rigs, mobile and fixed platforms and drilling ships. Sources include Lloyd's casualty reports, oil industry technical publications and government reports.
Producer: Institut Francais du Petrole (IFP)
Host: Institut Francais du Petrole (IFP)
Updated: approximately 50 reports per annum
File length: 1955 to date

C060

VEGA

Covers international law and statutory requirements governing the maritime industry, including marine pollution prevention, MARPOL - International Convention for the Prevention of Pollution from Ships, Chemical Code and Gas Code.
Producer: Det Norske Veritas
Host: Det Norske Veritas
Updated: periodically
File length: 1966 to date

b. Chemical and physical data

C061

Crude Oil Analysis Database

Analyses of worldwide crude oil deposits since 1920. Information on location, specific gravity, sulphur and nitrogen content, viscosity, colour and distillation is provided for each oil.
Producer: US Department of Energy
Host: US Department of Energy
Updated: periodically
File length: 1920 to date

C062

NBS FLUIDS

Consists of calculational programs which can be used to determine thermophysical and transport properties for butane, ethane, ethylene and propane, over a wide range of temperatures and pressures.
Producer: National Institute of Standards and Technology (NIST)
Host: STN International

C063

Physical Properties Data System

A modelling system based on several groups of physical property data. Estimates of thermo-dynamic properties of petroleum fractions can be calculated given the mean boiling points and a measure of density, and petroleum fractions can be combined with other data from the system. Also available on magnetic tape.
Producer: Institution of Chemical Engineers (
Host: Institution of Chemical Engineers (ICE)
Updated: annually
File length: current information

C064

Petroleum Chemicals Package Library

A floppy disk of spectra for over 20 classes of commercially available products used in identifying and improving petroleum products.
Producer: Bio-Rad Laboratories
Host: Bio-Rad Laboratories
Updated: periodically

Wells, drilling, exploration and production data

C065

Active Well Data Online

Provides information on the current status of offshore and onshore wells throughout the US. Test information, treatments, volumes, total depth and final status data is given for each well. The database is also available on floppy disk.
Producer: Petroleum Information Corporation
Host: Petroleum Information Corporation
Updated: daily
File length: data on each well available for 13 months following completion or last reported activity.

C066

Advanced Recovery Week

Full text of a weekly newsletter on tertiary oil recovery which provides technical and economic analyses of current recovery projects.
Producer: Pasha Publications Inc
Host: Data-Star; Dialog (both as part of PTS Newsletter Database).
Updated: weekly
File length: 1988 to date
Hard copy version: Advanced Recovery Week

C067

API Joint Association Survey

Contains data on US oil and gas drilling expenditure including wells, footage and related costs. This database is only available on magnetic tape.
Producer: American Petroleum Institute; Independent Petroleum Association of America; Mid-Continent Oil and Gas Association
Host: not available online. Available on magnetic tape from American Petroleum Institute
Updated: annually
File length: 1985 to date
Hard copy version: API Joint Association Survey

C068

API Monthly Completion Report

Monthly and annual time series on well completions and footage drilled for oil and gas wells in the USA.
Producer: American Petroleum Institute
Host: I P Sharp Associates
Updated: monthly
File length: 1970 to date

C069

AXSES INFOATLAS: Fundy, Maine & Georges Bank Coastal Zone

Contains digital maps and information on the coastal and ocean environment between Cape Cod, Massachusetts and Halifax, including data on geology, sediments, oceanography, oil and gas, and the environment. This database is available on floppy disk only.
Producer: Axses Information Systems; Environment Canada
Host: not available online. Available on floppy disk from Axses Information Systems
Updated: irregularly
File length: 1961 to date

C070

British Geological Survey Gravity & Magnetics Databanks

The following categories of data are stored on disk under the ORACLE database management system and are available for purchase on magnetic tape; 1600 or 6250 bpi, ASCII or EBCDIC, sequential files:
- Marine (shipboard) magnetics
- Marine (shipboard) gravity:
    unadjusted and adjusted
- Gridded gravity
- Marine bathymetry
Producer: British Geological Survey (Edinburgh)
Host: not available online. Available on floppy disk or magnetic tape from British Geological Survey (Edinburgh)

C071

British Geological Survey Offshore Database

During the routine geological mapping of the UK Continental Shelf carried out by BGS from 1969 to 1990, data from over 30,000 sample sites was collected. Digital location, descriptive and analysis data are held at Keyworth under the ORACLE database management system and may be purchased on tape or floppy disk, usually restricted to a small area of specific interest. The digital data comprise the following:

- Sample station location data
- Descriptive data
- Particle size analysis
- Geochemical data
- Geotechnical data
- Seismic track data

Producer: British Geological Survey (Keyworth)
Host: not available online. Available on floppy disk or magnetic tape from British Geological Survey (Keyworth)

C072

COREFINDER

References to boxes of petroleum, stratigraphic and hydrological diamond drill cores stored at the Londonderry Core Library. Mine or prospect name, company name, title information, location, depth and year are provided for each core.
Producer: New South Wales Department of Minerals and Energy
Host: INFO-ONE International
Updated: periodically

C073

Drilling Analysis Data Online

Provides information on oil and gas drilling activity in the US including well permit, spud and completion data. The database allows users to perform statistical analyses of geographical areas, geological formations, drilling depth and costs to generate a variety of reports. Also available on floppy disk and magnetic tape.
Producer: Petroleum Information Corporation
Host: Petroleum Information Corporation
Updated: weekly
File length: 1970 to date

C074

Drilling Information Services

Contains current and historical drilling information for the US Continental and offshore petroleum and gas industries, collected from bit, mud, log records and scout tickets. The information can be used to study drilling conditions in specific geographic areas and to calculate production measurements.
Producer: Adams Engineering Inc
Host: PSI Energy Software; SPENET
Updated: continuously
File length: 1950 to date

C075

Dwight's Hotline Energy Reports

Dwight's Oil & Gas Reports

Comprises several files of oil and gas information which form part of the Dwight's/Softsearch Petroleum Network:
Production Data
- monthly, annual and cumulative production data for over 1 million gas wells and oil leases in the US, Pacific and Gulf Coast Federal Offshore areas, and Canadian province of Alberta.
Reservoir and Field Data
- geologic and production data, including offshore lease and bid data for over 150,000 fields and reservoirs in the US and Canada.
Well Data
- new permit, drilling activity and completion data for oil wells in the Rocky Mountains, Mid-Continent, Gulf Coast, West Coast and Texas.
Petroleum Registry
- listings of producing properties offered for sale or sought for purchase
Producer: Dwight's Energydata Inc
Host: Softsearch Inc; Dwight's On-Line System (DOLS)
Updating: varies by file
File length: varies by file

C076

GEOBANQUE

Information on sub-soil exploration in France, including oil and gas exploration, boreholes, wells, mines and quarries.
Producer: Bureau des Recherches Geologiques et Minieres
Host: Questel
Updated: twice per annum
File length: 19th century to date

C077

Gower Federal Service

Contains the full text of adjudicative decisions by the Interior Board of Land Appeals concerning federal onshore and offshore oil and gas leases.
Producer: Rocky Mountain Mineral Law Foundation
Host: West Publishing Company
Updated: periodically
File length: 1971 to date
Hard copy version: Gower Federal Service

C078

Historical Well Data Online

Companion database to Active Well Data Online. Over 2 million completed gas, dry, service and re-entry wells are included, with detailed information on core, drillstem and production tests, treatments, volumes, depth and final status provided for each. This database is also available on floppy disk and magnetic tape.
Producer: Petroleum Information Corporation
Host: Petroleum Information Corporation
Updated: monthly
File length: from original recorded activity to date

C079

Hughes International Rig Count

Provides tallies of exploratory drilling rigs throughout the world.
Producer: Hughes Tool Company
Host: I P Sharp Associates
Updated: weekly for USA and Canada; monthly for rest of world (98 countries)
File length: 1973 to date for USA and Canada; 1981 to date for rest of world.

C080

Infield Database

Information about the world's existing and planned offshore oil and gas field developments. The database comprises 12 files: Fields; Platforms; Subsea Completions; Pipelines; Pigg-Back Lines; Bundles Lines; Wells; Control Lines; SPM; Operators; Manufacturers/Fabricators; Terminals.
Producer: Infield Systems Ltd
Host: Infield is not available online. The database can be supplied on disk in its entirety or in sections, or searches can be ordered from the producers.

C081

GRID DATA:
Petroleum Information
Cartographic Database

Contains digital cartographic data for the entire US. Available on floppy disk and magnetic tape only.
Producer: Petroleum Information Corporation
Host: not available online. Available on floppy disk or magnetic tape from Petroleum Information Corporation
Updated: monthly
Hard copy version: US Geological Survey Quadrangle Maps

C082

Marine Information & Advisory Service (MIAS)

MIAS acts as an agent for the British Oceanographic Data Centre, (BODC) and provides clients with information from the BODC database of numerical data on waves, currents, tides, sea level, temperature, salinity and other oceanographic parameters for UK waters and the Northeast Atlantic.
Producer: Natural Environment Research Council, Institute of Oceanographic Sciences, Marine Information and Advisory Service
Host: not available online. MIAS staff will undertake searches on behalf of commercial organisations, government agencies and research establishments

C083

Mobile Satellite Reports

Contains the full text of Mobile Satellite Reports, a monthly newsletter covering aeronautical, maritime and mobile satellites and radio determination services within the aerospace, aviation, shipping and offshore oil industries.
Producer: Warren Publishing Inc
Host: NewsNet Inc
Updated: every 2 weeks
File length: 1989 to date
Hard copy version: Mobile Satellite Reports

C084

North Sea Letter
Monthly Rig Market Forecast

Provides the full text of a newsletter covering usage, demand, operator requirements and prices for semi-submersible and jack- up oil and gas drilling rigs in the North Sea.
Producer: Financial Times Electronic Publishing
Host: Data-Star; Profile (both as part of Financial Times Business Reports: Energy); Mead Data Central (as a NEXIS database).
Updated: monthly
File length: Data-Star and Profile 1987 to date with 3 month forecasts; Mead Data Central 1989 to date
Hard copy version: North Sea Letter Monthly Rig Market Forecast

C085

Offshore Data Services

Offers a range of services including the telex-based Intelex *Wire Service,* a daily newswire for the offshore oil industry; *Telex Construction Wire,* covering the offshore construction industry; and *Telex Drilling Wire,* covering the offshore drilling industry. Several databases are also available either on floppy disk and/or magnetic tape:
Gulf of Mexico Construction Locator
- tabulates all platforms, pipelines and subsea installations under study, design or construction. Available on floppy disk, updated monthly
Gulf of Mexico Drilling Report
- provides information on current and planned drilling in the region. Supplied on floppy disk or magnetic tape, updated weekly, biweekly, or monthly

Gulf of Mexico Rig Locator
- an account of all drilling units operating in the Gulf, including mobile and platform rigs. Supplied on floppy disk, - updated weekly, biweekly or monthly
Ocean Construction Locator
- lists international offshore field developments. Available on floppy disk or magnetic tape, updated monthly
Rig Locator
- reports on offshore drilling units with current drilling locations and contract status for mobile and fixed platform rigs worldwide. Supplied on floppy disk or magnetic tape, updated monthly
Producer: Offshore Data Services
Host: not available online. Databases are available on floppy disk/magnetic tape, or searches can be ordered from the producers.

C086

Offshore Gas Report

Offshore Gas Report is a biweekly newsletter focusing on the production, marketing and transportation of natural gas produced offshore in US state waters and in the federal domain of the Outer Continental Shelf (OCS). The full text of this newsletter is available online, providing information on OCS well determinations, transportation policies and operations of OCS pipelines, gathering lines, transmission projects, pipeline capacity, and the availability of gas freed up for sale in the spot market.
Producer: Atlantic Information Services Inc
Host: NewsNet
Updated: biweekly
Hard copy version: Offshore Gas Report

C087

Offshore Rig Report

Worldwide data on mobile drilling rigs that are active or under construction, including name, owner, type, operator, location and design. Also provides an equipment listing service for new, surplus and used marine rigs, land rigs, drilling equipment and tubulars.
Producer: Oceandril Inc, from data supplied by Pennwell Publishing Company
Host: Oceandril Inc
Updated: daily
File length: current information
Hard copy version: Offshore Magazine Rig Locator

C088

Oil Cargo Management Service

One of a number of Lloyd's Maritime multi-subscriber services, supplied with PC software, to enable clients to perform analyses and prepare reports. The Oil Cargo Management Service simulates the impact on oil distribution and tanker fleet disposition resulting from disruption at a number of strategic points.
Producer: Lloyd's Maritime
Host: Lloyd's Maritime

C089

Online Resource Exchange (PSI Listing of Oil & Gas Opportunities)

Describes oil and gas properties offered by independent and major oil companies, brokers and banks. Covers on- and offshore properties, royalty and working interests, mature and developing fields. Details of location, size, price, reserves, production history and operating costs are provided for each property.
Producer: Online Resource Exchange Inc
Host: Petroleum Information Corporation
Updated: daily
File length: current information

C090

Permit Data Online

Daily record of all new oil well drilling permits granted in the USA. Information for each permit includes operator details, location, classification and projected depth. The database is also available on floppy disk.
Producer: Petroleum Information Corporation
Host: Petroleum Information Corporation
Updated: daily
File length: 30 days

C091

Petroleum Data System

Contains geological, engineering and production data on over 100,000 oil and gas fields in the USA and Canada, and on the outer continental shelf. Records include location, production status, depth, reservoir pressure and analysis data. The database is also available on floppy disk and magnetic tape.
Producer: Petroleum Information Corporation
Host: Petroleum Information Corporation
Updated: periodically
File length: varies by file

C092

Production Data Online

Provides current and historical monthly production volumes and cumulative totals, test data, and field and reservoir summary data for US gas and oil wells. Location, operator and lease details are given for each well. Covers Michigan, Texas, Rocky Mountains, Mid-Continent, West Coast and Southeast USA. This database is also available on floppy disk and magnetic tape.
Producer: Petroleum Information Corporation
Host: Petroleum Information Corporation
Updated: monthly
File length: varies by well, earliest data from 1960

C093

Seismic Crew Count

Contains monthly figures for US and Canadian land crews and marine vessels engaged in seismic oil and gas exploration, with subtotals by onshore/offshore and by sponsorship. Figures are compiled from data supplied by companies that own, operate or contract seismic exploration crews and vessels.
Producer: Society of Exploration Geophysicists
Host: I P Sharp Associates
Updated: monthly
File length: 1974 to date

C094

SPENET

An online service operated by the Society of Petroleum Engineers which acts as a gateway to a range of databases including Permit Data Online, Active Well Data, Historical Well Data, Drilling Analysis Data Online, Petroleum Data System, Hotline Energy Reports, Drilling Information Services, and Newport Associates Energy Database. SPENET also provides information on over 200 software programs, SPE publications and events.
Producer: Society of Petroleum Engineers
Host: BT Tymnet Dialcom Service
Updating: varies by file
File length: varies by file

C095

TITLEFINDER

Contains approximately 8900 references to licenses for offshore drilling for oil, minerals and coal, issued by the New South Wales Department of Minerals and Energy.
Producer: New South Wales Department of Minerals and Energy
Host: INFO-ONE International
Updated: monthly
File length: 1973 to date

C096

Well History Control System

Provides over 2 million completion histories for wells in continental US, Alaska, offshore state and federal waters. All wells drilled since 1972, plus selected pre-1972 wildcats and development wells are included. Data includes production, drillstem and wireline tests, and may be used to produce reports, statistics, maps and graphic displays. The database is also available on floppy disk and magnetic tape.
Producer: Petroleum Information
Host: Petroleum Information
Updated: continuously
File length: 1972 to date, with selected earlier data

C097

WISER (Western Information System for Energy Resources)

Comprises 3 files of drilling and production data for Alberta, British Columbia, Manitoba and Saskatchewan.
Rapid
- monthly production and injection data on producing wells in the Canadian Provinces , including details of name, location, type, number, operator and status code.
Drill Hole Status
- general information on the status of holes drilled in Alberta.
Land Lease
- specifications and details on each provincial mineral lease in Alberta.
Producer: Dataline Inc
Host: Dataline Inc
Updated: monthly
File length: 1963 to date

Research projects

C098

ENERLINKS The Australian Energy Research, Development & Demonstration Projects Database

Provides summaries of energy research projects in the areas of oil and gas, oil shale and supporting technology. Covers current projects and those completed in the past 5 years.
Producer: Department of Primary Industries and Energy
Host: CSIRO Australis
Updated: annually
File length: 5 years
Hard copy version: Compendium of Australian Energy Research, Development and Demonstration Projects

C099
INFOIL-SESAME Database

This merged database contains part of the SESAME database developed by the Directorate-General for Energy (DGXVII) from 1982, and INFOIL developed by the UK Health and Safety Executive, the Royal Norwegian Council for Scientific and Industrial Research, and the Norwegian Petroleum Directorate from 1980. This new database provides information on hydrocarbon technology, offshore oil and gas, and other petroleum subject areas. For each project, details given include the title, people involved, sponsors, start and finish dates, and an abstract of the aims/ work of the project.
Producer: Directorate-General for Energy (DGXVII), Commission of the European Communities; Health and Safety Executive (HSE); Royal Norwegian Council for Scientific and Industrial Research (NTNF); Norwegian Petroleum Directorate (NPD)
Host: Marine Technology Support Unit; Fabritius A.S.
Updated: periodically
File length: 1981 to date

BUSINESS & COMMERCIAL INFORMATION

Industry news

C100
AGA GASNET

Covers the US natural gas industry and developments in state and federal legislation. Provides daily summaries of industry-related articles in the Wall Street Journal, New York Times, Washington Post, USA Today and Oil Daily, plus abstracts of articles in over 160 gas industry and business periodicals.
Producer: Information Inc, in co-operation with the American Gas Association
Host: Knowledge Systems Inc
Updated: daily
File length: current information

C101
API Energy Business News Index (APIBIZ)

Formerly P/E NEWS
Worldwide coverage of commercial, financial, marketing and regulatory information affecting the petroleum and energy industries. Sources include Oil Daily, Petroleum Economist, Petroleum Intelligence Weekly, Petroleum Times Business Review, Petroleum Market Intelligence and World Oil.
Producer: American Petroleum Institute (Central Abstracting & Indexing Services)
Host: Data-Star; Dialog; Orbit
Updated: weekly
File length: 1975 to date
Hard copy version: Petroleum/Energy Business News Index; Guide to Petroleum Statistical Information

C102

Chemical Age Project File

Provides details of oil refining, petrochemical and mineral processing plants completed since 1980, together with information on those under construction, planned or undergoing feasibility studies.
Producer: MBC Information Services
Host: ESA-IRS; Pergamon Financial Data Services
Updated: daily
File length: 1980 to date

C103

Chemical Business Newsbase

A major business database with emphasis on European news relating to the chemical and allied industries, including petrochemicals.
Producer: Royal Society of Chemistry
Host: Dialog; ESA-IRS
Updated: weekly
File length: 1984 to date

C104

Chemical Market Associates - Petrochemical Markets Reports

Contains the full text of 3 petrochemical market reports. Chemicals covered include butadiene, butylenes, ethylene, benzene, propylene, styrene, vinyl chloride, acrylic, nylon and polyester fibre intermediates.
Producer: Chemical Market Associates Inc
Host: I P Sharp Associates
Updated: monthly
File length: current 2 isssues of reports; current quarter for forecasts
Hard copy version: C4 Market Report; Monomers Market Report; Fiber Intermediate Market Report

C105

CHEM-INTELL Chemical Plant Database

Provides information on manufacturing plants, plus trade and production figures for over 100 organic and inorganic chemicals including petrochemicals.
Producer: Chemical Intelligence Services
Host: Dialog; ESA-IRS; Pergamon Financial Data Services
Updated: monthly
File length: current information

C106

Conferences In Energy, Physics, Mathematics & Chemistry

Announcements of past and forthcoming conferences, with notes on related publications.
Producer: FIZ Karlsruhe; FIZ Chemie
Host: STN International
Updated: weekly
File length: 1976 to date

C107

Daily Oil Bulletin

Provides the full text of a daily newsletter on oil and gas exploration, production and distribution in Canada. Coverage includes international oil and gas markets, production figures and prices.
Producer: Southam Business Information and Communications Group
Host: Infomart Online; News$ource
Updated: daily
File length: June 1989 to date
Hard copy version: Daily Oil Bulletin

C108

Datatimes Corporation

Supplies full text news and feature articles from several US newspapers, including the Houston Chronicle and Oklahoma Journal Record, which emphasise regional coverage of the oil and gas industries.
Producer: varies by newspaper
Host: Datatimes Corporation
Updated: daily
File length: 1985 to date for Houston Chronicle & Oklahoma Journal Record

C109
East European Chemical Monitor

Company and market information on the East European chemical industry, including oil, gas, basic petrochemicals and downstream products. *Producer: Business International Corporation*
Host: Data-Star
Updated: monthly File length: 1984 to date

C110
The Edmonton Journal

Full text of The Edmonton Journal which includes regional coverage of the oil industry primarily in Alberta, northern Saskatchewan, northern British Columbia, and the Northwest and Yukon Territories.
Producer: Southam Business Information and Communications Group
Host: Infomart Online; News$ource
Updated: daily
File length: March 1989 to date
Hard copy version: The Edmonton Journal

C111
Energy Alert

The full text of a daily newsletter covering major news items and Securities and Exchange Commission filings in the oil and gas industries. Includes Baker and Hughes Rotary Rig Count, API activities, and Williams Act filings.
Producer: Vestlash Group Inc
Host: Data-Star; Dialog (both as part of PTS Newsletter Database)
Updated: daily on Dialog; weekly on Data-Star
File length: 1990 to date
Hard copy version: Energy Alert

C112
The Energy Report

Full text of a weekly newsletter on US energy policy relating to oil, natural gas and other energy sources.
Producer: Pasha Publications
Host: Data-Star; Dialog (both as part of PTS Newsletter Database); NewsNet Inc
Updated: weekly
File length: Data-Star and Dialog 1988 to date; NewsNet Aug 1989 to date
Hard copy version: The Energy Report

C113
European Energy Report

A full text newsletter on energy-related developments in Western Europe, covering oil and gas production, legislation and trade agreements, prices, demand and contracts for major projects.
Producer: FT Electronic Publishing
Host: Data-Star; FT Profile (both as part of FT Business Reports: Energy); Mead Data Central (as part of NEXIS)
Updated: twice monthly
File length: Data-Star and Profile 1987 to date; Mead Data Central 1989 to date
Hard copy version: European Energy Report

C114
FAIRBASE

A guide to past and forthcoming trade fairs, exhibitions, conferences and meetings worldwide in all industries, including oil drilling.
Producer: Fairbase Database Ltd
Host: BRS; Data-Star; FIZ Technik; German Business Information; National Center of Scientific and Technological Information
Updated: monthly
File length: 1986 to 2010

C115
Financial Times Business Reports: ENERGY

A current awareness service providing news, comment and market analysis of the coal, gas and oil industries, and the world energy market. The database is enhanced with summaries from Financial Times conference papers and reports, and full text specialist newsletters including European Energy Report, International Gas Report, North Sea Letter and North Sea Letter Monthly Rig Market Forecast.
Producer: Financial Times Business Information
Host: Data-Star; ESA-IRS; FT Profile
Updated: weekly
File length: Data-Star and FT Profile 1985 to date; ESA-IRS 1987 to date

C116

Foster Natural Gas Report

Full text of a newsletter covering US federal and state legislative, regulatory and judicial developments related to the natural gas industry. Includes major rate filings, import and export developments, briefs and decisions in major cases, actions of the Federal Energy Regulatory Commission, Economic Regulatory Administration, US Department of the Interior, and relevant state agencies.
Producer: Foster Associates Inc
Host: Mead Data Central (as part of LEXIS Public Utilities Law Library and as a NEXIS database)
Updated: weekly
File length: 1981 to date
Hard copy version: Foster Natural Gas Report

C117

Gas Daily

Contains full text of Gas Daily, a newsletter on the US natural gas industry covering exploration, production, capacity, transportation, regulatory issues, leases and prices. This database is also available on floppy disk from the producer.
Producer: Pasha Publications Inc
Host: Data-Star; Dialog (both as part of PTS Newsletter Database)
Updated: daily on Dialog; weekly on Data-Star; monthly on disk
File length: online Aug 1989 to date; disk 1986 to date
Hard copy version: Gas Daily

C118

International Gas Report

A full text newsletter on the natural gas and gas liquids industries worldwide. Coverage includes exploration and production, reserves, prices, business developments, technology and gases as feedstocks.
Producer: Financial Times Electronic Publishing
Host: Data-Star; FT Profile (both as part of Financial Times Business Reports: Energy); Mead Data Central (as part of NEXIS)
Updated: twice monthly
File length: 1987 to date
Hard copy version: International Gas Report

C119

LEXIS

The Lexis databases offer oil industry reports and full text coverage of several energy-related newsletters including Foster Natural Gas Report, Oil & Gas Journal, Platt's Oilgram News and Platt's Oilgram Price Report.
Producer: Mead Data Central
Host: Mead Data Central
Updated: varies by title
File length: varies by title

C120

McGraw-Hill GASWIRE

Contains news and analyses of the US natural gas market, including business, marketing and transportation developments, legal and regulatory actions, and spot market prices.
Producer: DRI/McGraw-Hill
Host: DRI/McGraw-Hill
Updated: weekly
File length: current information

C121

McGraw-Hill Publications Online

Provides the full text of many major McGraw-Hill publications including Industrial Energy Bulletin, Inside Energy/With Federal Lands, Inside F.E.R.C.'s Gas Market Reports, and Platt's International Petrochemical Report, Oilgram News and Oilgram Price Report.
Producer: McGraw-Hill Inc
Host: Dialog; Dow Jones News/Retrieval; Mead Data Central (as part of NEXIS)
Updated: weekly
File length: 1985 to date

C122

Mid-east Business Digest

Provides the full text of Mid-East Business
Digest, a weekly newsletter covering business
developments in the Middle East, including oil
and mineral discoveries, pipeline and refinery
expansion.
Producer: Hopewell Agency
Host: Data-Star; Dialog (both as part of PTS
Newsletter Database); MCI International;
NewsNet
Updated: weekly
File length: 1984 to date, except Dialog - 1988
to date
Hard copy version: Mid-East Business Digest

C123

NEXIS

The Nexis databases provide coverage of
news, EC legislation, business and financial
information from a wide range of sources,
including energy industry reports and
newsletters. International news sources
include: International Petrochemical Report,
International Gas Report, Oilweek, North Sea
Letter, Offshore, and Petroleum Times Price
Report.
Producer: Mead Data Central
Host: Mead Data Central
Updated: varies by title
File length: varies by title

C124

North Sea Letter

Full text of the weekly newsletter covering
technical and commercial developments in
North Sea offshore oil and gas drilling.
Producer: Financial Times Electronic
Publishing
Host: Data-Star; FT Profile (both as part of
Financial Times Business Reports: Energy);
Mead Data Central (as part of Nexis)
Updated: weekly
File length: Data-Star and FT Profile 1987 to
date; Mead Data Central 1989 to date
Hard copy version: North Sea Letter

C125

Oil & Gas Journal Energy Database

An extensive range of energy software and
data products on floppy disk is available from
the Oil & Gas Journal Energy Database.
Included are financial and company data,
industry indicators and forecasts, prices and
capital spending, offshore and onshore
exploration and drilling, production and
reserves, pipelines, refining, petrochemicals
and gas processing.
Producer: Oil & Gas Journal
Host: Available on disk from Oil & Gas Journal
Updated: varies by database
File length: varies by database

C126

Platt's Oilgram News

Full text of a newsletter on the oil and gas
industry worldwide, including coverage of
exploration and development, crude oil
supplies and prices, legislation, trade
agreements, and OPEC and IEA meetings.
Producer: McGraw-Hill
Host: Dialog (as part of McGraw-Hill
Publications Online database); Dow Jones;
Mead Data Central (as part of NEXIS)
Updated: daily
Hard copy version: Platt's Oilgram News

C127

PTS Newsletter Database

Contains the full text of articles from over 400
business and trade newsletters, including APS
Review: Oil Market Trends, Worldwide Energy,
and Energy Alert which covers major news
stories and Securities and Exchange
Commission filings in the oil and gas industry.
Producer: Predicasts
Host: Data-Star; Dialog
Updated: daily
File length: 1988 to date (with earlier coverage
for selected titles)

C128

Tecnon Petrochemical Newswires

Provides full text of 3 monthly newsletters covering the petrochemical industry: Ethpro (covering ethylene and propylene); Octane (covering BTX, gasoline and oxygenates); and Tecmeth (covering methanol).
Producer: Tecnon (UK) Ltd
Host: McGraw-Hill Inc
Updated: monthly
File length: current year
Hard copy version: Ethpro; Octane; Tecmeth

C129

Tecnon Petrochemical Product Summaries/ Newsletters

Full text coverage of 16 newsletters and product summary reports on current marketing developments and future prospects for petrochemical products.
Producer: Tecnon (UK) Ltd Host: I P Sharp Associates
Updated: monthly
File length: 1988 to date
Hard copy version: Tecnon Petrochemical Newsletters

C130

Trade & Industry Index

Includes abstracts of key trade journals in the oil and gas industries, covering companies, financial and economic information, products and technology etc. This database is also available on magnetic tape from Information Access Co.
Producer: Information Access Co
Host: Carl Systems; Data-Star; Dialog
Updated: varies by host from weekly to monthly
File length: 1981 to date

C131

US Oil Week

A full text newsletter for petroleum wholesalers and retailers, providing coverage of industry issues, news of major companies, marketing, prices and market share trends.
Producer: Capitol Publications Inc
Host: NewsNet Inc
Updated: weekly
File length: current and previous year
Hard copy version: US Oil Week

C132

Worldwide Energy

The full text of Worldwide Energy, a monthly newsletter covering energy sources and applications, including the exploration and operation of oil and gas, research and development, and alternative energy sources.
Producer: Worldwide Videotex
Host: Data-Star; Dialog; Newsnet
Updated: monthly
File length: Nov 1990 to date
Hard copy version: Worldwide Energy

Company information

C133

Canadian Oils Corporate Database (CANOILS)

Provides financial data on oil and gas producers listed on Canadian stock exchanges. Covers sources and uses of funds, drilling activity, production and reserves.
Producer: Woodside Research Ltd
Host: I P Sharp Associates
Updated: periodically
File length: annual data 1980 to date; quarterly data 1985 to date

C134
DRI Natural Gas Pipeline

Gives financial and operating statistics for the 21 largest US interstate natural gas pipeline companies. Data is taken from returns to the Federal Energy Regulatory Commission and includes assets and debits, liabilities and credits, natural gas sales, revenues and reserves, operation and maintenance expenses, number of customers, and gas delivered and received.
Producer: DRI/McGraw-Hill
Host: DRI/McGraw-Hill
Updated: continuously
File length: 1978 to date for annual figures; 1982 to date for monthly information series

C135
Newport Associates Energy Database

Financial data and operating statistics for over 500 publicly traded companies primarily or secondarily involved in the exploration, production, refining, or marketing of crude oil, natural gas, coal and uranium resources.
Producer: Newport Associates Ltd
Host: Newport Associates Ltd
Updated: weekly
File length: 350 annual and quarterly data series, plus 4 years of restated information

C136
Oil & Gas Journal Financial Database

Provides financial and operating data for 67 of the largest US oil and gas companies.
Producer: Oil & Gas Journal
Host: General Electric Information Services (GEIS)
Updated: varies by series
File length: 1981 to date

Statistics

C137
Analysis of Petroleum Exports (APEX)

An oil export monitoring service which analyses liftings of crude and refined petroleum products from 52 major oil producing countries by individual vessel. Covers all tankers and combined carriers over 10,000 metric dead weight tons. The database is also available on floppy disk, and is supplied with PC software to enable users to produce analyses.
Producer: Lloyd's Maritime Information Services Ltd
Host: Lloyd's Maritime Information Services Ltd
Updated: fortnightly
File length: 1986 to date

C138
API Inventories of Natural Gas Liquids & Liquefied Refinery Gases

Inventories of liquefied petroleum and refinery gases in plants, refineries and underground storage.
Producer: American Petroleum Institute
Host: American Petroleum Institute
File length: current information

C139
API Monthly Statistical Report

Analyses and comment on the significance of trends reflected in the weekly data.
Producer: American Petroleum Institute
Host: American Petroleum Institute
Updated: monthly
File length: current information

C140

API Weekly Statistical Bulletin

Contains time series on crude oil and petroleum products, and refinery operations in the USA. The database is also available on floppy disk and magnetic tape as the Weekly Data Historical File (covering 1982-87) which is updated annually with the next year of historical data.
Producer: American Petroleum Institute
Host: American Petroleum Institute; I P Sharp Associates
Updated: weekly
Hard copy version: corresponds to tables 1-6 in API Weekly Statistical Bulletin

C141

Australian Major Energy Statistics

Time series on energy consumption and production in Australia, with emphasis on petroleum products. Covers production, sales, stocks, imports and exports by fuel and by state.
Producer: Department of Primary Industries and Energy
Host: I P Sharp Associates
Updated: monthly, with a lag of 6 weeks
File length: 1980 to date

C142

CPASTATS

Provides weekly, monthly, quarterly and annual time series on Canadian oil and natural gas production including exploration and drilling, reserves, refining, imports and exports, prices, transportation and consumption.
Producer: Canadian Petroleum Association (CPA)
Host: General Electric Information Services (GEIS)
Updating: varies by series
File length: varies by series, earliest data 1947 to date

C143

DRI Canadian Energy

Details sources and uses of major fuels analysed by province and sector. Data covers demand, production and fuel use, prices, costs and taxes for oil and gas.
Producer: DRI/McGraw-Hill
Host: DRI/McGraw-Hill
Updated: continuously
File length: 1951 to date

C144

DRI Canadian Energy Forecast

Forecasts of energy supply, demand and prices for Canada, both at national level and for the individual provinces of Alberta, Atlantic, British Columbia, Manitoba, Ontario, Quebec and Saskatchewan. Includes data on energy consumption and prices analysed by fuel and sector, and production, trade and stock levels for crude oil, petroleum products and natural gas.
Producer: DRI/McGraw-Hill
Host: DRI/McGraw-Hill
Updated: 3 times a year
File length: 25 year forecasts, with historical series from 1978 to date

C145

DRI Energy Database

Large database profiling all aspects of the US energy sector. Provides an extensive collection of statistics for crude oil, motor gasoline, petroleum products and natural gas, including exploration, drilling, production, refinery operations, stocks, prices, sales and consumption.
Producer: DRI/McGraw-Hill
Host: DRI/McGraw-Hill
Updated: varies by series
File length: 1960 to date

C146

DRI European Energy Forecast

A forecasting service providing data on projected energy supply, demand and prices for crude oil, petroleum products and natural gas in 15 European countries. Forecasts for 15 to 20 years ahead.
Producer: DRI/McGraw-Hill
Host: DRI/McGraw-Hill
Updated: quarterly
File length: 1971 to date for most series

C147

DRI Middle East & African Forecast

Annual historical and forecast time series for 10 Middle Eastern and African economies. Forecasts for 10-15 years ahead include data on oil production. *Producer: DRI/McGraw-Hill Host: DRI/McGraw-Hill; Information Plus Updated: quarterly File length: 1950 to date*

C148

DRI Natural Gas Forecast

Contains detailed forecasts of oil and natural gas supply, prices and production costs in the US for 15-20 years ahead.
Producer: DRI/McGraw-Hill
Host: DRI/McGraw-Hill; Forecast Plus
Updated: quarterly
File length: 1970 to date

C149

DRI US Long-term Energy Forecast

Projections of energy demand, supply and prices in the US, at national and regional level, for the next 20-25 years. Includes production, trade and stock figures for crude oil and natural gas.
Producer: DRI/McGraw-Hill
Host: DRI/McGraw-Hill; Forecast Plus
Updated: quarterly
File length: data available from 1960s, 1970s or 1980s depending on series

C150

DRI US Short-term Energy Forecast

2 to 3 year forecasts for fuel supply, demand and prices in the US, including production, trade and stock figures for natural gas, crude oil and products.
Producer: DRI/McGraw-Hill
Host: DRI/McGraw-Hill; Forecast Plus
Updated: monthly
File length: 1981 to date

C151

DRI World Oil Forecast

Provides the 15- to 20-year outlook for supply, demand and prices of crude oil. Based on the DRI World Oil Simulation Model which balances the effects of world-wide economic growth, exchange rates and inventories, against OPEC and non-OPEC supply policies in specific countries.
Producer: DRI/McGraw-Hill
Host: DRI/McGraw-Hill Updated: quarterly
File length: 1975 to date

C152

EDMC Energy Databank

Approximately 34,000 weekly, monthly, quarterly and annual time series on energy production, consumption, imports, exports and prices in major industrialised nations. Database use restricted to Japan.
Producer: The Institute of Energy Economics Energy Data and Modelling Centre (EDMC)
Host: Kaihatsu Computing Centre
Updated: varies by series
File length: 1945 to date for annual data; 1984 to date for monthly data

C153

ENERGY

Weekly, monthly, quarterly and annual time series of detailed supply and demand statistics for oil, petroleum and natural gas.
Producer: The WEFA Group (Wharton Econometric Forecasting Associates)
Host: The WEFA Group (Wharton Econometric Forecasting Associates)
Updated: varies by series
File length: 1940s to date

C154

GASLINE

Contains time series on the world energy industry including production, reserves, imports, exports, and prices for natural gas, gas liquids and oil in over 70 countries.
Producer: Institute of Gas Technology (IGT)
Host: IGT Net
Updated: every two months
File length: varies by series
Hard copy version: Energy Statistics

C155

IEA International Energy

Provides country-specific information on production, consumption, stocks, imports and exports of energy by fuel type for member nations of the International Energy Agency and the OECD, and their trading partners. Includes indicators from a range of reports including Energy Balances, Energy Prices and Taxes, Energy Statistics and Monthly Oil Statistics.
Producer: DRI/McGraw-Hill
Host: DRI/McGraw-Hill
Updated: monthly
File length: annual data from 1960; quarterly data from 1978; monthly data from 1980

C156

Imports of Crude Oil & Petroleum Products

Time series on shipments into the US of over 100 petroleum and petroleum-related products. For each shipment, details of country of origin, port of entry, destination, viscosity and gravity are given. The database is also available on floppy disk and magnetic tape from API.
Producers: American Petroleum Institute; US Department of Energy, Office of Oil Imports
Host: I P Sharp Associates
Updated: monthly, with a lag of 2 to 3 months
File length: 1977 to date
Hard copy version: Imported Crude Oil and Petroleum Products

C157

International Energy Annual

Contains 11 files of annual time series covering production, supply, stocks, prices, imports and exports, and refining capacity for crude oil, petroleum products, dry and liquid natural gas. Data is compiled for over 190 countries and regions.
Producer: I P Sharp Associates; US Department of Energy, Energy Information Administration
Host: I P Sharp Associates
Updated: annually
File length: 1973 to date
Hard copy version: International Energy Annual

C158

Lundberg Survey Share of Market

Monthly data on gasoline and distillate wholesale sales volumes and market shares by brand, PAD, US national and regional.
Producer: Lundberg Survey Inc
Host: I P Sharp Associates
Updated: periodically
File length: 1978 to date

C159

Monthly Energy Review

Monthly data on US production and consumption of natural gas, exports, imports, wellhead, FOB and landed costs of crude oil.
Producer: I P Sharp Associates; US Department of Energy
Host: I P Sharp Associates
Updated: monthly, with a lag of 1 to 3 months
File length: 1977 to date
Hard copy version: corresponds to tables 10, 22, 26, 27, 28, 40, 41 and 42 in Monthly Energy Review

C160

OECD Quarterly Oil & Gas Statistics

Quarterly time series on the oil and gas industries of OECD member nations. Data provided for each country includes oil production, refinery output, stocks, imports, exports and international marine bunkers. A wide range of statistical data is also available on floppy disk or magnetic tape direct from OECD, as part of the International Oil Market Information System.
Producer: Organization for Economic Cooperation and Development (OECD)
Host: I P Sharp
Updated: quarterly
File length: 1974 to date
Hard copy version: OECD Quarterly Oil and Gas Statistics

C161

Oil & Gas Journal Energy Database

Current estimates and forecasts of supply, demand and prices including drilling and exploration, refining, operating data and revenue accounting for the US and international oil and gas industries.
Producer: Oil & Gas Journal
Host: General Electric Information Services (GEIS)
Updated: varies by series
File length: 1971 to date for most series
Hard copy version: Oil & Gas Journal

C162

Petroleum Supply Monthly

27 monthly time series on imports, exports, supply, disposition and stocks of crude oil and petroleum products in the US.
Producer: I P Sharp Associates; US Department of Energy, Energy Information Administration
Host: I P Sharp Associates; US Department of Energy (as part of the Energy Information Administration Electronic Publication System)
Updated: monthly, with a lag of 2 to 3 months
File length: 1971 to date

C163

Petroleum Supply Monthly State Stocks

Contains time series from the current months Petroleum Supply Monthly on refinery and bulk terminal stocks of selected petroleum products by state.
Producer: US Department of Energy, Energy Information Administration
Host: US Department of Energy (as part of the Energy Information Administration Electronic Publication System)
Updated: monthly
File length: current information

C164

State Energy Data System

Time series on energy consumption in each US state. Data is analysed by geographic area, type of fuel and by sector, and covers natural gas, petroleum and products.
Producer: US Department of Energy
Host: I P Sharp Associates
Updated: annually
File length: varies by series, earliest data 1960 to date

C165

SUPPLYLINE

Current and historical time series on oil and gas exploration, production, inventories, imports and exports worldwide. Areas covered include land and marine seismic crews and rigs, OPEC production estimates and petroleum stocks for OECD countries.
Producer: Reuters Ltd
Host: Reuters Ltd
Updated: varies by series
File length: 1971 to date

C166

US Energy Forecast

Time series of historical and forecast data on US petroleum and natural gas consumption by household, commercial and industrial sector.
Producer: The WEFA Group (Wharton Econometric Forecasting Associates)
Host: The WEFA Group
Updated: quarterly
File length: 1960 to date

C167

Weekly Petroleum Status Report

Provides data on production, imports, exports and prices for crude oil and petroleum.products.
Producer: US Department of Energy, Energy Information Administration
Host: I P Sharp Associates; US Department of Energy (as part of the Energy Information Administration Electronic Publication System)
Updated: weekly
File length: I P Sharp Assoc Dec 1988 to date; US Department of Energy current 5 weeks
Hard copy version: Weekly Petroleum Status Report

C168

World Forecast Database

Annual time series of historical and forecast economic data for OECD countries and S.Africa. Includes crude oil export data for 16 oil producing countries.
Producer: The WEFA Group (Wharton Econometric Forecasting Associates)
Host: The WEFA Group
Updated: quarterly
File length: annual data 1948 to date; quarterly data 1970 to date

Prices

C169

Business & Finance Report

Provides economic, financial & business news, including petroleum industry futures reports.
Producer: Dow Jones & Company Inc
Host: Dow Jones News/Retrieval
Updated: continuously
File length: current information

C170

Daily Petro Futures

Contains opening, closing and stop prices for No 2 heating oil, gasoline and propane futures traded at the New York Mercantile Exchange.
Producer: News-A-Tron Corporation
Host: General Videotex Corporation/Delphi; Western Union Telegraph Company
Updated: twice daily
File length: current information

C171

Datastream Financial & Commodity Futures

Covers "live" prices from the London International Petroleum Exchange, and futures contracts for oil and gas commodities traded in Chicago, New York and London.
Producer: Datastream
Host: Datastream
Updated: daily
File length: 1979 to date

C172

DeWitt Petrochemical Newsletters

Contains 9 newsletters providing analyses of recent market developments in supply, demand, price quotations and forecasts for a variety of petrochemical products including benzene, butadiene, butylenes, ethylene, hydrocarbons, methanol, propylene, toluene and xylenes.
Producer: DeWitt and Company
Host: I P Sharp Associates
Updated: newsletters: weekly; forecasts: monthly
File length: current 3 issues of newsletters; current issue of forecasts
Hard copy version: DeWitt Petrochemical Newsletters

C173

DRI European Oil Spot Market Forecast

Provides 6 month forecasts of average spot prices of OPEC and 15 representative crudes, together with prices for a range of petroleum products.
Producer: DRI/McGraw-Hill
Host: DRI/McGraw-Hill
Updated: weekly
File length: 6-month forecasts with 6 months historical data

C174
DRI Natural Gas Spot Prices

Quotes spot prices at each major US transaction point: wellhead, as delivered to pipelines, at the city-gate or burner-tip.
Producer: DRI/McGraw-Hill
Host: DRI/McGraw-Hill; Information Plus
Updated: weekly or monthly depending on series
File length: 1983 to date

C175
DRI/Platt's Oil Prices - Platt's Oilgram Price Report

Details spot and posted crude oil and petroleum product prices throughout the world. Includes rack, barge, pipeline and city prices of US markets, and extensive coverage of the European bulk, Mediterranean, Caribbean and Far Eastern markets.
Producer: DRI/McGraw-Hill
Host: Dialog (as part of McGraw-Hill Publications Online database); Dow Jones; DRI/McGraw-Hill; Mead Data Central (as part of NEXIS)
Updated: daily
File length: daily series from 1979; weekly from 1978; monthly from the mid 1970s
Hard copy version: Platt's Oilgram Price Report

C176
Electronic Markets & Information Systems (EMIS)

News on the petroleum and chemical industries, including commodities, spot and futures prices, shipping and tanker rates.
Producer: Tecnon (UK) Ltd
Host: McGraw-Hill Inc
Updated: monthly
File length: most recent year

C177
European Chemical News

Contains full text of the weekly journal providing current news on petrochemical prices, market trends, new plants and processes.
Producer: European Chemical News
Host: Data-Star
Updated: weekly
File length: 1984 to date
Hard copy version: European Chemical News

C178
FT Energy Economist

Full text of a monthly newsletter on world energy patterns, covering prices and trends in oil, gas and refined products, and market commentary on all major energy sources.
Producer: Financial Times Electronic Publishing
Host: Data-Star; FT Profile (as part of Financial Times Business Reports: Energy); Mead Data Central (as part of both Lexis and Nexis)
Updated: monthly
File length: 1988 to date
Hard copy version: FT Energy Economist

C179
Gas Price Report

Current and historical data on oil and gas prices, including spot and bid prices, Canadian border prices, current highest and average wellhead prices in Texas, and new contract prices for Texas, Louisiana and Oklahoma. Also covers interstate pipeline purchase costs and transportation revenue. Available only on floppy disk.
Producer: Parent Information Enterprises
Host: not available online. Available on floppy disk from Parent Information Enterprises
Updated: every two weeks
File length: 1986 to date
Hard copy version: Gas Price Report

C180

ICIS-LOR Oil & Chemical Reports

The ICIC-LOR Group produce reports and assessments on worldwide market activity and spot prices for all aspects of the oil and petrochemical industries. Their Oil and Chemical Reports database corresponds to the following printed reports:
- Chemical Tanker Market Report (weekly)
- Ethylene Cracker Report (monthly)
- Worldwide Product Report (daily)
- Petrochemicals and Chemicals (weekly)
- LPG Report (twice weekly)
- Netbacks (daily)
- Worldwide Crude Report (daily)
- European Aromatics Report (monthly)
- European Ethylene Derivatives Report (monthly)
Producer: ICIS-LOR Group
Host: Data-Star; I P Sharp Associates; McGraw-Hill's EMIS time-sharing networks
Updated: varies by report
File length: 1980 to date except Data-Star - 1987 to date
Hard copy version: all files correspond to a hard copy report

C181

Inside F.E.R.C.'s Gas Market Report

Full text coverage of a newsletter covering market growth, prices, futures trading and trends and developments in the US natural gas market.
Producer: McGraw-Hill Inc
Host: Dialog (as part of McGraw-Hill Publications Online database); Dow Jones News/Retrieval; Mead Data Central (as a Nexis database).
Updated: every two weeks
File length: 1988 to date except Mead Data Central - 1981 to date
Hard copy version: Inside F.E.R.C.'s Gas Market Report

C182

Lundberg Survey Spot Product Assessment

Contains biweekly assessments of petroleum product spot market prices for gasolines, jet kerosene, No 2 fuel, 180 CST and Bunker C at Chicago, Los Angeles, New York and Gulf Coast harbours.
Producer: Lundberg Survey Inc
Host: I P Sharp Associates
Updated: twice weekly
File length: 1984 to date

C183

Lundberg Survey Wholesale Prices & Moves

Current and historical wholesale price data for all grades of gasoline and No.2 distillate diesel fuel in over 200 US regions. Covers brands with a 2% or greater share of the market.
Producer: Lundberg Survey Inc
Host: I P Sharp Associates
Updated: continuously
File length: most recent 2 years

C184

PETROFLASH!

Subdivided into two daily oil price reports:
Crude Report
- market prices and gross product values of principal US domestic grade crudes and key world export crudes.
Products Report
- spot and contract prices for products in oil markets in the US, Canada, Caribbean and Europe.
Producer: Petroflash Inc
Host: I P Sharp Associates; Bonneville Market Information
Updated: varies by report from daily to weekly
File length: current information

C185
Petroleum Argus Crude Deals Done Database

Contains information on over 40,000 spot and forward deals for 147 types of crude oil reported since 1973. Includes daily quotations of main traded crudes since 1979.
Producer: Petroleum Argus Ltd
Host: most recent 3 months of the database are accessible online via I P Sharp Associates. The complete database (1973 -) is available on diskette from Petroleum Argus
Updated: daily
File length: 1973 to date

C186
Petroleum Argus Daily Market Report

Subdivided into three files:
Crude Report
- price assessments of key crudes at close of trading in Houston and London, together with information on other widely traded crude oils including the dated and 15 day Brent market, West Texas Intermediate and Dubai.
Atlantic Products Report
- daily coverage of Mediterranean, New York, and Northwest European cargo and barge markets, including forward prices and confirmed deals.
Asia Pacific Products Report
- end-of-day assessments for cargo and pipeline prices for products from the Arab Gulf, Japan, Singapore and US West Coast.
Producer: Petroleum Argus Ltd
Host: I P Sharp Associates
Updated: daily
File length: most recent 50 reports
Hard copy version: Crude Report; Atlantic Products Report; Asia Pacific Products Report

C187
Petroleum Argus Prices

Contains 3 files of time series for crude oil and petroleum products quoted in Petroleum Argus Daily Market Report. Includes first, second and third month prices of major traded crudes, spot cargo prices for products, and information on shipping freight costs.
Producer: Petroleum Argus Ltd
Host: I P Sharp Associates
Updated: crude oil and products daily; shipping freight costs weekly
Hard copy version: Weekly Petroleum Argus

C188
Petroleum Intelligence Weekly

Subdivided into 4 databases:
Crude Oil Production
- provides monthly time series on volume of crude oil produced by major producers and exporters, and totals for natural gas liquids produced for OPEC and the world
Key Crude Prices
- shows the broad trend of basic oil price indicators - official price, spot price and spot product value, for Mideast Light, Mideast Heavy and African Light/North Sea
Spot Product Prices
- representative monthly spot prices for the primary petroleum products in 6 key refining centres
Official Crude Prices
- official daily government selling prices and the equivalent term contract prices for 114 crude oils
Producer: Petroleum Intelligence Weekly
Host: I P Sharp Associates
The online versions of these databases are no longer updated and file lengths close in 1987. For current information, Petroleum Intelligence Weekly now offers a monthly disk service containing the weekly and monthly data on world oil production, crude oil and product prices.
Hard copy version: Petroleum Intelligence Weekly

C189

Petroleum Marketing Monthly

Contains monthly petroleum price, consumption and volume data analysed by US state and PAD district.
Producer: US Department of Energy
Host: I P Sharp Associates
Updated: monthly, with a lag of 3 months
File length: 1983 to date
Hard copy version: Petroleum Marketing Monthly

C190

PETRONET US

Canadian and Mexican data and product-trading information on gasolines, fuel oils, jet oils and propane. Provides spot market and composite prices, and daily wholesale and retail prices from suppliers in 400 US locations.
Producer: Computer Petroleum Corporation
Host: Computer Petroleum Corporation
Updated: continuously
File length: current information, with 9 years of historical data

C191

PETROSCAN

17 files of current US wholesale price and market information for petroleum futures and products, including No.2 oil, No.1 (kerosene), regular, un-leaded and premium grades of gasoline, LPG, propane, natural gas, residual and jet fuel.
Producer: United Communications Group
Host: Petroscan - a division of United Communications Group
Updated: varies from daily to monthly according to file
File length: most recent day or week
Hard copy version: API Supply and Production Data file; API Weekly Statistical Bulletin

C192

Platt's Chemical Prices

Provides spot and contract prices for petrochemical products as reported in Platt's Petrochemicalscan, Solventwire, Olefinscan and Polymerscan reports.
Producer: DRI/McGraw-Hill
Host: DRI/McGraw-Hill
Updated: weekly
File length: 1983 to date

C193

Reuter Monitor Oil Service

Offers a complete package of up-to-the-minute news reports and real-time spot and futures prices. May be combined with Reuter Monitor Graphics Service to produce instant comparisons between spot and futures prices, crude prices and netback values.
Producer: Reuters
Host: Reuters
Updated: daily
File length: current information

C194

Reuter Pipeline

Information on US and European crude oil and petroleum product markets, including spot market prices and confirmed sales for Arab Light, Brent, Dubai and West Texas intermediate crudes, and crudes from Libya, Egypt and Indonesia. Also covers netback calculations for topping and cracking refineries in the US, Gulf Coast, Singapore, NW Europe and the Mediterranean.
Producer: Reuters
Host: I P Sharp Associates; Reuters
Updated: daily
File length: current 2 months

C195

SPOTLINE

Current and historical spot and futures prices for crude oil, petroleum products, petrochemicals and natural gas.
Producer: Reuters
Host: Reuters
Updated: varies by series
File length: 1973 to date

C196

Telerate Energy Service

Provides real-time energy market information including futures and options, spot market prices for crude oil, petroleum products, natural gas and LPG, posted prices for key crudes around the world, supply/demand statistics, tanker market fixtures and shipping information.
Producer: Telerate; Associated Press; Dow Jones
Host: Telerate
Updated: daily
File length: current information

CD-ROM DATABASES

Applied Science & Technology Index

Includes energy resources and development, geology, marine technology, oceanography, petroleum and gas.
Producer: H.W. Wilson Company
Supplier: H.W. Wilson Company (Wilsondisc)
Updated: monthly
File length: October 1983 to date
Hard copy version: Applied Science and Technology Index

C198

Aquatic Sciences & Fisheries Abstracts

Worldwide literature on all aspects of freshwater and marine environments including oceanography, marine and offshore technology, ocean engineering, diving systems, submersibles and all related technologies.
Producer: Cambridge Scientific Abstracts
Supplier: Compact Cambridge; Microinfo Ltd
Updated: quarterly
File length: 1982 to date
Hard copy version: Aquatic Sciences and Fisheries Abstracts

C199

Arctic & Antarctic Regions

Comprises 8 databases of information on temporarily or permanently frozen areas, including the Arctic, Antarctica, the Antarctic Ocean and sub-Antarctic islands. Databases include the Arctic Science and Technology Information System (ASTIS), C-CORE from the Cold Ocean Resources Engineering Centre, and the library catalogues of the Boreal Institute for Northern Studies and the Scott Polar Research Institute.
Producer: Library of Congress, Science and Technology Division, with support from the National Science Foundation, Division of Polar Programs
Supplier: National Information Services Corporation (NISC)
Updated: twice per annum
File length: varies by database
Hard copy version: ASTIS Current Awareness Bulletin; ASTIS Bibliography; ASTIS Occasional Publications; Boreal Institute for Northern Studies Library Bulletin; Antarctic Bibliography; Bibliography on Cold Regions Science & Technology; Polar and Glaciological Abstracts

C200

AXSES INFO ATLAS

Contains digital maps and information on the coastal and ocean environment between Cape Cod, Massachusetts and Halifax, Nova Scotia including data on the environment, geology, sediments, oceanography, oil and gas.
Producer: Axses Information Systems; Environment Canada
Supplier: Axses Information Systems
Updated: irregularly
File length: 1961 to date

C201

CHEM-BANK

A collection of 4 databases covering potentially hazardous chemicals:- Registry of Toxic Effects of Chemical Substances (RTECS); Chemical Hazard Response Information System (CHRIS) Oil and Hazardous Materials Technical Assistance Data System (OHMTADS); and Hazardous Substances Databank (HSDB).
Producer: Silverplatter, with information supplied by US Dept of Health and Human Services, Centers for Disease Control; National Institute for Occupational Safety and Health; US Coast Guard; US Environmental Protection Agency Emergency Response Division; US Environmental Protection Agency Office of Pesticides and Toxic Substances
Supplier: Microinfo Ltd; Optech Ltd; Silverplatter
Updated: quarterly
File length: current information
Hard copy version: Registry of Toxic Effects of Chemical Substances; Toxic Substances Control Act Initial Inventory

C202

COMPENDEX PLUS

Covers a wide range of engineering disciplines including marine, petroleum and fuel engineering. Also includes literature from engineering-related areas of science and management including petroleum refining, transportation and environmental pollution. The database is also available as part of Ei CHEMDISC.
Producer: Engineering Information Inc
Supplier: Microinfo Ltd; Dialog:Faxon Europe
Updated: quarterly
Coverage: 1985 to date
Hard copy version: Engineering Index; Engineering Meetings

C203

Deep Sea Drilling Project

Contains marine geological and geophysical data obtained by the D/V Glomar Challenger from over 1000 holes drilled at 624 marine sites worldwide. Covers hardrock and sediment data, including age profiles, chemistry, paleomagnetism, paleontology and x-ray mineralogy, logging and geophysical data.
Producer: National Geophysical Data Center
Supplier: National Geophysical Data Center
File length: data collected from the 1960s through the 1980s

C204

Earth Sciences

Comprises 3 databases:
Earth Science Data Directory
- A catalogue of databases and sources of information on earth sciences and natural resources
Geoindex
- Citations to published geological maps of the US and its territories
US Geological Survey Library
- Bibliographic records providing core collection of materials pertaining to all areas of earth sciences. Includes all acquisitions since 1975
Publisher: OCLC Online Computer Library Center Inc
Supplier: Microinfo Ltd; OCLC Online Computer Library Center Inc
Updated: quarterly

C205

Ei CHEMDISC

Abstracts of world conference proceedings and journal articles. Extensive coverage of technology issues for chemical engineers in all industries - of special interest to chemical and petroleum companies.
Producer: Engineering Information Inc
Supplier: Dialog; Engineering Information Inc
Updated: quarterly
Coverage: 1980 to date

C206

Ei Energy/Environment Disc

A subset database of abstracts from Compendex Plus. Subject areas include petroleum refining, oil field equipment, operations and production, ocean engineering, gas fuels, pollution, hazardous materials and economic considerations in energy and the environment.
Producer: Engineering Information Inc
Supplier: Dialog; Engineering Information Inc; Microinfo Ltd
Updated: quarterly
File length: 1980 to date
Hard copy version: Energy Abstracts; Engineering Index

C207

Ei PAGE ONE

Fast access to citations of world engineering journals and conference proceedings with 30% wider coverage than Compendex Plus. May be used as an alerting service for upcoming abstracts in Compendex Plus, Ei ChemDisc or Ei Energy/Environment Disc.
Producer: Engineering Information Inc
Supplier: Engineering Information Inc
Updated: monthly
File length: latest 2 years

C208

Energy Analyst

Contains over 200,000 energy and macroeconomic time series from more than 40 sources worldwide. Covers oil and gas production, drilling, oil and gas prices and economic statistics. Has ability to download to spreadsheets.
Producer: Quick Source Inc
Supplier: Automated Sciences Group
Updated: monthly
File length: 1940s to date

C209

Energy Library

An OCLC database providing information on oil, coal, nuclear, solar, and other related energy sources.
Producer: OCLC Online Computer Library Center Inc
Supplier: OCLC Online Computer Library Center Inc; Doc-6
Updated: annually
File length: pre-1900 to date

C210

ENVIRO/ENERGYLINE Abstracts Plus

Comprises 3 databases of information on energy and the environment including US petroleum and natural gas resources, fuel transport etc.
Producer: Bowker A&I Publishing
Supplier: Bowker Electronic Publishing; Microinfo Ltd
Updated: quarterly
File length: varies by database
Hard copy version: Acid Rain Abstracts; Energy Index (1971-1976); Energy Information Abstracts (1976 -); Environment Abstracts

C211

EXPLORE!

Contains an index to over 1.5 million time series and data fundamentals from 11 DRI online databases of statistics on the US economy. Includes Platt's Oil Prices covering spot and posted crude oil prices by crude type and field location available in the Platt's Oil Prices online database.
Producer: DRI/McGraw-Hill Data Products Division
Supplier: DRI/McGraw-Hill Data Products Division
Updated: quarterly
File length: current information

C212

GEOBASE Alberta General Well File

Includes detailed data on over 150,000 petroleum wells in Alberta, Canada.
Producer: CD PubCo Inc
Supplier: CD PubCo Inc
Updated: monthly
File length: current information
Hard copy version: Alberta General Well File (microfiche)

C213

GEOBASE Alberta Production History File

Contains monthly time series of production data for over 150,000 petroleum wells in Alberta, Canada, including well header and yearly cumulative data.
Producer: CD PubCo Inc
Supplier: CD PubCo Inc
Updated: monthly
File length: 1962 to date

C214

GEOBASE British Columbia & Saskatchewan General Well File

Contains over 50 data items on petroleum wells in the Canadian provinces of British Columbia and Saskatchewan.
Producer: CD PubCo Inc
Supplier: CD PubCo Inc
Updated: monthly
File length: current information

C215

GEOBASE British Columbia & Saskatchewan Production History

Contains monthly time series of production data for petroleum wells in the Canadian provinces of British Columbia and Saskatchewan, including well header and yearly cumulative data.
Producer: CD PubCo Inc
Supplier: CD PubCo Inc
Updated: monthly
File length: late 1950's to date

C216

Geophysics of North America

Provides land and marine physical data for the Northern Hemisphere, including topograghy, bathymetry, gravity, magnetics, thermal aspects, stress data, and satellite images.
Producer: National Geophysical Data Center
Supplier: National Geophysical Data Center
Updated: this database is not updated
File length: current information

C217

GEOREF

Contains around 1.5 million citations, with some abstracts, to worldwide literature on earth sciences including marine geology and oceanography, sedimentary, igneous and metamorphic petrology, minerals and petroleum.
Producer: American Geological Institute (AGI)
Supplier: Microinfo Ltd; Silverplatter
Updated: quarterly
File length: North American material 1785 to date; international material 1933 to date
Hard copy version: Bibliography and Index of Geology (1969 -); Bibliography and Index of Micropaleontology (1972 -); Bibliography and Index of North American geology (1785-1970); Bibliography and Index of Geology exclusive of North America (1933-1968); Bibliography of Theses in Geology (1965-1966); Geophysical Abstracts (1966-1971)

C218

GLORIA Data on the Gulf of Mexico

Contains 3 sets of image data for the sea floor of the Gulf of Mexico: raw raster images, 2-degree square computer-generated mosaics, and composite image files for the eastern and western halves of the Gulf.
Producer: US Geological Survey, National Oceanic and Atmospheric Administration Joint Office for Mapping and Research
Supplier: National Geophysical Data Center; US Geological Survey, National Oceanic and Atmospheric Administration Joint Office for Mapping and Research

C219

MUNDOCART

Contains digital cartographic data for the entire world at a scale of 1:1,000,000 and for Antarctica at a scale of 1:2,500,000. Available as a complete database or as 8 regional subsets.
Producer: Petroconsultants (CES) Ltd
Supplier: Petroconsultants (CES) Ltd; Microinfo Ltd; Chadwyck-Healey Ltd
Hard copy version: US Defence Mapping Agency Operational Navigation Charts

C220

National Energy Research Seismic Library Prototype Disk

Contains 3 sets of processed multichannel seismic reflection data covering the Atlantic Outer Continental Shelf, the New Mexico Church Rock area and the southern Montana-north eastern Wyoming area.
Producer: US Geological Survey, National Energy Research Seismic Library
Supplier: US Geological Survey, National Energy Research Seismic Library This is a prototype product. NERSL plans to release additional seismic reflection data on CD-ROM.
Hard copy version: US Geological Survey NERSL

C221

NTIS

Important multidisciplinary sci-tech database, containing bibliographic citations and abstracts of technical reports from US and non-US government sponsored research, development, and engineering analyses. Coverage includes natural resources and ocean technology.
Producer: National Technical Information Service (NTIS)
Supplier: Dialog; Microinfo Ltd; NTIS; OCLC Online Computer Library Center Inc; Optech Ltd; Silverplatter
Updated: quarterly
File length: varies according to supplier, earliest data from 1980
Hard copy version: corresponds in part to Government Reports Announcements & Index; Abstract Newsletters

C222

OSH-ROM

A collection of 4 major databases on occupational health and safety information:-
NIOSHTIC - database of the US National Institute for Occupational Safety and Health;
HSELINE
- database of the UK Health and Safety Executive;
CISDOC
- database of the International Occupational Safety & Health Information Centre of the International Labour Organisation;
MHIDAS
- Major Hazards Incident Data Service developed by the Major Hazards Assessment Unit of the Health and Safety Executive.
Producer: Silverplatter, with information supplied by National Institute for Occupational Safety and Health; Health and Safety Executive; International Occupational Safety & Health Information Centre
Supplier: Microinfo Ltd; Optech Ltd; Silverplatter
Updated: quarterly
File length: current information

C223

PASCAL CD-ROM

French and English multidisciplinary database covering the world literature on science, technology and engineering. Includes *Pascal: Energie* covering energy economics, fuels and technology, and *Pascal: Geode* covering geochemistry, marine geology, stratigraphy and mineral economics.
Producer: Centre National de la Recherche Scientifique, Institut de l'Information Scientifique et Technique (CNRS/INIST)
Supplier: INIST (DIFFUSION)
Updated: every 6 months
File length: current disc 1990 to date; archival disc 1987-1989
Hard copy version: corresponds to Pascal Sigma and Pascal Thema titles

C224

Publications of the US Geological Survey

Produced from the GEOREF database. Includes references to US Geological Survey documents published from 1880-1988, non-survey publications and reports produced by the Hayden, King, Powell and Wheeler surveys.
Producer: American Geological Institute
Supplier: Americal Geological Institute
Updated: quarterly
File length: 1880 to date

C225

TOMES PLUS

Comprises 7 databases of chemical, medicinal and toxicological data including
HAZARDTEXT
- initial response protocols to incidents involving hazardous material;
OHM-TADS
- health and environmental effects of over 1,000 petroleum products and hazardous chemicals;

DOT
- Emergency Response Guides for Hazardous Incidents;
CHRIS
- Chemical Hazard Response Information System.
Producer: Micromedex Inc, with information supplied by National Library of Medicine; US Coast Guard; US Department of Transportation; US Environmental Protection Agency
Supplier: Micromedex Inc; Microinfo Ltd
Updated: quarterly
File length: 1989 to date

C226

UKCS Digital Well Logs/ Norwegian Digital Well Logs/ Western Australian Digital Well Logs

Contains information on petroleum wells on the UK Continental Shelf, Norway and Western Australia. Users can order data on separate CD-ROMs for each geographic area.
Producer: ERICO Petroleum Information Ltd
Supplier: ERICO Petroleum Information Ltd
Updated: approximately 150-200 wells per annum
File length: 1965-1984

C227

World Energy

Coverage includes the European Community's SESAME-INFOIL database of 2,700 hydrocarbon and related energy topics, information on OPEC, data on over 500 organisations involved in energy research, details of all major oil, gas and chemical companies involved in the industry, key personnel from the chemical, oil, gas and mining industries, production and market data.
Producer: Longman Cartermill Ltd
Supplier: Microinfo Ltd; Optech Ltd
Updated: annually
File length: current information

STOCKBROKER RESEARCH REPORTS

ARTHUR ANDERSEN

Petroleum Services
1 Surrey Street
London WC2R 2PS
Tel: 071-438 3866
Fax: 071-438 3881
Contact: Louise Baker

The Petroleum Services Group provides a wide range of upstream oil industry information services, supplied both as hard copy and as computer database packages. Initially the services concentrated on the North Sea, but have recently been expanded to cover Australasia and the Far East.
All prices given are for first copy subscriptions. Discounts for multiple copies and successive subscriptions are offered for most services. There is a surcharge for non-UK subscribers, and computer-based services are quoted exclusive of VAT.

Hard Copy Services

C228
Weekly Service
A subscription news service providing the latest information on drilling activity, rig movements and licence changes in the North Sea region and onshore UK and France.
Price: £1,500 pa for total North West Europe; £1,200 for offshore coverage only; £550 for onshore only.

C229
North West Europe Field and Production Report
A monthly scouting service providing up-to-date information on production figures and field development activities.
Price: £750 pa.

C230
Guide to European Oil and Gas Licensing Terms
An annual publication providing an outline of the current licensing terms and conditions applicable for new awards for onshore and offshore areas of 17 European countries, together with an outline of the fiscal regime. Summary tables provide comparisons between the different countries.
Price: details available on request.

C231
North West Europe Company Report
Gives a complete summary of the exploration and production interests of more than 60 of the most active companies with oil and gas interests in the North Sea region and onshore UK. A bi-annual publication produced in A3 spiral bound format.
Price: £1,000 pa.

C232
North West Europe Field Reports
Provide detailed geological, technical and financial information on oil and gas fields in Denmark, Ireland, The Netherlands, Norway, Germany and the UK. Covers fields in production and under development and certain potentially commercial fields for all offshore areas of the countries covered, together with some onshore UK fields.Published annually.
Price: £2,500 pa.

C233
North West Europe Deals Report
A bi-annual publication containing detailed information on all asset/corporate deals implemented in the UK since 1984, both onshore and offshore, and those in the remainder of North West Europe since 1987.
Price: £3,000 pa.

C234
Oil Share Listing
A daily fax service which provides subscribers with a list of 40 energy companies quoted on the London International Stock Exchange, giving their type of listing, the number of issued shares, the share price and market capitalisation. In addition the same information is given for 23 companies listed on the New York Stock Exchange, 10 companies on the Toronto Stock Exchange, 2 companies on the Paris Stock Exchange and one company each on the Madrid and Brussels Stock Exchanges. Finally, the current Brent Blend and WTI crude prices are quoted together with a selection of key Sterling exchange rates.
Price: £1,050 pa for UK client location.

C235
Australasian Upstream Petroleum Database
A subscription service providing data on companies, licences, wells and reserves, together with cash flow analyses of producing fields. The database comprises eight individual volumes, six for Australia and one each for New Zealand and Papua New Guinea. Published annually with over 1,000 pages of information.
Price: £2,500 pa.

C236
Indonesia Upstream Petroleum Database
Comprises six volumes of data on companies, licences, wells and reserves, together with cash flow analyses of producing fields. Produced bi-annually.
Price: £3,500 pa

C237
UK Upstream Petroleum Database
Provides full listings of companies, licences, wells and reserves, together with analyses of fields and companies for both onshore and offshore UK. The database comprises eight bound volumes and two A3 quadrant map books. Produced bi-annually.
Price: £1,500 pa.

C238
NW Europe Upstream Petroleum Database
Provides full listings of companies, licences, wells and reserves, together with analyses of fields and companies for offshore sectors of Norway, The Netherlands, Denmark, Ireland and Germany. Details of the onshore region of Denmark and Ireland are also included. The database comprises one A3 quadrant map book together with six volumes, one for each country plus one for total company interests. Produced bi-annually.
Price: 1,900 pa.

Computer-based Services:

C239
Database Management Service
A new international service due for completion in 1992. The database will provide information on company ownership, licences, wells and fields, which will be easily accessible through a series of menus, query screens and report facilities. The system is being developed in UNIFACE, an advanced fourth generation ; application environment which supports several database management systems including Sybase, Oracle, Ingres, Adabas and Informix. On completion of the first phase of development, the Petroleum Services Group will provide a service to customise the system to clients' specifications.
Price: details available on request.

C240
Financial Analysis Service
An analytical tool used to evaluate the cashflow and net present value of oil and gas fields, and companies. Separate services are produced for the UK, Norway, The Netherlands, Australia and Indonesia. The service is written using the LOTUS 123 spreadsheet package and consists of a suite of computer software models together with a full set of data for a large number of fields in the sector. These include every field in production, under development, or which has had a significant amount of appraisal work. The service is updated bi-annually with the exception of Australasia.
Price: Australasia, UK, Indonesia, Netherlands, Norway - £7,000 pa each for the first year; £4,500 pa each for subsequent years.

Arthur Andersen also produce two further computer-based services:- **Historic Production Figures** and **Licence & Well Co-ordinates.** Details of these services are available on request.

COUNTY NATWEST
WOODMAC

Kintore House
74-77 Queen Street
Edinburgh EH2 4NS
Tel: 031 225 8525
Fax: 031 243 4434
Contact: Paul Gregory

County NatWest WoodMac has developed a series of business publications and consultancy services for the oil and gas industry. The establishment of their North Sea Service in 1973 was followed by the launch of similar services covering the Far East, North West Europe and West Africa. All of the hard copy services, with the exception of North Sea Valuation, use the same Reports and Reference Section format detailed under the entry for the North Sea Service. Subscription rates quoted are for first copies. In most cases there is a 50% discount for additional copies.

C241
North Sea Service
A primary reference source covering the UK upstream oil and gas industry. Comprises two main parts:-
(a) Reports
- monthly reports combine topical research studies with current information on exploration activity, production levels, new developments, asset sales and purchase.
(b) Reference Section
- regularly updated loose-leaf section provides comprehensive coverage of all aspects from licence details to estimates of reserves, production profiles and costs for fields in production, under development or likely to be developed in the near future. Field analyses are supported by computer models that provide cash flow calculations with present values and rates of return.
Price: £2,750 pa.

C242
North West Europe Oil Service
Established in 1986 the service provides an objective view of developments in the European oil and gas industry. Covers Denmark, France, Germany, Ireland, The Netherlands, and Norway.
Price: £2,750 pa.

C243
Far East Oil Services
(SE Asia & Australasia)
Launched in 1984 the Far East Oil Service has recently been restructured to form two smaller, complementary services covering SE Asia and Australasia. SE Asia includes coverage of Brunei, China, Indonesia, Malaysia, Philippines, Thailand, and Vietnam. Australasia covers Australia, New Zealand, and Papua New Guinea.
Price: £3,500 pa for SE Asia; £1,750 pa for Australasia; £4,750 pa special price for subscribers to both services.

C244
West Africa Oil Service:
a complete guide to the upstream oil and gas industry in Cameroon, Equatorial Guinea, Gabon, Congo, Zaire, and Angola.
Price: £2,750 pa.

C245
North Sea Valuation Service:
includes every company which participates in at least one UK offshore field in production, under development, or likely to be developed in the next two years. Updated quarterly. Similar valuation products for NW Europe, the Far East and West Africa are currently under development.
Price: £3,300 pa.

C246
CNWM Computer-based Services:
CNWM Economic Models: County NatWest WoodMac Economic Models and Field and Company Databases are available for all countries covered by the hard copy services. Data is provided for all producing fields, fields under development and those likely to be developed in the near future. Company data files are included for all companies with an interest in the fields supplied. In addition, oil and gas price and exchange rate files contain the latest CNWM forecasts. Also included in the package is the cost and production data used in-house to produce the cash flows shown in the hard copy services.
The models incorporate a customised Lotus 1-2-3 menu structure and are flexible in allowing changes to all key economic variables.
Price: £200 pa (plus VAT) per country.

Please note that the models are only available to subscribers of the companion hard copy information service.

C247
Integrated Mapping System
(& Well/Licence Database):
the IMS has the ability to produce customised maps for all regions covered by the hard copy services. Included in the package is the Well & Licence Database providing details of all offshore wells and licences in the UK. This database may also be purchased separately. The Integrated Mapping System is only available to subscribers to CNWM's existing services.
Price: on application.

DAVY STOCKBROKERS

Davy House
49 Dawson Street
Dublin 2
Ireland
Tel: 6797788
Fax: 712704
Contact: Mary Looney, Information Officer
C248
Davy Stockbroker Updates

Several publications covering the oil and gas exploration sector are produced by the company's analyst. These include regular company updates, **Annual Energy Review**, and **Resource Quarterly.**
Prices on application.

GIROZENTRALE GILBERT ELIOTT

Salisbury House
London Wall
London EC2M 5SB
Tel: 071-628 6782 ext 336
Fax: 071-628 3500
Contact: John Hartley
C249
Girozentrale Gilbert Eliott Publications

The company produces a range of publications on the oil and gas industry, all available on subscription with discount available on multiple copies.
Drilling Weekly:
£550 pa and its companion publication
Production Drilling Monthly, available only to subscribers to Drilling Weekly.
International Explorer: published monthly, *price £500 pa.*
North Sea Notes: published monthly, *price £550 pa.*
Offshore Results Service and **Onshore Results Service**: two weekly publications covering NW Europe, each available at a subscription *price of £300 pa.*
Onshore Weekly: *price £450 pa.*

ALPHABETICAL INDEX

PRODUCER INDEX

A

Adams Engineering Inc - C074

Alberta Oil Sands Technology and Research
C001, C028

American Geological Institute
C024, C217, C224

American Meteorological Society - C034

American Petroleum Institute
C002, C003, C048, C067, C068, C101,
C138, C139, C140, C156

Arctic Institute of North America - C005

Arthur Andersen - C228–C240

Asian Institute of Technology -C025

Atlantic Information Services Inc - C086

Australian Energy Research, Development
& Demonstration Projects Database
C098

Australian Mineral Foundation - C006

Axses Information Systems; Environment Canada
C069, C200

B

BHRA - C017

Bio-Rad Laboratories - C064

BMT Cortec Ltd - C007

Bowker - C015, C210

British Geological Survey - C070, C071

Bureau des Recherches Geologiques et Minieres
C011, C076

Business International Corporation - C109

C

Cambridge Scientific Abstracts
C004, C039, C198

Canadian Petroleum Association (CPA) - C142

Capitol Publications Inc - C131

CD PubCo Inc
C212, C213, C214, C215

Center for Short-Lived Phenomena - C056

Centre National de la Recherche Scientifique
C223, C018

Chemical Intelligence Services - C105

Chemical Market Associates Inc - C104

Cold Region Research and Engineering Laboratory
of the US Army Corps of Engineers - C008

Computer Petroleum Corporation - C190

Cutter Information Group - C057

D

Dataline Inc - C097

Datastream - C171

Datatimes Corporation - C108

Davy Stockbrokers - C248

Department of Primary Industries and Energy
C098, C141

Derwent Publications Ltd - C048

Det Norske Veritas - C060

DeWitt and Company - C172

Directorate-General for Energy (DGXVII),
Commission of the European Communities
C099

Dow Jones & Company Inc - C169, C196

DRI/McGraw-Hill
C147, C174, C120, C134, C143, C144,
C145, C146, C148, C149, C150, C151,
C155,C173, C175, C211,

Dwight's Energydata Inc - C075

E

Engineering Information Inc
C009, C202, C205, C206, C207

Environment Canada, National Environmental
Emergency Centre - C053

ERICO Petroleum Information Ltd - C226

European Association for Grey Literature
Exploitation (EAGLE) - C044

European Chemical News - C177

F

Fairbase Database Ltd - C114

Federal Institute for Geosciences and Natural
Resources - C032

Financial Times Business Information - C115

O

Oceandril Inc - C087

Oceanroutes Inc - C054

OCLC Online Computer Library Center
C204, C209

Offshore Data Services - C085

Oil & Gas Journal - C125, C136, C161

Online Resource Exchange Inc - C089

P

Parent Information Enterprises - C179

Pasha Publications - C066, C112, C117

PennWell Publishing Company - C087

Petroconsultants (CES) Ltd - C219

Petroflash Inc - C184

Petroleum Abstracts at the University of Tulsa
C042

Petroleum Argus Ltd - C185, C186, C187

Petroleum Information Corporation
C065, C073, C078, C081, C090,
C091, C092, C096

Petroleum Intelligence Weekly - C188

Predicasts - C127

Q

Quick Source Inc - C208

R

Reuters - C165, C193, C194, C195

Rock Mechanics Information Service - C023

Rocky Mountain Mineral Law Foundation
C077

Royal Norwegian Council for Scientific and
Industrial Research (NTNF)- C099

Royal Society of Chemistry - C103

S

Silverplatter - C201, C222

Society of Exploration Geophysicists - C093

Society of Petroleum Engineers - C094South
Australian Department of Mines and
Energy (SADME) - C045

Southam Business Information and
Communications Group - C107, C110

T

Tecnon (UK) Ltd - C128, C129, C176

Telerate; Associated Press - C196

Texas A & M University Library - C014

Tsintichimneftemash - C029

U

United Communications Group - C191

Universite des Sciences Sociales de Grenoble
018

US Coast Guard - C049, C201, C225

US Department of Energy
C013, C016, C043, C061, C156, C157,
C159, C162, C163, C164, C167, C189

US Department of Transportation - C047, C225

US Dept of Health and Human Services,
Centers for Disease Control - C201

US Environmental Protection Agency
C055, C201, C225

US Geological Survey - C218, C220

US National Oceanic and Atmospheric
Administration - C004

V

Vestlash Group Inc - C111

W

Warren Publishing Inc - C083

The WEFA Group - C153, C166, C168

H W Wilson Company (Wilsondisc) - C197

Woodside Research Ltd - C133

World Information Systems - C050

Worldwide Videotex - C132

New titles and classics

for petroleum geologists

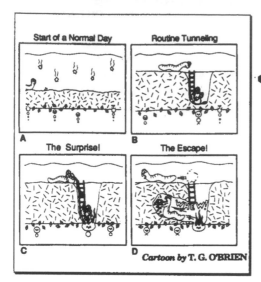

Start of a Normal Day | Routine Tunneling

A | B

The Surprise! | The Escape!

C | D

Cartoon by T. G. O'BRIEN

J.H. Barwis, J.G. McPherson, J.R.J. Studlick (Eds.)
Sandstone Petroleum Reservoirs

1990. XV, 583 pp. 448 figs. in 683 parts including 48 color
plates. (Casebooks in Earth Sciences) Hardcover £ 64.50
ISBN 3-540-97217-X

R.H. Bennett, W.R. Bryant, M.H. Hulbert (Eds.)
Microstructure of Fine-Grained Sediments
From Mud to Shale

1991. XXII, 582 pp. 458 figs. in 708 parts (Frontiers in
Sedimentary Geology) Hardcover £ 70.50
ISBN 3-540-97339-7

K.O. Bjørlykke
Sedimentology and Petroleum Geology

1989. XII, 363 pp. 186 figs. Softcover £ 28.00
ISBN 3-540-17691-8

G. Einsele, W. Ricken, A. Seilacher (Eds.)
Cycles and Events in Stratigraphy
1991. XIX, 955 pp. 393 figs. 11 tabs. Hardcover £ 45.50
ISBN 3-540-52784-2

H. Ibbeken, R. Schleyer
Source and Sediment
**A Case Study of Provenance and Mass Balance at an
Active Plate Margin (Calabria, Southern Italy)**

1991. X, 286 pp. 161 figs. Hardcover £ 81.50
ISBN 3-540-53282-X

Springer-Verlag
Berlin
Heidelberg
New York
London
Paris
Tokyo
Hong Kong
Barcelona
Budapest

8 Alexandra Rd., London SW 19 7JZ, England

dtp 30352/A/1

Section D

DIRECTORY

Addresses relevant to Petroleum and Marine Technology

Tony and Sybil Richardson

INTRODUCTION

This Directory aims to assist its users by identifying "addresses relevant to petroleum and marine technology."

This remit has been interpreted widely and the following pages include: government departments, agencies and laboratories concerned with research, development and standards; professional and trade associations and societies; publishers and bureaux; database producers and vendors; and conference organisers. As well as the concerns of administration their interests (like yours) range from the normal activities of petroleum, chemical, mechanical, electrical or civil engineering to more specialised areas such as marine biology, seismology, firefighting or oil prices.

Every effort has been made to locate the publishers and others mentioned in Section A; naturally the organisations covered in Sections B and C are included in the Directory (marked ¶ and § respectively).

In response to reactions from users, this volume reverts to normal alphabetical order where *of* and *for* appear in the names of organisations, and if you can't find *University of X*, try *X University*.

Entries generally appear under an organisation's full name but, to help users, a list of commonly-used acronyms is provided separately. One of our aims has been to minimise the duplication of addresses (which will streamline your amendment and annotation of information) and consequently there are many cross-references. A summary listing of these, and the list of acronyms, appear immediately before the main alphabetical entries.

We have done all we can to make this Guide comprehensive, accurate and helpful so that you can locate the information you need, from whatever source, as easily as possible. If we can do more, please let us know.

Sybil Richardson
(signing also for Tony, who died during the preparation of this book and in whose memory it has been completed)

Lancaster, August 1992

TELEPHONE & FAX NUMBERS

(246)   indicates a regional code only:
precede with appropriate country code
       USA and Canada = +1
       UK = +44

+   indicates that the country code follows:
replace "+" with your own country's
international telephone exchange number

Note   the first digit, 0, in a UK regional
code is needed within the UK only:
If you are calling from outside UK
you should NOT include it.

SECTION D CONTENTS

Familiar Acronyms

for organisations listed in this Directory

A

AAAF	Assn Aeronautique & Astronautique de France
AAPG	American Association of Petroleum Geologists
ABOI	Assn of British Offshore Industries
ABS	American Bureau of Shipping
ACE	Association of Consulting Engineers
ACOPS	Advisory Committee on Pollution
ADC	Association of Diving Contractors
AEMS	Ass Amicale des Anciens Eleves des Ecole de Maitre-Sondeurs
AERE	Atomic Energy Research Establishment
AFIP	Federation Francaise des Petroliers Independants
AGCAS	Assn of Graduate Careers Advisory Serv
AGI	American Geological Institute
AMF	Australian Mineral Foundation
AMTE	now Admiralty Research Establishments
ANSI	American National Standards Institute
AODC	Assn of Offshore Diving Contractors
AOSTRA	Alberta Oil Sands Technology & Research Authority
AOTC	Arctic Offshore Technology Conference
ARE	Admiralty Research Establishment
ASCOPE	Asean Council for Petroleum
ASLIB	Assn for Information Management
ASTM	American Society for Testing Materials
AWES	Asscn of West European Shipbuilders

B

BETC	Crude Oil Analysis Data Bank - National Institute for Petroleum & Energy Research
BGS	British Geological Survey
BRGM	Bureau des Recherches Geologiques et Minieres

BRINDEX	Assn of British Independent Oil Exploration Companies
BSRA	now British Maritime Technology

C

C-CORE	Centre for Cold Ocean Resources Engineering
CEC	Commission of the European Communities
CEDIGAZ	International Information Centre for Natural Gas & Hydrocarbon Gases
CEDOCAR	Centre de Documentation de l'Armement
CEFRACOR	Centre Francais de la Corrosion
CEPM	Comite d'Etudes Petrolieres Marines
CIRIA	Construction Industry Res & Inf Assn
CIS	Chemical Information Service
CISTI	Canada Institute for Scientific & Technical Information
CMCI	Agence Regionale de l'Energie
CNEXO	Centre National pour 1'Exploitation des Oceans : IFREMER
CNRS	Centre National de la Recherche Scientifique
CNRS/CDSH	Centre de Doc Sciences Humaines
CONSTRADO	Constructional Steel Res & Dev Organisation
CPA	Canadian Petroleum Association
CPPI	Canadian Petroleum Products Institute

D

DEMINEX	Deutsche Erdolversorgungs ges'ft mbH
DGEMP	Direction Generale de l'Energies & Matieres Premieres
DONG	Danish Oil & Natural Gas Corp
DSCR	Decompression Sickness Central Registry
DTO	Dansk Teknisk Oplysningstjeneste

E

E & P Forum   Oil Industry International
Exploration & Production Forum

EAEG   European Association of Exploration
Geophysicists

EAGLE   European Association for Grey Literature
Exploitation

ECOR   Engin'ng Committee on Oceanic Research

EDTC   European Diving Technology Committtee

EEZ   Exclusive Economic Zone

EFIC   European Fuel Information Centre

EIC   Environment Information Centre

EMIS   Electronic Markets & Information Systems

ENI   National Hydrocarbons Authority

ENIEPSA   Empressa Nacional de Investigacion y
Exploracion de Petroleo SA

EUBS   European Underwater Biological Society

F

FAO   Food & Agriculture Organisation

FEDEX   National Energy Information Centre

FEPEM   Federation of European Petroleum & Gas
Equipment Mfters

FFPI   Fed Francaise des Petroliers Independants

G

GERMINAL   c/o Bureau de Recherches
Geologiques et Minieres

GESAMP   Group of Experts on the Scientific Aspects
of Marine Pollution FAO

GSA   Geological Society of America

I

IABSE   Int'l Assn for Bridge & Structural Eng'g

IADC   Int'l Assn of Drilling Contractors

IAEA   Int'l Atomic Energy Agency

IAHR   Norwegian Institute of Technology

IAWPRC   UK Committee - c/o Marine Biological Assn

ICE   Institution of Civil Engineers

IChemE   Institution of Chemical Engineers

ICIS   Independent Chemical Inform'n Services

ICMRD   International Center for Marine Resource
Development : University of Rhode Island

ICSTI   International Centre for Scientific &
Technical Information

IEA   International Energy Agency

IEE   Institution of Electrical Engineers

IEEE   Institution of Electrical & Electronic Eng'rs

IFEG   Information for Energy Group :
c/o Institute of Petroleum

IFIP   International Federation for Information
Processing

IFP   Institut Francais du Petrole

IHRDC   International Human Resources
Development Corp

IIS Inform   Inst of Information Scientists

ILO   International Labour Office

IMER   Inst for Marine Environmental Research

IMO   Int'l Maritime Organization

INIST   Institute de l'Information Scientifique et
Technique : Centre National de la
Recherche Scientifique

INPP   Inst National de Plongee Professionnelle &
d'Intervention Milieu Aquatiques

IOS   Institute Of Oceanographic Sciences

IPE   International Petroleum Exchange of London

IPIECA   Intl Petroleum Industry Environmental
Conservatiom Assn

IRO   Netherlands Industrial Council for
Oceanology

ISI   Institute for Scientific Information

ITOPF   International Tanker Owners Pollution
Federation

L

LPGITA   Liquefied Petroleum Gas Ind Tech Assn

M

MARIN   Maritime Research Institute Netherlands

MaTSU   Marine Technology Support Unit

MIAS   Marine Information Advisory Service : Institute
of Oceanographic Sciences

MIT   Massachusetts Institute of Technology

MOLARS   Meteorological Office Library Accession
System : Meteorological Office

MRC   Medical Research Council

MWV   Mineralolwirtschaftsverband ev

N

NACE   Natl Assocn of Corrosion Engineers

NATES   Natl Environmental Emergency Centre :
Environment Canada

NAVSEA   Naval Sea Systems Command :
US Dept of the Navy

NEDO	National Economic Development Office
NEIC	National Earthquake Information Center : US Geological Survey
NERC	Natural Environment Research Council
NESDA	North East Scotland Devel Authority
NHL	Norwegian Hydrodynamic Laboratories
NIVA	Norwegian Inst for Water Research
NNI	Netherlands Standards Institute
NOAA	US National Oceanic & Atmospheric Admin
NOIA	National Ocean Industries Assn
NPD	Norwegian Petroleum Directorate
NPF	Norwegian Petroleum Society
NSI	Norsk Senter for Informatik A/s
NTH	Norwegian Institue of Technology
NTIS	National Technical Information Service : US Dept of Commerce
NTNF	Royal Norwegian Council for Science & Technology Research
NUTEC	Norwegian Underwater Technology Centre

O

OAPEC	Organization of Arab Petroleum Exporting Countries
OCSEAP	Outer Continental Shelf Environmental Assessment Pogram : US National Oceanic & Atmospheric Administration
OCSIP	Outer Continental Shelf Information Prog
OECD	Organisation for Economic Coop & Devel
OPEC	Organization of the Petroleum Exporting Countries
OPITB	Offshore Petroleum Industry Training Board
OXERA	Oxford Economic Research Associates

P

PADI	Professional Assocn of Diving Instructors
PETANS	Petroleum Training Assn North Sea
PETEX	Petroleum Extension Service : University of Texas
PFDS	Pergamon Financial Data Services

R

RGIT	now the Robert Gordon University
RINA	Royal Institution of Naval Architects
RVS	Research Vessel Service (NERC)

S

SADME	South Australian Dept of Mines & Energy
SBM	Single Buoy Moorings Inc
SCOTA	Scottish Offshore Training Assn
SEG	Society of Exploration Geophysicists
SERC	Science & Engineering Research Council
SNEIL	Secretariat for the Nordic Energy Libraries
SPE	Society of Petroleum Engineers
SPWLA	Society of Professional Well Log Analysts
STF	State Pollution Control Authority
SUT	Society for Underwater Technology

T

Tapir	Technical University of Norway
TNO	Netherlands Organisation for Applied Scientific Research
TPAO	Turkieye Petrolleri Anonim Ortakligi

U

UEG	Underwater Engineering Group : now MTD Ltd
UFG	Union Francaise du Geologues
UKAEA	Atomic Energy Research Establishment
UKOOA	UK Offshore Operators Ass
UMIST	University of Manchester Istitute of Science & Technology

V

VTT	Technical Research Centre of Finland

W

WEGEMT	West European Graduate Education in Marine Technology : MTD Ltd

Cross References

For information on these organisations please
refer to the entry indicated

A

Academic Press - Harcourt Brace Jovanovich

Arctic Biological Station - Environment Canada

Arctic Environmental Information & Data Centre
- University of Alaska

Assn Europeenee des Gaz de Petroles Liquefies
- European Liquefied Petroleum Gas Assn

Association of Earth Science Editors
- c/o American Geological Institute

B

BETC Crude Oil Analysis Data Bank - National
Institute for Petroleum & Energy Research

Bidston Observatory
- Institute of Oceanographic Sciences

Blaise-Line
- British Library

British Nuclear Fuels
- BNF Metals Technology Centre

British Ship Research Association
- British Maritime Technology

C

Canadian Nationional Energy Board
- Environment Canada

Centre National pour l'Exploitation des Oceans
- IFREMER

City of Birmingham Polytechnic
- University of Central England in Birmingham

Cold Regions Research & Engineering Laboratory
- US Army Corps of Engineers

Continental Shelf Institute - IKU

Current Awareness Bulletin
- Assn for Information Management

D

Deep Sea Research & Oceanographic Science, A
- Pergamon Press

Deep Sea Research & Oceanographic Science, B
- Wood's Hole Oceanographic Institute

Diving Inspectorate - Department of Energy

E

Earthquake Engineering Research Center
- University of California Berkeley Campus

Electrical Research Assn - ERA Technology Ltd

Engineering Sciences Data Unit
- ESDU International Ltd

Ente Nazionale Idrocarburi - National Hydrocarbons
Authority

Enviro-Marine Club - c/o Marinetech North West

Environment & Natural Resources Institute
- University of Alaska

Environmental Protection Service
- Canadian Dept of the Environment

Estuarial & Coastal Marine Science
- Institute of Oceanographic Sciences

European Diving Technology Committee
- Commission of the European Communities

F

Federal Interagency Cttee on the Health & Safety
Effects of Energy Technology
- US Dept of Commerce, NTIS

Fluid Engineering Centre - BHR Group Ltd

E & P Forum - Oil Industry International Exploration
& Production Forum

French Assn of Independent Petroleum Companies -
Federation Francaise des Petroliers
Independants

G

General Directorate of Environmental Protection
- Department of the Environment

Geophysics Directorate - British Geological Survey

Grafton Books / Granada Publishing
- HarperCollins Publishers

Group of Experts on the Scientific Aspects of Marine
Pollution - GESAMP :
Food & Agriculture Organisation

H

Heinemann - Reed International Books

Her Majesty's Stationery Office - HMSO

Hydrographic Society - c/o NELondon Polytechnic

I

I AWPRC, UK Committee
- c/o Marine Biological Assn

Industriele Raad voor de On- en Offshore
- Nethelrands Industrial Council for
Oceanology

Infoil II - MTD/Norwegian Petroleum Directorate

Institute de l'Information Scientifique et Technique
- Centre National de la Recherche
Scientifique

Institut Francaise de Recherche pour l'Exploitation
de la Mer - IFREMER

Institute of Metals - Institute of Materials

Institute of Sound & Vibration Research
- University of Southampton

Institution of Metallurgists - Institute of Materials

International Center for Marine Resource
Development - University of Rhode Island

L

Law of the Sea Institute - University of Rhode Island

Lloyd's List - Lloyd's of London Press Ltd

London Marine Technology Centre
- Imperial College

M

Marine Directorate - Department of Transport

Marine Information Advisory Service
- Inst of Oceanographic Sciences

Marine Sciences & Information Directorate
- Canadian Dept of Fisheries & Environment

Marine Surveys Directorate
- British Geological Survey

Marine Technology Centre, Newcastle
- Univ of Newcastle upon Tyne

Marine Technology Directorate - MTD Ltd

Mechanical Engineering Publications Ltd - MEP Ltd

Metals Soceity - Institute of Materials

Middlesex Polytechnic - now Middlesex University

N

Nansen Institute - Fridtjof Nansen Institute

National Centre of Scientific & Technological
Information - Ministry of Energy &
Infrastructure

National Geophysical Data Center
- US Natl Oceanic & Atmospheric Admin

National Maritime Institute
- now British Maritime Technology

National Meteorological Library
- Meteorological Office

National Technology Information Service (NTIS)
- US Dept of Commerce

Naval Sea Systems Command
- US Dept of the Navy

Nederlands Normalisatie Instituut
- Netherlands Standards Institute

Newcastle upon Tyne Polytechnic
University of Northumbria in Newcastle

Norges Geologiske Undersokelse
- Geological Survey of Norway

Norges Geotekniske Institutt
- Norwegian Geotechnical Inst

Norges Skippsforskningsinstitutt
- Norwegian Ship Research Inst

Norges Standardiseringsforbund
- Norwegian Standards Assn

Norges Teknisk Naturvitenskapelige Forskningsgrad
- Royal Norwegian Council for Scientific &
Industrial Res

Norsk Petroleumsforening
- Norwegian Petroleum Society

Norske Olje Revy
- Norwegian Information Publishers AS

Norske Tekniske Hogskole
- Norwegian Institute of Technology

Norwegian Oil Review
- Norwegian Information Publishers AS

O

Ocean Engineering - Pergamon Press

Office of Minerals Management Information
- US Department of the Interior

Office Of Oil Imports - US Department of Energy

Offshore Energy Technology Board
- Offshore Supplies Office

Offshore Minerals Group - Warren Spring Laboratory

Oil & Gas Consultants
- c/o Society of Petroleum Engineers

Oil Division - Department of Energy

Olje og Energidepartementet
- see Royal Ministry of Oil & Energy

Oljedirektoratet - Norwegian Petroleum Directorate

Outer Continental Shelf Oil & Gas Information Prog -
US National Oceanic & Atmospheric
Administration

P

Peches et Oceans - Institut Maurice-Lamontagne

Pentech Press - John Wiley & Sons

Petroleum Abstracts - University of Tulsa

Petroleum Directorate - Canada Newfoundland
Offshore Petroleum Board

Petroleum Equipment Institute - Gulf Publishing

Petroleum Extension Service
- University Of Texas at Austin

Pipelines Inspectorate - Dept of Energy

Plymouth Polytechnic - University of Plymouth

Plastics & Rubber Institute - Institute of Materials
(from January 1993)

Polar Continental Shelf Project - Canadian Dept of
Energy, Mines & Resources

Q

Queen Mary College - University of London

R

Royal School of Mines - Imperial College

Rubber & Plastics Research Association
- now RAPRA Technology Ltd

Routledge - International Thomson

S

Sea Fisheries Inspectorate
- Ministry of Agriculture, Fisheries & Food

Shipbuilding & Electrical Engineering Division - Dept
of Trade & industry

Skibsteknisk Laboratorium
- Danish Ship Research Labs

Societe National Elf-Aquitatine - Elf-Aquitaine

Spill Control Newsletter - Environment Canada

Statens Dykkerskole
- Norwegian State Diving School

T

Technische Hogeschool Eindhoven
- Einhoven University of Technology

Technische Hogeschool te Delft
- Delft University of Technology

Technische Hogeschool Twente
- Twente University of Technology

Technology Reports Centre
- British Library Document Supply Centre

U

UEG (Underwater Engineering Group)
- now MTD LTd

Ufficio Nazionale Minerario pergl
- Natl Mining Office for Hydrocarbons

UK Online User Group
- Institute of Information Scientists

UN Ocean Economics & Technology Branch
- publications: Kluwer / Graham & Trotman

Underwater Engineering Group
- now part of MTD Ltd

Underwater Technology
- Society for Underwater Technology

V

Verband Fur Schiffbau & Meerestechnik Ev
- German Shipbuilding & Ocean Industries
Association

Vereniging van de Nederlandse Aardolie Industrie
- Assn of the Dutch Petroleum Ind

W

Welding Institute - TWI

European Petroleum Year Book
Annuaire Européen du Pétrole
Jahrbuch der Europäischen Erdölindustrie

ANEP

Handbook and Reference Work for the Oil and Natural Gas Industries.
400 pages of comprehensive information and documentation.

Oil and Gas Fields in Western Europe. North Sea Oil and Natural Gas. Crude Oil, Product and Natural Gas Pipelines in Western Europe. West European Refinery Survey. Oil and Natural Gas in the East European Countries. Europe's Oil and Natural Gas Statistics in a Worldwide Comparison: Primary Energy Consumption, Petroleum Reserves, Crude Oil Production, Petroleum Cunsumption, Crude Oil Imports, North Sea Oil within the West European Oil Supply, Natural Gas Reserves, Natural Gas Production, Natural Gas Consumption, North Sea Gas within the West European Natural Gas Supply, Import Dependence of Western Europe's Primary Energy Supply.

Country-by-Country Oil Statistics (Austria, Belgium, Switzerland, Germany, Denmark, Spain, France, Great Britain, Greece, Italy, Luxemburg, Norway, Netherlands, Portugal, Sweden, Finland, Eastern Europe): Primary Energy Consumption, Crude Oil Production, Natural Gas Production, Imports of Crude Oil, Imports of Petroleum Products, Imports of Natural Gas, Refinery Output, Consumption of Petroleum Products, Consumption of Natural Gas, Exports of Petroleum Products.

Full Information About More than 2000 Companies in All European Countries
Companies engaged in oil and gas exploration, refining, marketing, trading and distribution, storage and transportation. Associations, government offices, institutes, etc.

Directory of Suppliers and Buyers' Guide
An offer of the European supply and service industry with about 600 products and services.

**URBAN-VERLAG
Hamburg/Wien GmbH**

**P.O.Box 70 16 06
D-2000 Hamburg 70
Germany**

**Phone +49-40-656 70 71
Fax +49-40-656 70 75**

ANEP 1993, 26th edition,
400 pages, size A4 (210 x 297 mm),
paperback. Price DM 168,
postage and packing included.
Additional air mail charges:
Europe DM 8, Americas DM 20,
Australasia DM 26.

DIRECTORY

There is further information about organisations marked "¶" in Section B
Those marked "§" are producers, suppliers or hosts for databases included in Section C

A

AAA Technology &
Specialities Company Inc
3000 Rodgerdale Road
Houston TX 77042-4189
USA
Tel:(713) 789 6200
Tlx: 910 881 2424

Aangstrom Precision
Corporation
5805E Pickard
One Energy Place, Suite 160
Mount Pleasant MI 48858
USA
Tel:(517) 772 2232

AB Svenska Shell
171 78 Solna
Sweden

ABS Europe Ltd
- see American Bureau of
Shipping

AS Veritas Research (see
also Det Norske Veritas)
Veritasveien 1
1322 Hovik
Norway
Tel:+41 2 477500
Tlx:76192 verit n
Fax:+41 2 479871

ABC - Clio
2040 Alameda Padre Serra
PO Box 4297
Riviera Campus
Santa Barbara CA 93103
USA
Tel:(805) 963 4221

Aberdeen City Libraries
Business & Technical Dept
Rosemount Viaduct
Aberdeen AB9 1GU
Scotland   ¶
Tel:(0224) 634622
Tlx:739196 ABD 019
Fax:(0224) 641985

Aberdeen Conference Centre
Foveran House
Newburgh
Aberdeen AB4 0AP
Scotland
Tel:(03586) 398

Aberdeen District Council
St Nicholas House
Broad Street
Aberdeen AB9 1DE
Scotland
Tel:(0224) 642121
Tlx: 73360

Aberdeen Harbour Board
16 Regent Quay
Aberdeen AB9 1SS
Scotland

Aberdeen Petroleum
Publications Ltd
35-37 Huntly Street
Aberdeen AB1 1TJ
Scotland
Tel:(0224) 644725
Tlx: 73663

Aberdeen University
Old Aberdeen
Aberdeen AB9 2UF
Scotland
Tel:(0224) 40241
Tlx: 837330

Aberdeen University Press
Farmer's Hall
Aberdeen AB9 2XT
Scotland
Tel:(0224) 630724
Tlx: 739477
Fax:(0224) 643286

Academic Press
- see Harcourt Brace
Jovanovich Ltd

ACI Computer Services
PO Box 42
310 Ferntree Gully Road
Clayton North
Victoria 3168
Australia
Tel:(03) 544 8433
Tlx: 33852

ACS
23272 Del Lago Drive
Laguna Hills CA 92653
USA
Tel:(714) 472 8186

Action Instrument Inc
8601 Aero Drive
San Diego CA 92123
USA
Tel:(619) 279 5726
Tlx: 910 335 2030

Adage Inc
1 Fortune Drive
Billerica MA 01821
USA
Tel:(617) 667 7070
Tlx: 710 347 1594

Adam Hilger
Techno House
Redcliffe Way
Bristol BS1 6NX
UK
Tel: (0272) 297481
Tlx: 449149
Fax:(0272) 294318

Adams Engineering Inc
8484 Breen
Houston TX 77064
USA   §
Tel:(713) 937 8320
Fax:(713) 937 6503

Adaptive Systems Inc
8807 Brummel Drive
Houston TX 77099
USA
Tel:(713) 530 9403

Administration des Mines Service Geologique de Belgique
Ministere des Affaires Economisques
rue J A de Mot 30
1040 Brussels
Belgium
Tel: +32 2 647 64 00

Admiralty Research Establishment
Ministry of Defence
ARE (Dunfermline)
St Leonard's Hill
Dunfermline, Fife KY11 5PW
Scotland
Tel:(0383) 721346
Tlx: 72363 NCRE DF
and
ARE (Haslar)
Haslar
Gosport
Hampshire PO12 2AG
UK
Tel:(0705) 22351
Extension 41400
and
ARE (Holton Heath)
Holton Heath
Poole
Dorset BH16 6JU
UK
Tel:(0202) 22711
and
ARE (Portland)
Southwell
Portland
Dorset DT5 2JS
UK
Tel:(0305) 820381
and
ARE (Portsdown)
Portsdown Cosham
Portsmouth
Hampshire PO6 4AA
UK
Tel:(0705) 379411

Advisory Committee on Pollution of the Sea
57 Duke Street
Grosvenor Square
London W1N 5DH
UK
Tel:(071) 499 0704
Fax:(071) 493 3092

Agence Regionale de l'Energie
rue Henri-Barbusse
13241 Marseilles Cedex 1
France
Tel:+33 91 91 36 7

AGIP SpA
Box 12069 GESA
20120 Milan
Italy
Tel:+39 2 5201
Tlx: 310246 ENI

AGS Management Systems Inc
880 First Avenue
King of Prussia
PA 19406
USA
Tel:(215) 265 1550
Tlx: 510 660 3320

AIMM
- see Cooperative Group Marine Science & Marine Technology

Aker Engineering A/S
Tjuvholmen
0250 Oslo 2
Norway

Alaska Division of Oil & Gas
Pouch 7034
Anchorage AK 99510
USA
Tel:(907) 762 4277

Alaska Oil & Gas Association
505 W Northern Lights Blvd
Suite 219
Anchorage AK 99503
USA
Tel:(907) 272 1481

Alaska State Library
PO Box G
Juneau AK 99811
USA
Tel:(907) 465 2920

Alaska Update
c/o Ray Piper Co Inc

Alba International Ltd
Leading Light Building
Sinclair Road
Torry
Aberdeen AB1 3PR
Scotland
Tel:(0224) 878188
Tlx:73337
Fax:(0224) 879781

Alberta Oil Sands Technology & Research Authority
600 Highfield Place
10010-106 Street, 6th Floor
Edmonton
Alberta T5J 3L8
Canada   ¶§
Tel:(403) 427 8382
Tlx: 037-3519
Fax: (403) 422 9112

Alena Enterprises
PO Box 1779
Cornwall
Ontario K6H 5V7
Canada

Alexanders Laing & Cruikshank
7 Copthall Avenue
London EC2R 7BE
UK
Tel:(071) 588 2800
Tlx:888397
Fax:(071) 256 9545

Allegro Technology Corp
618 Baronne street
Box 2211
New Orleans LA 70176
USA
Tel:(504) 524 8505

Ambersoft Oilfield Applications
188 Warwick Road
Basingstoke
Hants RG23 8EB
UK
Tel:(0256) 4638

American Association for the Advancement of Science
1333 H Street NW
Washington DC 20005
USA
Tel:(202) 326 6400

American Association of Engineering Societies
345 East 47th Street
New York NY 10017
USA
Tel:(212) 705 7338

American Association of Petroleum Geologists
1444 S Boulder Ave
PO Box 979
Tulsa OK 74101-0979
USA   ¶
Tel:(918) 584 2555
Fax:(918) 584 0469

American Bureau of Shipping
45 Eisenhower Drive
Paramus, NJ 07653-0528
USA
Tlx: ITT 421966
Fax:(201) 368 0255
and
ABS Europe Ltd
1 Frying Pan Alley
London E1 7HR
UK   ¶
Tel:(071) 247 3255
Tlx:88562.8811205
Fax:(071) 377 2453

American Chemical Society
1155 16th Street NW
Washington DC 20036
USA
Tel:(202) 872 4600

American Concrete Institute
22400 W Seven Mile Road
PO Box 19150
Redford Station Detroit
Detroit MI 48219
USA
Tel:(313) 532 2600

American Enterprise Inst for Public Policy Research
1150 NW 17th Street
Washington DC 20036
USA
Tel:(202) 862 5800

American Gas Association
Industry Information
Services
1515 Wilson Boulevard
Arlington VA 22209
USA   ¶§
Tel:(703) 841 8400
Fax:(703) 841 8406

American Geological Institute
GeoRef Information System
4220 King Street
Alexandria VA 22302
USA   §
Tel:(703) 379 2480
Fax:(703) 379 7563

American Geophysical Union
200 Florida Avenue
Washington DC 20009
USA
Tel:(202) 462 6903

American Independent Refiners Association
114 3rd Street, SE
Washington DC 20003
USA
Tel:(202) 543 8811

American Institute of Aeronautics & Astronautics
Technical Information
Service
1290 Avenue of the
Americas
New York NY 10019
USA
Tel:(212) 581 4300

American Institute of Chemical Engineers
345 East 47th Street
New York NY 10017
USA
Tel:(212) 705 7338

American Institute of Professional Geologists
7828 Vance Drive
Suite 103
Arvada CO 80003
USA
Tel:(303) 431 0831

American Journal of Science
Kline Geology Laboratory
Yale University
Box 6666
New Haven CT 06511
USA
Tel:(203) 436 3827

American Management Systems Inc
Suite 950
9800 Center Parkway
Houston TX 77036
USA
Tel:(713) 270 9095

American Meteorological Society
45 Beacon Street
Boston MA 02108
USA   §
Tel:(617) 227 2425

American National Standards Institute
1430 Broadway
New York NY 10018
USA

American Petroleum Institute
1220 L Street NW
Washington DC 20005
USA
Tel:(202) 682 8128
Fax:(202) 682 8537
Finance, Accounting & Statistics Department
Tel: (202) 682 8505   §
Fax: (202) 682 8036
and
Central Abstracting/Index
Services
275 7th Avenue, 9th Floor
New York NY 10001
USA   §
Tel: (212) 366 4040
Fax: (212) 366 4298
and
Production Division
25-35 1 Main Place
Dallas TX 75202-3904
USA
Tel:(214) 748 3841

American Society for Non-Destructive Testing
4153 Arlington Plaza
Caller # 28518
Columbus OH 43228
USA

American Society for Photogrammetry
210 Little Falls Street
Falls Church VA 22046
4398
USA
Tel:(203) 534 6617

American Society for Testing & Materials
1916 Race Street
Philadelphia PA 19103
USA   ¶
Tel:(215) 299 5400
Fax:(215) 977 9679

American Society of Civil Engineers
United Engineering Center
345 East 47th Street
New York NY 10017
USA   ¶
Tel:(212) 705 7496
Tlx: 422847 ASCE UI
Fax:(212) 705 7712

American Society of Lubrication Engineers
833 Busse Highway
Park Ridge IL 60068
USA
Tel:(312) 825 5536

American Society of Mechanical Engineers
United Engineering Center
345 East 47th Street
New York NY 10017
USA
Tel:(212) 705 7722
Fax:(212) 705 7674

American Society of Petroleum Operations Engineers
PO Box 956
Richmond VA 23207
USA
Tel:(804)271 0794

American Technology Laboratories
5420 Old Orchard Road
Skokie IL 60077
USA

American Welding Society
2501 NW Seventh Street
Miami FL 33125
USA
Tel:(305) 642 7090

Ametek Inc
Offshore Research & Engineering Division
41 Aero Camino (Goleta CA)
Box 6447
Santa Barbara CA 93160
USA
Tel:(805) 683 2151
Tlx: 65 8483

AMS Press Inc
56 East 13th Street
New York NY 10003
USA
Tel:(212) 777 4700

Analytical & Computational Research Inc
3106 Inglewood Boulevard
Los Angeles CA 90066
USA
Tel:(213) 398 0956

Anchor Management Systems Inc
Box 19247
Oklahoma City OK 73144-0247
USA
Tel:(405) 947 1458

AnchorPress / Doubleday
501 Franklin Avenue
Garden City NY 11530
USA
Tel:(516) 294 4561

Ann Arbor Science (Publishers) Inc
230 Collingwood Avenue
Ann Arbor MI 48106
USA
Tel:(313) 761 5010

ANSLICS
Robert Gordon's Institute of Technology Library
School Hill
Aberdeen AB9 1FR
Scotland
Tel:(0224) 633611

Applied Automation Inc
Box 9999
Bartlesville OK 74005
USA
Tel:(918) 661 6141
Tlx: 49 2455

AP/DJ Telerate
Interfinet UK Ltd
12-15 Fetter Lane
London EC4A 1BR
UK
Tel:(01) 583 1144
Tlx: 943117

Arab Petroleum Research Centre
7 Avenue Ingres
75016 Paris
France
Tel:+33 1 45 243310
Tlx: 613497 OIL

Arbeitsgemeinschaft Erdol-Gewinnung und Verabeitng
Association of Oil Extractors & Producers
Steindamm 71
2000 Hamburg 1
Germany
Tel:+49 40 2802279

Arctic Biological Station
- see Environment Canada

Arctic Environmental Information & Data Center
- University of Alaska, Anchorage

Arctic Institute of North America
University of Calgary
2500 University Drive, NW
Calgary
Alberta T2N 1N4
Canada
Tel: (403) 220 7515
Fax: (403) 282 7298
and (403) 282 4609

Arctic Offshore Technology Conference
101-3009-23 Avenue SW
Calgary
Alberta T3E 0J3
Canada
Tel:(403) 382 5727

Arnold (Edward) Publisher
- see Edward Arnold

Arthur Andersen & Co
Petroleum Services
1 Surrey Street
London WC2R 2PS
UK
Tel:(071) 438 3866
Fax:(971) 438 3881

ASEAN Council for Petroleum
c/o Pertamina Pusat
Jalan Merdeka Timur 1A
Jakarta 10110
Indonesia
Tel:+62 21 343 710
Tlx: 44 347

ASEA-ATOM
Box 53
72104 Vasteras 1
Sweden
Tel:+46 21 10 70 0
Tlx: 40629 ATOMA S

Asian Institute of Technology
Library and Regional Documentation Centre
PO Box 2754
Bangkok 10501
Thailand
Tel:+66 2 529 0100
Fax: 529 0374
Asian Geotechnical Eng'ing Information Centre

Asociacion de Geologos y Geofisicos Espanoles del Petroleo
Santa Feliciana 14-1
28010 Madrid
Spain
Tel:+34 91 445 0139

Assn Europeenee des Gaz de Petroles Liquefies
- see European Liquefied Petroleum Gas Assn

Associacao Portuguesa dos Gases Combustiveis
Rua a Paricular
Quinta do Figo Maduro
2685 Sacavem
Portugal
Tel:+351 1 94 17428
Fax:+351 1 94 11954

Associated Press
50 Rockefeller Plaza
New York NY 10020
USA   §
Tel: (212) 621 1500

Association Aeronautique et Astronautique de France
80 rue Lauriston
75116 Paris
France
Tel:+33 1 47 048068

Association Amicale des Anciens Eleves des Ecolesde Maitres-Sondeurs (AEMS-IFP)
4 ave de Bois-Preau, BP 10
92506 Rueil-Malmaison
France
Tel:+33 1 47 323692

Association Amicale des Ingenieurs Diplomes de L'Ecole Nationale Superieure de Petrole et des Moteurs (AAID-ENSPM)
1-4 Ave Bois-Preau BP 311
92506 Rueil-Malmaison
France
Tel:+33 1 47 323692

Association for Information Management
Information House
20-24 Old Street
London EC1V 9AP
UK
Tel:(071) 253 4988
Tlx: 23667

Association Francaise des Techniciens du Petrole
16 avenue Kleber
75116 Paris
France
Tel:+33 1 45 006450

Association of British Directory Publishers
Imperial House
17 Kingsway
London WC2B 6UN
UK

Association of British Independent Oil Exploration Companies
c/o Premier Consolidated Oilfields
23 Lower Belgrave Street
London SW1W 0NR
UK
Tel:(071) 730 0752

Association of British Offshore Industries
4th Floor
30 Great Guildford Street
London SE1 0HS
UK
Tel:(071) 928 9199
Fax:(071) 928 6599

Association of Consulting Engineers
Alliance House
12 Caxton Street
London SW1H 0QL
UK
Tel:(071) 222 6557

Association of Consulting Engineers of Canada
Suite 616
130 Albert Street
Ottawa
Canada   ¶
Tel:(613) 236 0569
Fax:(613) 236 6193

Association of Crude Oil & Gas Producers
Wirtschaftsverband Erdol und Erdgasgewinnung
Bruhlstrasse 9
3000 Hannover
Germany
Tel:+49 511 326016

Association of Diving Contractors
1799 Strumpf Boulevard
Building 7, Suite 4
Gretna LA 70053
USA
and
AODC, UK   ¶
- see Hollobone (T A) & Co
Ltd   ¶

Association of Earth Science Editors
c/o American Geological Institute

Association of Energy Professionals
521 5th Avenue
New York NY 10175
USA
Tel:(212) 757 6454

Association of German Petroleum Industry
Mineraloewirtschaftsverband
e.v.
Steindamm 71
D-2000 Hamburg 1
Germany
Tel:+49 40 28541

Association of Graduate Careers Advisory Service
CSU Crawford House
Precinct Centre
Manchester M13 9EP
UK

Association of the Dutch Petroleum Industry
Balistraat 97
2585 XR The Hague
The Netherlands
Tel:+31 70 469444

Association of United Kingdom Oil Independents
Windsor and Regent House
83/89 Kingsway
London WC2B 6SD
UK
Tel:(081) 404 5161

Association of West European Shipbuilders
Frederikq 1
DK 1265 Copenhagen K
Denmark

Association pour la Prevention dans les Transports d'Hydrocarbures
4 Avenue Hoche
75008 Paris
France
Tel:+33 1 47 639894

Association Technique de l'Industrie du Gaz en France
62 rue de Courcelles
75008 Paris
France
Tel:+33 1 47 543434
Tlx:642621

Atkins Research & Development
Woodcote Grove
Ashley Road
Epsom
Surrey KT18 5BW
UK
Tel:(0372) 726140
Tlx:266701 GAtk
Fax:(0372) 740055

Atlantic Information Services Inc
1050 17th St NW, Suite 480
Washington DC 20036
USA   §
Tel: (202) 775 9008

Atomic Energy Research Establishment
Harwell - Building 10.28
Didcot
Oxfordshire OX11 0RB
UK
Tel:(0235) 24141
Extension 2039
- seeMarine Technology
Support Unit

Ausbildungszentrum fur Armeetaucher
Brugg
Switzerland
Tel:+41 56 410311

Ausbildungszentrum fur Polozei
Taucher
Oberrieden
Switzerland
Tel:+46 1 720 7021

Australian Department of Primary Industries and Energy
GPO Box 858
Canberra City ACT 2601
Australia   §
Tel:+61 62 724751
Fax:+61 62 731232

Australian Institute of Engineering Associates
PO Box 408
North Sydney, NSW 2000
Australia

Australian Institute of Petroleum Ltd
227 Collins Street
Melbourne
Victoria 3000
Australia
Tel:+61 3 63 2756

Australian Mineral Foundation
63 Conyngham Street
Glenside
South Australia 5065
Australia   §
Tel:+61 8 379 0444
Fax:+61 8 379 4634

Australian Petroleum & Exploration Association
London Assurance House
20 Bridge Street
Sydney NSW 2000
Australia
Tel:+61 2 27 6718

Automated Science Group
700 Roeder Road
Silver Springs MD
20910-4405
USA   §

Avesta Projects AB
PO Box 557
651 09 Karlstad
Sweden
Tel:+46 541 102770
Tlx:66108 apab s
Fax:+46 545 188254

Avia International
Badenerstrasse 329
8040 Zurich
Switzerland
Tel:+41 1 491 4343

Axses Information Systems
Boutiliers Point
Halifax
Nova Scotia B0J 1G0
Canada   §
Tel: (902) 826 2440
Fax: (902) 826 7274

B

Balch Institute
18 South 7th Street
Philadelphia PA 19106
USA
Tel:(215) 925 8090

Balkema (A A) Publishers
Vijverweg 8
PO Box 1675
3000 Rotterdam
The Netherlands
Tel:+31 10 4145822
Tlx:41605 tkomnlba
Fax:+31 10 2523 7478
and
Old Post Road
Brookfield VT 05036
USA
Tel:(802) 276 3162
Tlx:759615

Ballinger Publishing Company
59 Church Street, PO Box 281
Harvard Square
Cambridge, MA 02136
USA
Tel:(617) 492 0670

Banff Centre
PO Box 1020
Banff
Alberta T0L 0CO
Canada

Bank of Scotland (Oil Division)
The Mound
Edinburgh EH1 1YZ
Scotland
Fax:(031) 243 5436

Barbour Index Ltd
New Lodge
Drift Road
Windsor
Berkshire SL4 4RQ
UK
Tel:(0344) 884121

Barclays de Zoete Wedd Securities Ltd
Ebbgate House
2 Swan Lane
London EC4R 3TS
UK
Tel:(071) 623 2323
Tlx: 88567
Fax:(071) 623 6075

Battelle Memorial Institute
7 Route de Drize
1227 Garouge
Geneva
Switzerland
Tel:+41 22 423472

Battelle Memorial Institute
15 Hanover Square
London W1R 9AJ
UK
Tel:(071) 493 0184/5
Tlx:23773
Fax:(071) 629 9705
and
Battelle Columbus
Laboratories
505 King Avenue
Columbus OH 43201
USA
Tel:(614) 424 6424

Bedford Institute of Oceanography
Dept of Fisheries & Oceans
Canada
PO Box 1006
Dartmouth
Nova Scotia B2Y 4A2
Canada   ¶
Tel:(902) 426 3675
Tlx:019-31552
Fax:(902) 426 7827

Bell & Howell Micro Photo Division
Old Mansfield Road
Wooster OH 44691
USA
Tel:(216) 264 6666

Benn Business Information Services Ltd
PO Box 20
Sovereign Way
Tonbridge
Kent TN9 1RQ
UK
Tel:(0732) 362666
Tlx:95454

Benn Electronics Publications Ltd
Chiltern House
146 Midland Road
Luton
Bedfordshire LU2 0BL
UK
Tel:(0582) 421981
Tlx:827648 BENNLU
Fax:(0582) 29691

Benn Technical Books Ltd
Tolley House
17 Scarbrook Road
Croydon
Surrey CR0 1SQ
UK
Tel:(081) 686 9141

BETC Crude Oil Analysis Data Bank
- see National Institute for Petroleum & Energy Research

BGS Maps Database
- see British Geological Survey

BGS MINSEARCH
- see British Geological Survey
Mineral Intelligence, Statistics & Economics Subgroup

BHR Group Ltd
Cranfield
Bedford MK43 0AJ
UK   ¶§
Tel:(0234) 750422
Tlx: 825059
Fax:(0234) 750074
- Fluid Engineering Centre

Bibliographisches Institut AG
Dudenstrasse 6
Postfach 311
D6800 Mannheim 1
Germany
Tel:+49 621 39011
Tlx:462107

Bidston Observatory
- see Institute of Oceanographic Science

Biochemical Society
7 Warwick Court
London WC1R 5DP
UK
Tel:(071) 424 1076

Bio-Rad Laboratories
Sadtler Division
3316 Spring Garden Street
Philadelphia PA 19104
USA   §
Tel: (215) 382 7800
Fax: (215) 662 0585

Blackie and Son Limited
Bishopbriggs
Glasgow G64 2NZ
Scotland
Tel:(041) 772 2311
Tlx:777283 BLACKI
Fax:(041) 762 0897

Blackwell Scientific Publications Ltd
Osney Nead
Oxford OX2 0EL
UK
Tel:(0865) 240201
Tlx:83355 837022
Fax:(0865) 721205

Blaise-Line
- see British Library   §

BMT Cortec Ltd
Wallsend Research Station
Wallsend
Tyne & Wear NE28 6DY
UK   §
Tel: (091) 262 5242
Fax: (091) 263 8754

BNF Metals Technology Centre
Grove Laboratories
Duckworth Road
Wantage
Oxfordshire OX12 9BJ
UK
Tel:(0235) 72882
Tlx:837166 BNF MTC
Fax:(0235) 73428

Bohai Oil Corporation of CNOOC
Tanggu, PO Box 501
Tianjin
China
Tel:+86 22 2128
Tlx:23163 OBPCC CN

Bonneville Market Information
19 W South Temple, # 200
Salt Lake City UT
84101-1503
USA   §
Tel:(801) 532 3400
Fax:(801) 532 3202

Bowker A&I Publishing / Bowker Electronic Pub'ing
245 West 17th Street
New York NY 10011
USA   §
Tel: (212) 337 6989
Fax: (212) 645 0475

Bowker Saur (UK)
60 Grosvenor Street
London W1X 9DA
UK   §
Tel: (071) 493 5841
Fax: (071) 499 1590

Bowker (R R) Company
Borough Green
Sevenoaks
Kent TN15 8PH
UK
Tel:(0732) 884567
Tlx:95678
Fax:(0732) 884079

BP Norway Limited U A
Forusbeen 35
PO Box 197
4033 Forus
Norway
Tel:+47 4 803000
Tlx:73965
Fax:+47 4 570405

BP Research Centre
Chertsey Road
Sunbury on Thames TW16 7LN
UK
Fax:(09327) 62999

BPS Exhibitions Ltd
28 Marine Drive
Rottingdean
Brighton
East Sussex BN2 7HQ
UK
Tel:(0273) 32431

Brady (R J) Publisher
RTES 197
Bowie, MD 20715
USA
Tel:(301) 262 6300

British Cement Association
Wexham Springs
Slough
Bucks SL3 6PL
UK   ¶
Tel:(0753) 662727
Fax:(0753) 660499

British Coal
Hobart House
14-15 Lower Grosvenor Place
London SW1W 0EX
UK
Tel:(071) 235 2020
Fax:(071) 235 5632

British Constructional Steelwork Association
4 Whitehall Court
Westminster
London SW1A 2ES
UK
Tel:(071) 222 2254
Fax:(071) 1634

British Electrical & Allied Mfrs Association
8 Leicester Street
London WC2H 7BN
UK
Tel:(071) 437 0678
Tlx:263536

British Gas plc
59 Bryanston Street
London W1A 2AZ
UK
Tel:(071) 723 7030
Tlx:261710
Fax:(071) 402 9554
and
Exploration & Production
100 Thames Valley Park Drive
Reading RG6 1PT
UK
Tel: (0734) 353222
Tlx:846231
Fax:(0734) 353484

British Geological Survey
Nicker Hill
Keyworth
Nottingham NG12 5GG
UK   ¶§
Tel: (06077) 6111
Tlx:8812180 Geosci
Fax: (06077) 6602
- *Marine Surveys Directorate*
- *Geophysics Directorate*
Scotland Regional Office
Murchison House
West Mains Road
Edinburgh EH9 3LA
Scotland   ¶§
Tel: (031) 667 1000
Tlx:727343 SeisEd
Fax: (031) 668 4140

British Institute of Management
2 Savoy Court, 3rd Floor
Strand
London WC2R 0EZ
UK   ¶
Tel:(071) 497 0580
Fax:(071) 497 0463

British Library - move to new HQ site in London between 1991-1996
-Helpline for Science,   ¶
Technology and Industry
Collections:
Tel: (071) 323 7915
- European Biotechnology
Information Project
- BLAISE Help Desk
Tel:(071) 323 7070   §
Document Supply Centre
Boston Spa
Wetherby
West Yorkshire LS23 7BQ
UK   ¶
Tel:(0937) 843434
Tlx:557381
Fax:(0937) 546586
- Blaise-Line
Tel: (0937) 546585
Fax: (0937) 546586
and
Science Reference & Information Service
25 Southampton Buildings
London WC2A 1AW
UK   ¶
Tel:(071) 323 7494
Tlx:266959 ScirefG

British Library Board
2 Sheraton Street
London W1V 4BH
UK
Tel:(071) 323 7262

British Marine Equipment Council
32-38 Leman Street
London E1 8EW
UK
Tel:(071) 488 0171
Tlx:886593 BMEC
Fax:(071) 480 7632

British Maritime Technology Ltd
Wallsend Research Station
Tyne and Wear NE28 6UY
UK
Tel:(091) 262 5242
Tlx:53476
Fax:(091) 263 8574
and
Research & Technology
1 Waldegrave Road
Teddington
Middlesex TW11 0SJ
UK
Tel:(081) 943 5544
Tlx:263118
Fax:(081) 943 5347

British Nuclear Fuels
- see BNF Metals
Technology Centre

British Petroleum Co plc
Britannic House
Moor Lane
London EC2Y 9BU
UK
Tel:(071) 920 8000
Fax:(071) 920 8263

British Ship Research Assn
- see British Maritime
Technology

British Standards Institution
Linford Wood
Milton Keynes MK14 6LE
UK
Tel:(0908) 220022
Tlx:825777 BSIMK G
Fax:(0908) 320856
and
2 Park Street
London W1A 2BS
UK
Tel:(071) 629 9000
Tlx:266933 BSICON
and
Testing Services Division
Maylands Avenue
Hemel Hempstead
Herts HP2 4SQ
UK
Tel:(0442) 31111
Tlx:827682

British Steel
  BS Stainless HQ
  PO Box 161
  Shepcote Lane
  Sheffield S9 1TR
  UK
  Tel:(0742) 44331
  Tlx: 547025
  Fax:(0742) 448280
  and
  BS Technical HQ
  Swinden House
  Moorgate
  Rotherham S60 3AC
  UK
  Fax:(0709) 829415

British Technology Group
  101 Newington Causeway
  London SE1 6BU
  UK
  Tel:(071) 403 6666
  Tlx:894397

Brown, Son & Ferguson
  52 Darnley Street
  Glasgow G41 2SG
  Scotland
  Tel:(041) 429 1234

BRS Information Technologies
  8000 Westpark Drive
  McLean VA 22102
  USA   §
  Tel: (703) 442 0900
  Fax: (703) 893 4632

Brunel University
  Dept of Mechanical
  Engineering
  Uxbridge
  Middlesex UB8 3PH
  UK
  Tel:(0895) 37188

BT Tymnet Inc
  BT Tymnet Dialcom Service
  2560 North 1st Street
  PO Box 49019
  San Jose CA 95161-9019
  USA   §
  Tel: (408) 922 0250
  Fax: (408) 911 8015

Building Research Establishment
  Bucknall's Lane
  Garston
  Watford WD2 7JR
  UK   ¶
  Tel:(0923) 894040
  Tlx:923220
  Fax:(0923) 664010

Bundesanstalt fur Geowissenschaften & Rohstoffe
Fed Inst for Geoscience &
Natural Resources
PO Box 510153
Stilleweg 2
3000 Hannover 51
Germany
Fax:+49 511 643 2819
Tlx:9 23730 bgr ha

Bundesministerium fur Wirtschaft
Villemombler Strasse 76
5300 Bonn-Duisdorf
Germany
Tel:+49 228 6 15-1
Tlx:886747
Fax:+49 228 6254436

Bureau de Normalisation de Petrole
16 avenue Kleber
75116 Paris
France
Tel:+33 1 45 021120
Tlx:630545 ucsip

Bureau des Recherches Geologiques et Minieres (BRGM)
39-43 Quai Andre Citroen
75739 Paris Cedex 15
France
Tel:+33 1 45 783333
and
Dept Doc et Information
Geologiques
BP 6009
45060 Orleans Cedex
France   §
Tel:+33 38 643168
Tlx:780258
Fax:+33 38 663518

Bureau for Oil Exploration & Exploitation
Rua Braamcamp 11-4
1200 Lisbon
Portugal
Tel:+351 1 542544

Bureau of Economic Geology
University of Texas at Austin
Box X
University Station
Austin TX 78713-7508
USA
Tel:(512) 471 1534
and 471 7721

Bureau Veritas
Ocean Engineering Division
17 Bis Place des Reflets
Cedex 44
92077 Paris la Defense
France
Tel:+33 1 42 915291
Tlx:614149/611135
and
42 Weston Street
London SE1 3OL
UK
Tel:(071) 403 6266
Fax:(071) 403 1590

Burmass Publishing Company
PO Box 1768
Midland TX 79702
USA
Tel:(915) 682 1782

Busby Associates
576 South 23rd Street
Arlington, VA 22202
USA
Tel:(703) 892 2888

Business Briefings Ltd
565 Fulham Road
London SW6 1ES
UK
Tel:(071) 281 1284

Business International Corp
215 Park Avenue South
New York NY 10003
USA   §
Tel: (212) 460 0600
Fax: (212) 995 8837

Butterworth Publishers Inc
10 Tower Office Park
Woburn MA 01801
USA
Tel:(617)933 8260

**Butterworth & Co
(Publishers) Ltd**
Borough Green
Sevenoaks
Kent TN15 8PH
UK
Tel:(0732) 884567
Tlx:95678
Fax:(0732) 884079

Byram (R W) & Company
P O Drawer 1867
Austin TX 78767
USA
Tel:(512) 478 2551

C

CAES
Southampton University
125 High Street
Southampton SO1 0AA
UK
Tel:(0703) 221397

**Cahners Books International
Inc**
221 Columbus Avenue
Boston, MA 02116
USA
Tel:(617) 536 7780

California State Library
Regional Depository
Government Publications
Service
PO Box 2037
Sacramento CA 95809
USA

California State University
Library Documents
2000 Jed Smith Drive
Sacramento CA 95819
USA
Tel:(916) 278 6466

**Cambridge Scientific
Abstracts**
7200 Wisconsin Avenue,
Suite 601
Bethesda MA 20814
USA   §
Tel: (301) 961 6737
Tlx:898452
Fax: (301) 961 6720

Cambridge University Press
Edinburgh Building
Shaftesbury Road
Cambridge CB2 2RU
UK
Tel:(0223) 312393
Tlx:817256
Fax:(0223) 315052

**Canada Centre for Inland
Waters**
867 Lakeshore Road
PO Box 5050
Burlington
L7R 4A6
Canada

Canada Gas Association
55 Scarsdale Road
Don Mills
Toronto
Ontario M3B 2R3
Canada
Tel:(416) 447 6455
Tlx:06966824

**Canada Institute for
Scientific & Technical
Information (CISTI)**
National Research Council
Canada
Ottawa
Ontario K1A 0S2
Canada   ¶
Tel:(613) 993 1600
Fax:(613) 952 9112
and
Marine Dynamics Branch
PO Box 12093, Stn A
St John's
Newfoundland A1B 3T5
Canada   ¶
Tel:(709) 772 2468
Fax:(709) 772 2462

**Canada Newfoundland
Offshore Petroleum Board**
Suite 500, TD Place
140 Water Street
St John's
Newfoundland A1C 6H6
Canada   ¶
Tel:(709) 778 1400
Fax:(709) 778 1473

**Canada Oil & Gas Lands
Administration**
Environmental Protection
Branch
355 River Road
Ottawa
Ontario K1A 0E4
Canada
Tel:(613) 993 3760

**Canadian Association of
Oilwell Drilling
Contractors**
800 540 - 5th Avenue SW
Calgary
Alberta T2P 0M2
Canada   ¶
Tel:(403) 264 4311
Fax:(403) 263 3796

Canadian Centre for Marine Communications
PO Box 8454
St John's
Newfoundland A1B 3N9
Canada   ¶
Tel:(709) 579 4872
Fax:(709) 579 0495

Canadian Defense Research Establishment Pacific
Victoria
British Columbia VO5 1BO
Canada

Canadian Dept of Energy Mines & Resources
Polar Continental Shelf
Project
City Centre Tower
880 Wellington Street
Ottawa, Ontario K1A 0E4
Canada
and
Resource Management &
Conservation Bureau
Sir Willilam Lojan Building
580 Booth Street
Ottawa, Ontario K1A 0E4
Canada
and
Centre for Remote Sensing
2464 Sheffield Road
Ottawa
Ontario K1A 0Y7
Canada
Tel:(613) 993 0121
Tlx:0543 3777
and
- see Geological Survey of
Canada

Canadian Dept of Fisheries & Oceans
Marine Science &
Information Directorate
Ocean & Aquatic Science
Affairs Br
Ottawa
Ontario K1A 0H3
Canada   ¶
and
Physical & Chemical
Sciences Directorate
12th Floor, 200 Kent Street
Ottawa
Ontario K1A 0E6
Canada
Tel:(613) 990 0298
Fax:(613) 990 5510
and

Canadian Committee of Oceanography
200 Kent Street
Ottawa
Ontario K1A 0E6
Canada
and
580 Broth Street
Ottawa
Ontario K1A 0H5
Canada

Canadian Dept of the Environment
Environmental Protection
Service
Ottawa
Ontario K1A 1C8
Canada
and
Fisheries & Marine Service
Biological Station Library
St Andrew's
New Brunswick E0G 2X0
Canada
Tel:(506) 529 8854

Canadian Gas Research Institute
55 Scarsdale Road
Don Mills
Ontario M3B 2R3
Canada

Canadian Geotechnical Society
Suite 602
170 Attwell Drive
Rexdale
Ontario M9W 5Z5
Canada   ¶
Tel:(416) 674 0366
Fax:(416) 674 9507

Canadian Government Ocean Industry Devel Centre
PO Box 6026
St John's
Newfoundland A1C 5X8
Canada
Tel:(709) 576 2457

Canadian Government Publishing Centre
Supply Service
Hull
Quebec K1A 0S9
Canada
Tel:(418) 994 2085

Canadian Institute of Mining & Metallurgy
400-1130 Sherbrooke
Street W
Montreal, Quebec H3A 2M8
Canada

Canadian Library Association
200 Elgin Street
Ottawa
Ontario K2P 1L5
Canada   ¶
Tel:(613) 232 9625
Fax:(613) 563 9895

Canadian Maritime Industries Association
PO Box 1429, Stn B
801-100 Sparks Street
Ottawa
Ontario K1P 5R4
Canada   ¶
Tel:(613) 232 7127
Fax:(613) 232 2490

Canadian National Energy Board
- see Environment
Directorate

Canadian Petroleum Association
Frontier Division
800, TD Place
140 Water Street
St John's
Newfoundland A1C 6H6
Canada   ¶
Tel:(709) 726 7270
Fax:(709) 726 0232
and
140 Water Street
St John's
Newfoundland
Canada
Tlx:163250
and
Suite 3800
150 6th Avenue SW
Calgary
Alberta T2P 3Y7
Canada   ¶§
Tel: (403) 269 6721
Fax: (403) 261 4622

Canadian Petroleum Products Institute
Suite 1000
175 Slater Street
Ottawa
Ontario K1P 5H9
Canada   ¶
Tel:(613) 232 3709
Fax:(613) 236 4280

**Canadian Society of
Petroleum Geologists**
505,206 7th Avenue SW
Calgary, Alberta T2P 0W7
Canada   ¶
Tel:(403) 264 5010

**Canadian Standards
Association**
178 Rexdale Boulevard
Rexdale
Ontario M9W 1R3
Canada   ¶
Tel:(416) 737 4098
Tlx: 06-989344
Fax:(416) 737 2473

Canner (J S) & Company
49-65 Lansdowne Street
Boston MA 02115
USA
Fax:(617) 437 1923

Capital Planning Information
6 Castle Street
Edinburgh EH2 3AT
Scotland
Tel:(031) 226 4367

Capitol Publications Inc
1101 King Street, Suite 444
Alexandria VA 22314
USA   §
Tel: (703) 683 4100
Fax: (703) 739 6490

Carl Systems Inc
777 Grant, Suite 304
Denver CO 80203
USA   §
Tel: (303) 861 5319
Fax: (303) 830 0103

CEFI
13 Rue de Temara
78100 St Germain en Laye
France
Tel:+33 1 30 61088
Tlx:699467

**Centre de Documentation de
l'Armement (CEDOCAR)**
00460 Armees
France   §
Tel:+33 45 42 60 48
Fax:+33 45 52 45 74

**Centre de Documentation
Sciences Humaines**
54 Boulevard Raspail
75260 Paris Cedex 06
France
Tel:+33 1 4544 3849
Tlx:203104

Centre de la Mer et des Eaux
Albert 1er Prince de Monaco
195 rue St Jacques
75005 Paris
France

**Centre de Recherches &
d'Etudes Oceanographiques**
73-77 rue de Sevres
92100 Boulogne
France

**Centre Europeen de
Documentation et
d'Information Mer**
24 Quai de la Fosse
4400 Nantes
France
Tel:+33 40 73 98 79

**Centre for Cold Ocean
Resources Engineering**
Memorial University of
Newfoundland
St John's
Newfoundland AlB 3X5
Canada
Tel:(709) 737 8354
Tlx:016-4101
Fax:(709) 737 4706

**Centre for Petroleum &
Mineral Law Studies**
University of Dundee
Dundee DD1 4HN
Scotland
Tel:(0382) 23181

**Centre for Professional
Advancement**
Palestrinastraat 1
1061 LC Amsterdam
The Netherlands
Tel+31 20 :10662

**Centre for Short-Lived
Phenomena**
PO Box 199
Harvard Square Station
Cambridge MA 02238
USA   §
Tel: (617) 492 3310
Fax: (617) 492 3312
- *Oil Spill Intelligence Report*

**Centre Francais de la
Corrosion**
Maison de la Chemie
28 rue Saint Dominique
75007 Paris
France
Tel:+33 1 47 051073

**Centre National de la
Recherche Scientifique
(CNRS)**
2 Allee du Parc de Brabois
54514Vandoeuvre-les-Nancy
Cedex
France   §
Tel:+33 83 53 29 00
Fax:+33 83 57 63 09

**Centre National pour
l'Exploitation des Oceans**
- see IFREMER

**Centre Oceanologique de
Bretagne**
Boite postale 337
Sainte Anne du Porzic
29273 Brest CEDEX
France
Tel:+33 98 804650

**Centro Professionale Di
Immersione**
Subsea Oil Services
Via Venezia 3
Zingonia (BG)
Italy

CEP Consultants
26 Albany Street
Edinburgh EH1 3QH
Scotland
Tel:(031) 557 2478

Chadwyck-Healey Ltd
Cambridge Place
Cambridge CB2 1NR
UK   §
Tel: (0223) 311479
Tel: (0223) 66440

**Chalmers University of
Technology**
Sven Hultins GT
412 96 Gothenburg
Sweden
Tel+46 31 810100

**Chambre Syndicale du
Reraffinage**
44 rue La Boetie
75008 Paris
France
Tel:+33 1 45 634179
Tlx:660875

Chapman & Hall
2-6 Boundary Row
London SE1 8HN
UK
Tel:(071) 865 0066
Tlx: 290164
Fax:(071) 522 8623

Charles Knight Publishing
Tolley House
2 Addiscombe Road
Croydon, Surrey CR9 5AF
UK
Tel:(081) 686 9141
Fax:(081) 686 3155

**Chartered Institute of
Building Services Engin'rs**
Balham High Road
London SW12 9BS
UK   ¶
Tel:(081 675 5211

Chase Econometrics
150 Monument Road
Bala Cynwyd
PA 19004-1780
USA
Tel:(215) 667 6000

Chemical Abstracts Service
2540 Olentangy River Road
PO Box 3012
Columbus OH 43210
USA
Tel:(614) 421 3600

**Chemical Industries
Association Ltd**
Kings Buildings
Smith Square
London SW1P 3JJ
UK
Tel:(071) 834 3399
Fax:(071 834 4469

**Chemical Information
System**
7215 York Road
Baltimore MD 21212
USA   §
Tel:(301) 321 8440
Fax:(301) 296 0712
and
- UK, see Fraser Williams

Chemical Market Associates
11757 Katy Freeway, # 750
Houston TX 77079
USA   §
Tel:(713) 531 4660
Tlx:792318
Fax:(713) 531 9966

**CHEM-INTELL Chemical
Intelligence Service**
39a Bowling Green Lane
London EC1R 0BJ
UK   §
Tel: (071) 833 3812
Tlx:924015
Fax: (071) 833 1563

**China Bohai Racal
Positioning & Survey Co
Ltd**
9 Hebei Road
PO Box 557
Tanggu
Tianjin
China
Tel:+86 22 4817/8/9
Tlx:23212 BRJVC CN
and
1430 Hongqiao Road
Shanghai
China
Tel:+86 21 3219325
Tlx:33011 BTHJJ CN

**China France Bohai
Geoservices Co Ltd**
PO Box 548
Tianggu
Tianjin
China
Tel:+86 22 3640-642
Tlx:23181 CFBGC CN

**China Nanhai Racal
Positioning & Survey Co
Ltd**
PO Box 11
Potou
Zhanjiang
Guangdong Province
China
Tel:+86 24 3111
Tlx:44086COSSB CN

**China National Offshore Oil
Corporation**
31 Dong Chang An Jie
PO Box 1519
Beijing
China
Tel:+86 1 55-5 225
Tlx:22611 CNOOC CN

**China Offshore Oil
Development &
Engineering Corp**
20 Xue Yuan Road
PO Box 943
Beijing
China
Tel:+86 1 514 277731
Tlx:22611 CNOOC CN

**China Petroleum
Development Corporation**
Ministry of Petroleum
Industry
Liu Pu Kang
PO Box 766
Beijing
China
Tel:+86 1 444313
Tlx:22312 PCPRC CN
and
China Petroleum
Engineering Construction
Corp
Tel:+86 1 444404
Tlx:20047 CPECC CN

Chr.Michelsen Institute
N-5036 Fantoft - Bergen
Norway
Tel:+47 5 28 44 10
Tlx:40006 cmi n
Fax:+47 5 285613
COST-43
Tel:+47 5 40 006

**CISTI, CAN-OLE (Canadian
Online Enquiry Service)**
National Research Council
of Canada
Ottawa
Ontario K1A 0S2
Canada   §
Tel:(613) 993 1210
Fax:(613) 952 8244

**City Exhibitions &
Conferences**
8 Duke's Close
Alton, Hampshire
UK
Tel:(0420) 87303

**City of Birmingham
Polytechnic**
- now University of Central
England in Birmingham

City University
Northampton Square
London EC1V 0HB
UK
Tel:(071) 253 4399

Civil Aviation Authority
Greville House
37 Gratton Road
Cheltenham
GL50 2BN
UK
Tel:(0242) 235151
Fax:(0242) 584139

Clarkson (H) & Co Ltd
12 Camomile Street
London EC3A 7BP
UK
Tel:(071) 283 8955
Tlx:881297
Fax:(071) 626 3763

Clay Minerals Society
c/o Box 595
Clarkson
New York NY 14430
USA
Tel:(716) 395 2334

Claygate Services Ltd
Unit 9, Mitcham Industrial
Estate
Streatham Road
Mitcham, Surrey CR4 2AP
UK
Tel: (081) 640 7127
Fax: (081) 640 6875

Clio Press Ltd
55 St Thomas Street
Oxford OX1 1JG
UK
Tel:(0865) 250333

**CMAS World Underwater
Federation**
34 Rue du Colisee
75008 Paris
France

CML Publications
18 Spring Crescent
Ashurst
Southampton SO4 2AA
UK
Tel:(042) 129 3223

Coastal Diving Academy
108 West Main street
Bay Shore NY 11706
USA
Tel:(516) 666 3127
Tlx:144599

**Coastal Engineering
Research Council (ASCE)**
c/o Cubit Engineering Ltd
207 E Bay Street, Suite 311
Charleston, SC 29401
USA

**Coastal Zone Management
Institute**
PO Box 221
Sandwich MA 02563
USA

**Cold Regions Research &
Engineering Laboratory**
- see US Army Corps of
Engineers
§

College of Petroleum Studies
New Inn Hall Street
Oxford OX1 2QD
UK
Tel:(0865) 750181
Fax:(0865) 791474

Colorado School of Mines
1500 Illinois Street
Golden CO 80401
USA   ¶
Tel:(303) 273 3687
Fax:(303) 273 3199

**Comite Consultatif de
l'Utilisation de l'Energie**
97 rue de Grenelle
75700 Paris Cedex
France
Tel:+33 1 45 563152

**Comite d'Etudes Petrolieres
Marine**
28 Boulevard de Grenelle
75015 Paris
France
Tel:+33 1 578 1907

Comite National de l'Energie
rue du Commerce 44
1040 Brussels
Belgium
Tel:+32 2 511 81 34

**Comite Professionnel du
Petrole**
51 Boulevard de Courcelles
75008 Paris
France
Tel:+33 1 47 660382
Tlx:650436 CPDP Pa

Commercial Diving Center
272 S.Fries Avenue
Wilmington CA 90744
USA
Tel:(213) 834 2501
Tlx:684839/I:188235

**Commission of the European
Communities**
European Diving
Technology Committee
D-G Employment/Social
Affairs
Jean Monet Building
Kirchberg
Luxembourg

**Commission of the European
Communities: DG XIII-C**
SPRINT: Strategic
Programme for Innovation
& Technology Transfer
Jan Monnet Building
2920 Luxembourg
Luxembourg
Tel:+35 2 4301 2918

**Commission of the European
Communities: DG XII-E**
MAST: Marine Science &
Technology
rue de la Loi 200
B 1049 Bruxelles
Belgium
Tel:+32 2 2356787

Compact Cambridge
7200 Wisconsin Avenue
Suite 601
Bethesda MD 20814
USA   §
Tel: (301) 962 6737
Fax: (301) 961 6720

**Companie Francaise des
Petroles TOTAL**
5 rue Michel-Ange
75781 Paris Cedex 16
France
Tel:+33 1 743 8000
Tlx:611992

Compass Publications Inc
Suite 1000
1117 North 19th Street
Arlington VA 22209
USA
Tel:(703) 524 3136
Fax:(703) 841 0852

**Compuserve Business
Information Service**
5000 Arlington Centre
Boulevard
Columbus OH 43220
USA
Tel:(614) 457 8600
Tlx:(810) 482 1709

Computational Mechanics
- see CML Publications

**Computer Petroleum
Corporation**
6949 Valley Creek Road
St Paul MN 55125
USA   §
Tel:(612) 738 1088
Fax: (612) 738 1952

CONCAWE
Madouplein 1
B-1030 Brussels
Belgium ¶
Tel:+32 2 2203111
Tlx:20308
Fax:+32 2 2194646

Confederation of British Industry
Centre Point
London WC1A 1DU
UK
Tel:(071) 379 7400
Tlx:21332
Fax:(071) 240 1578

Conoco Norway Inc
PO Box 488
4001 Stavanger
Norway

Conoco (UK) Ltd
Rubislaw House
Anderson Drive
Aberdeen AB2 4AZ
Scotland
Fax:(0224) 205222
and
116 Park Street
London W1Y 4NN
UK
Tel:(071) 493 1235
Tlx:27307/24734
Fax:(071) 408 6660

Consiglio Nazionale della Richerche
Piazzale A Moro 7
Rome I-00185
Italy
Tel:+39 6 4923 107

Construction Industry Directorate
- see Dept of the Environment UK

Construction Industry Research & Information Assn
6 Storey's Gate
London SW1P 3AU
UK
Tel:071 222 8891
Tlx:262433
Fax:(071) 222 1708

Construction Press Ltd
Longman House
Burnt Mill
Harlow
Essex CM20 2JE
UK
Tel:(2079) 26721

Constructional Steel Research & Development Org
NLA Tower
12 Addiscombe Road
Croydon
Surrey CR9 3JH
UK
Tel:(081) 688 2688

Continental Shelf Institute
- see IKU

Cooperative Group Marine Research & Marine Technology Information
Ross Strasse 126
4000 Dusseldorf 30
Germany

COPPE/UFRJ - Federal University of Rio de Janeiro
Programma de Engeharia Civil
CP 68506
21945 Rio de Janeiro
Brazil ¶
Tel:+55 21 280 9993
Tlx: 2133817
Fax+ 55 21 290 6626

Copyright Licensing Agency Ltd
33-34 Alfred Place
London WC1E 7DP
UK
Tel:(071) 580 9729

Corad Technology Ltd
GPO Box 8310
Room 1201, East Town Building
41 Lockhart Road
Wanchai
Hong Kong
Tel+852 5 8610103
Tlx:62268

Cornell Maritime Press Inc
POB 456
Centreville MD 21613
USA
Tel:(301) 228 3850

Cornell University Press
124 Roberts Place
Ithaca NY 14850
USA
Tel:(607) 257 7000
Tlx:937478

Corporation Estatal Petrolera Ecuatoriana
Avenue Orellana y Juan Leon Mera
Quito
Ecuador

Corrosion & Protection Centre, CAPCIS
UMIST
Bainbridge House
Granby Row
Manchester M1 2PW
UK
Tel:(061) 236 6573
Tlx: 668810 Capcis
Fax:(061) 228 7846

County Natwest Woodmac
Kintore House
74-77 Queen Street
Edinburgh EH2 4NS
Scotland
Tel:(031) 225 8525
Tlx:72555
Fax:(031) 243 4434/5

Crane Limited
107 High Street
Chipping Ongar
Essex CM5 9DX
UK

Crane Russak & Co Inc
3E 44th Street
New York NY 10017
USA
Tel:(212) 867 1490

Cranfield Institute of Technology
Marine Technology Centre
Cranfield
Bedford MK43 0AL
UK
Tel:(0234) 750111
Tlx:825072

CRC Press Inc
2000 Corporate Boulevard NW
Boca Raton FL 33431
USA
Tel:(305) 994 0555
Tlx:568689

Croner Publications Ltd
173 Kingston Road
New Malden
Surrey KT3 3SS
UK
Tel:(081) 942 8966
Tlx:267778
Fax:(081) 650 6056

Croom Helm Ltd
North Way
Andover
Hants SP10 5BE
UK
Tel:(0264) 332424
Tlx:47214

CSIRO Australia
Information Resources
Branch
314 Albert Street
East Melbourne
Victoria 3002
Australia   §
Tel:+61 3 418 7333
Fax:+61 3 419 0459

Current Awareness Bulletin
- see Assn for Information
Management

**Cushman Foundation for
Foraminiferal Research**
E 501 US National Museum
Washington DC 20560
USA
Tel:(202) 357 2405

Cutter Information Group
37 Broadway
Arlington VA 02174-5539
USA   §
Tel: (617) 648 8700

D

Danish Energy Agency
Ministry of Energy
11 Landemaerket
DK 1119 Copenhagen
Denmark
Tel:+45 1 92 67 00
Tlx:22 450 denerg dk
Fax:+45 1 11 47 43

**Danish Geotechnical
Institute**
Maglebjergevej 1
DK-2800 Lyngby
Denmark
Tel:+45 2 88 44 44
Tlx:27230 geotecdk
Fax:+45 2881240

Danish Hydraulic Institute
Agern Alle 5
2970 Horsholm
Denmark
Tel:+45 2 86 80 33
Tlx:0063 37402 DHI
Fax:+45 2 867951

Danish Maritime Institute
Hjortekjaersvej 99
2800 Lyngby
Denmark   ¶
Tel:+45 2 87 93 25
Tlx:37223 shilab d
Fax:+45 2 879333

Danish Ministry for Energy
Strandgade 29
1401 Copenhagen K
Denmark
Tel:+45 1 543611

Danish Offshore School
AP Mollersvej 37
5700 Svendborg
Denmark
Fax:+45 9 215100
and
Vangled
6720 Nordby, Fano
Denmark
Tel:+45 5 162655
Tlx:9257313 dos

**Danish Oil & Natural Gas
Corp**
Agern Allee 24-26
2970 Horsholm
Denmark
Tel:+45 2 571022

**Danish Petroleum Industry
Association**
Vogmagergade 7
Postboks 120
DK-1004 Copenhagen K
Denmark
Tel:+45 33 11 3077
Fax:+45 33 32 1618

**Danish Ship Research
Laboratory**
Hjortekaersvej 99
DK 2800 Lyngby
Copenhagen
Denmark
Tel:+45 2 879325
Tlx:0063 37223 SHI
Fax:+45 2879333

Danish Welding Institute
345 Park Allee
2605 Brondby
Denmark
Tel:+45 2 968800
Tlx:33388 svc dk
Fax:+45 2 962636

Dansk Boreselskab A/S
50 Esplanaden
1263 Copenhagen K
Denmark
Tel:+45 1 114676
Tlx:16260 DABOR

**Dansk Teknisk
Oplysningstjeneste**
Ornevej 30
2400 Copenhagen NV
Denmark
Tel:+45 1 109195

Data Resources Inc
1750 K Street, NW
9th Floor
Washington DC 20006
USA
Tel:(202) 862 3700

Datacentralen
Retoretvej 6-8
DK 2500 Valby
Copenhagen
Denmark   §
Tel:+45 31 468122
Tlx:27122
Fax:+45 31 168805

Dataline Inc
67 Richmond Street West
Suite 700
Toronto
Ontario M5H 1ZS
Canada   §
Tel: (416) 365 1515

Datastream International Ltd
Monmouth House
58/64 City Road
London EC1Y 2AL
UK   §
Tel:(071) 250 3000
Tlx:884230
Fax:(017) 251 0454

Datatimes Corporation
Suite 450
14000 Quail Springs
Parkway
Oklahoma City OK 73134
USA   §
Tel: (405) 751 6400
Fax: (405) 755 8028

Data-Star Marketing Ltd
Plaza Suite
114 Jermyn Street
London SW1Y 6HJ
UK   §
Tel: (071) 930 5503
Fax: (071) 910 2581

David & Charles Ltd
Brunel House
Newton Abbott
Devon TQ12 2DW
UK
Tel:(0626) 61121
Tlx:42904

Davy Stockbrokers
Davy House
49 Dawson Street
Dublin 2
Eire
Tel:+353 6797788
Fax:+353 712704

**Dawson Microfiche - Wm
Dawson Ltd**
Cannon House
Parkfarm Road
Folkestone
Kent CT19 5EE
UK
Tel:(0303) 57421

**Decompression Sickness
Central Registry**
University of Newcastle
upon Tyne
21 Claremont Place
Newcastle upon Tyne
NE2 4AA
UK
Tel:(0632) 324987

**Deep Offshore Technology
bv1**
Museumplein 11
DJ Amsterdam
The Netherlands

**Deep Sea Research &
Oceanographic Science, A**
- see Pergamon Press

**Deep Sea Research &
Oceanographic Science, B**
- see Wood's Hole
Oceanographic Institute

**Defence Research
Information Centre**
St Mary Cray
Orpington
Kent
UK

DeGolyer & MacNaughton
One Energy Square
Dallas TX 75206
USA
Tel:(214) 368 6391

Delco Electronics
General Motors Corp
Santa Barbara CA
USA

Delft Hydraulics Laboratory
Information/Documentation
Section
PO Box 177
2600 MH Delft
The Netherlands
Tel:+31 15 569353
Tlx:38176.32770

**Delft University of
Technology**
134 Julianalaan
POB 5
Delft
The Netherlands
Tel:+31 15 789111
- *Dept of Geodetics*
- *Dept of Naval Architecture,*
- *Shipbuilding & Marine
Engineers*

**Department of Agriculture &
Fisheries for Scotland**
Scottish Office
Chesser House
500 Gorgie Road
Edinburgh EH11 3AW
Scotland
Tel:(031) 443 4020
Tlx:72162/727478
and
Marine Laboratory
PO Box 101
Victoria Road
Aberdeen AB9 8DB
Scotland
Tel:(0224) 876544
Tlx:73587
Fax:(0224) 879156

**Department of Community &
Regional Affairs**
PO Box B
Juneau AK 99811
USA
Tel:(907) 465 4700

Department of Energy
1 Palace Street
London SW1E 5HE
UK   ¶
Tel:(071) 238 3000
Tlx:918777
Fax:(071) 834 3771
- *Diving Inspectorate*
- *Gas Division*
- *Pipelines Inspectorate*
- *Energy Technology Division*
- *Petroleum Engineering Div'n*
and
Gas & Oil Measurement
Branch
Tigers Road (off Saffron Rd)
Wigston
Leicester LE8 2US
UK
Tel:(0533) 785354
and
Oil Division
Thames House South
Millbank
London SW1P 4QJ
UK
Tel:(071) 211 3000
Tlx:918777
and
-see **Offshore Supplies
Office**
-*Offshore Energy*
-*Technology Board*
-*Subsea Projects*
- see Marine Technology
Support Unit

Department of Energy
Petroleum Affairs Division
Beggars Busy
Haddington Road
Dublin 4
Eire
Tel:+353 1 715233
Tlx:90870
Fax:+353 1 604462

Dept of Mines & Energy
18 Jalan Merdeka Selatan
PO Box 344
Jakarta
Indonesia
Tlx: 44363

**Department of Primary
Industries & Energy**
- see under Australia

Department of the Environment
2 Marsham Stret
London SWIP 3EB
UK
Tel:(071) 212 3434
Tlx:22221
- *General Directorate of Environmental Protection*
- *Construction Industry Directorate*
- *Environmental Protection Service*
- see Fire Research Station

Department of Trade & Industry
1-19 Victoria Street
London SW1H 0ET
UK   ¶
Tel:(071) 215 7877
Tlx:918777
- *ESA/IRS*
and
Shipbuilding & Electrical
Engineering Division
Ashdown House
123 Victoria Street
London SW1E 6RB
UK
Tel:(071) 215 5000
Tlx:8813148
Fax:(071) 828 3258

Department of Transport
Marine Library
Sunley House
90 High Holbron
London WC1V 6LP
UK   ¶
Tel:(071) 405 6911
Tlx:264084
Fax:(071) 831 2508

Dept of Fisheries & Oceans Canada
Physical & Chemical
Services Directorate
200 Kent Street, 12 Floor
Ottawa
Ontario K1A 0E6
Canada
Tel:(613) 990 0298
Fax:(613) 990 5510
- see Bedford Institute of Oceanography

Derwent Publications Ltd
Rochdale House
128 Theobalds Road
London WC1X 8RP
UK   §
Tel:(071) 242 5823
Fax: (071) 405 3630

Det Norske Creditbank
21 Kirkegaten
Oslo 1
Norway

Det Norske Veritas/Veritec
Veritasveien
PO Box 300
1322 Hovik
Norway   ¶§
Tel:+47 2 47 7250
Tlx:76192
Fax:47 2 47 7474
and
Veritas House
112 Station Road
Sidcup
Kent DA15 7BU
UK
Tel:(081) 309 7477
Tlx:896526
Fax:(081) 309 1834

Deutsche Erdolversorgungs gesellschaft mbH
PO Box 332
Dorotheenstrasse 1
4300 Essen 1
Germany
Tel:+49 201 7261
Tlx:8571141

DeWitt & Company
16800 Greens Point Park
North Atrium, Suite 120
Houston TX 77060
USA   §
Tel:(713) 875 5525
Tlx:762854 HOU
Fax: (713) 875 0175

Dialcom Inc
6120 Executive Boulevard
Rockville MD 20852
USA
Tel:(301) 881 9020
Tlx:710/825 0435

DIALOG Information Services
PO Box 188
Oxford OX1 5AX
UK   §
Tel:(0865) 730275
Tlx:837704
Fax:(0865) 736354
and
3460 Hillview Avenue
Palo Alto CA 94304
USA   §
Tel:(415) 858 3785
Fax:(415) 858 7069

Direction des Hydrocarbures
Ministere de l'Industrie
3-5 rue Barbet de Jouy
75007 Paris
France
Tel:+33 1 45 563636
Tlx:200802

Direction Generale de l'Energies et des Matieres Premieres
99 rue de Grenelle
75700 Paris Cedex
France
Tel:+33 1 45 563636
Tlx:270257

Directorate of Fisheries Research
Ministry of Agriculture,
Fisheries & Food
Pakefield Road
Lowestoft
Suffolk NR33 0HT
UK
Tel:(0502) 62244
Tlx:97470
Fax:(0502) 513865

Dirrecion General de Energia y Combustibles
Serrauo 37
Madrid 1
Spain

Diver Equipment Information Centre
Battelle Columbus
Laboratories
505 King Avenue
Columbus, OH 43201
USA

Divers Institute of Technology Inc
4601 Shilshole Avenue NW
PO Box 70312
Seattle. WA 98107
USA
Tel:(206) 783 5542

Divers Training Academy Inc
RFD 1
Box 193-C
Link Port, Fort Pierce
Florida
FL 33450
USA
Tel:(305) 465 1994

Diving Inspectorate
- see Department of Energy

**Diving Medical Advisory
Committee**
28/30 Little Russell Street
London WC1A 2HN
UK

DOC-6
Tuset, 21, 6e.3a
Barcelona 08006
Spain   §
Tel:+34 3 414 0679

Don Mor Productions Ltd
69 Dee Street
Aberdeen AB1 2EE
Scotland
Fax:(0224) 210126

DOT bv
Muaseumplein 11
1071 DJ Amsterdam
The Netherlands
Tel:+31 20 18709

Douglas W Hilchie Inc
PO Box 75
Boulder CO 80403
USA
Tel:(303) 642 7318

Dow Jones & Co Inc
PO Box 300
Princeton NJ 08543-0300
USA   §
Tel: (609) 520 4000
Fax: (609) 520 4660

**Drewry (H P) (Shipping
Consultants) Ltd**
34 Brook Street
London W1Y 2LL
UK
Fax:(071) 629 5362
Tlx:21167 HPDLDN G

DRI/McGraw Hill
Data Products Division
1750 K Street NW, 10th Flr
Washington DC 20006
USA   §
Tel: (202) 663 7720
- *Forecast Plus*   §
Tel:(202) 862 3720
DRI Euroope Ltd
Wimbledon Bridge House
1 Hartfield Road
Wimbledon
London SW19 3RU
UK   §
Tel:(081) 543 1234
Fax:(081) 545 6248

DTI Energy Library
1 Palace Street
London SW1E 5HE
UK
Tel:(071) 238 3038
Tlx:918777 Energy G
Fax:(071) 834 3771

DTS Singapore Pte Ltd
Delta House
2 Alexandra Road
No 03-01B
Singapore 0315
Singapore
Tel:65 2788566
Tlx:51057

Duke University
Marine Laboratory
Pivers Island
Beaufort NC 28516-9721
USA   ¶
Tel:(919) 728 2111
Fax:(919) 728 2514

Dun & Bradstreet
26-32 Clifton Street
London EC2P 2LY
UK
Tel:(071) 377 4439
Tlx:934627
Fax:(071) 247 3836

**Dunstaffnage Marine
Research Laboratories**
PO Box 3
Oban
Argyle PA34 4AD
Scotland
Tel:(0631) 62244/6
Tlx:776216

Dwight's Energydata Inc
1633 Firman Drive
Richardson TX 75081
USA   §
Tel:(214) 783 8002
Fax:(214) 783 0058

E

E & P Forum
- see Oil Industry
International Exploration &
Production Forum

**Earthquake Engineering
Research Center**
- see Univ of California
Berkeley Campus   §

Easams Ltd
Lyon Way
Frimley Road
Camberley
Surrey GU6 5EX

**Eastern Petroleum Directory
Inc**
370 Morris Avenue
Trenton, NJ 08611
USA
Tel:(609) 393 4694

**Economic Geology
Publishing Company**
c/o 91-A Yale Station
New Haven CT 06520
USA
Tel:(203) 436 2499

Economist Intelligence Unit
Spencer House
27 St James's Place
London SW1A 1NT
UK
Tel:(071) 493 6711
Tlx:266353

Editions Technip
27 rue Ginoux
75737 Paris, Cedex 15
France
Tel:+33 1 45 771108
Tlx:200375
Fax:+33 1 45 753711

Editorial Garsi
Londres 17
Madrid 28
Spain
Tel:+34 24 2256809

**Edward Arnold (Publishers)
Ltd**
Mill Road
Dunton Green
Sevenoaks, Kent TN13 2YA
UK
Tel:(0732) 450111
Tlx:95122
Fax:(0732) 461321

**Eidg Verkehrs und Energie-
Wirtschaftsdepartement**
Eidg Rohrleitunsinspektorat
Freiestrasse 88
8030 Zurich
Switzerland

Eindhoven University of Technology
Den Dolech 2, PB 513
5600 MB Eindhoven
The Netherlands
Tel:+31 40 47 9111

Electrical Research Assn
- see ERA Technology Ltd

Electricity Association
30 Millbank
London SW1P 4RD
UK
Tel:(071) 834 2333
Fax:(071) 834 7606

Electronic Markets and Information Systems Inc
1221 Ave of the Americas
42nd Street
New York NY 10020
USA   §
Tel: (212) 512 6119

Elf Aquitaine Editions
Centre Micoulau
64018 Pau Cedex
France
Tel:+33 59 836237

Elf Aquitaine Norge AS
PO Box 168
4001 Stavanger
Norway

Elf-Aquitaine: Societe Nationale Elf-Aquitaine
Tour Elf
Cedex 45
92078 Paris La Defense
France
Tel:+33 1 47444546
Tlx:615400 ELFA
Fax:+33 147447373

Ellis Horwood Ltd
Market Cross House
Cooper Street
Chichester
West Sussex PO19 1GB
UK
Tel:(0243) 789942
Tlx:86290

Elsevier Applied Science Publishers
Crown House, Linton Road
Barking
Essex IG11 8JU
UK
Tel:(081) 594 7272
Tlx:896950
Fax:(081) 594 5942

Elsevier Science Publishers
PO Box 211
335 Jan van Galenstraat
1000 AE Amsterdam
The Netherlands
Tel:+31 20 5803.911
Tlx:18582 espa nl
Fax:+31 20 5803.769
and
  52 Vanderbilt Avenue
New York, NY 10017
USA
Tel:(212) 867 9040
Tlx:420643
Fax:(212) 916 1288

Empresa National del Petroleo
Ahumala 341 - 3x piso
Santiago
Chile   ¶
Tel:+56 2 381845
Fax:+56 2 380164

Empressa Nacional de Investigacion y Exploracion de Petroleo SA
Pez Volator 2
Madrid 30
Spain
Tel:+34 1 2747200
Tlx:46606 EMPE

Energy Business Centre
Bright's Meadow
Ham
Marlborough
Wiltshire SN8 3RB
UK
Tel:(0488) 4412

Energy Communications Inc
PO Box 1589
Dallas TX 75221
USA
Tel:(214) 691 3911
Tlx:910 861 9306

Energy Consultancy
24 Elm Close
Bedford MK41 8BZ
UK

Energy Datasearch
2810 Glenda Avenue
Fort Worth TX 76117
USA
Tel:(817) 831 0761

Energy Economics Research Ltd
7/9 Queen Victoria Street
Reading RG1 1SY
UK
Tel:(0734) 587689

Energy Enterprises
1580 Lincoln
Suite 1000
Denver CO 80203
USA
Tel:(303) 832 7111

Energy Industries Council
Newcombe House
45 Notting Hill Gate
London W11 3LQ
UK
Tel:(071) 221 7372
Tlx:27273
Fax:(071) 221 8813

Energy Information Administration
National Energy
Information Center
1000 Independence Ave SW
Washington DC 20585
USA   ¶
Tel:(202) 586 8800
for the deaf)

Energy Information Resource Inventory
Office of Scientific &
Technical Information
US Dept of Energy
PO Box 62
Oak Ridge TN 37831
USA
Tel:(615) 576 1222

Energy Publications Ltd
PO Box 147
Grosvenor House
Newmarket
Cambs CB8 9AL
UK
Tel:(0638) 663030

Energy Publishing Company
1412 SW H Avenue
Winter Haven FL 33880
USA
Tel:(813) 293 2576

Energy Systems Trade Association
OI Vix 16
Stroud G15 5EB
UK
Tel: (0453) 873668

Energy, Mines & Resources Canada
580 Booth Street
Ottawa
Quebec K1A 0E4
Canada

Engineering Committee on Oceanic Research
British Committee for ECOR
76 Mark Lane
London EC3R 7JN
UK
Tel:(071) 481 0750
Tlx:886841
Fax:(071) 481 4001

Engineering Equipment & Material User Assn
14 Belgrave Square
London SW1X 8PS
UK
Tel:(071) 235 5316

Engineering Industry Training Board
PO Box 176
54 Clarendon Road
Watford, Herts WD1 1LB
UK
Tel:(0923) 38441
Fax:(0923) 243025

Engineering Information Inc
345 East 47th Street
New York, NY 10017-2387
USA   §
Tel:(212) 705 7600
Fax:(212) 935 2927

Engineering Sciences Data Unit
- see ESDU Internat'l Ltd

Engineering Societies Library
United Engineering Centre
345 East 47th Street
New York, NY 10017
USA
Tel:(212) 705 7611
Fax:(212) 486 1086

Ente Nazionale Idrocarburi
- see National Hydrocarbons Authority

Environment Canada
National Environmental Emergency Centre (NATES)
Hazardous Waste Hall
Ottawa
Ontario K1A 0H3
Canada   §
Tel: (819) 997 3742
Fax: (819) 953 5261
and
45 Alderney Drive, 5th Floor
Queen Square
Dartmouth
Nova Scotia B2Y 2N6
Canada   §
Tel: (902) 426 8301
Fax: (902) 426 9709
and
Arctic Biological Station
Box 400 St Anne de Bellvue
Quebec H9X 3L6
Canada
Tel:(514) 457 3660
and
Atmospheric Environment Service
4905 Dufferin Street
Downsview
Ottawa, M3H 5TA
Canada

Environment Directorate
National Energy Board
311 6th Avenue SW
Calgary
Alberta T2P 3H2
Canada
Tel:(403) 299 3675
Fax:(403) 292 5503

Environment Information Centre Inc
292 Madison Avenue
New York NY 10017
USA
Tel:(212) 944 8500

Environment & Natural Resources Institute
- see University of Alaska, Anchorage   ¶

Environmental Conservation Executive
PO Box 740
North Sydney
New South Wales 2060
Australia

Environmental Data Services
Unit 24
Finsbury Business Centre
40 Bowling Green Lane
London EC1R 0NE
UK

Environmental Data & Information Service
3300 White Laven Street NW
Washington, DC 20235
USA
Tel:(202) 634 7305

Environmental Protection Agency
Office of the Administrator
401 M Street SW, A-100
Washington DC 20460
USA
Tel:(202) 382 2090

Environmental Protection Service
- Canadian Dept of the Environment

Environmental Research Institute
PO Box 8618
Ann Arbor, MI 48107
USA
Tel:(810) 223 7019

Enviro-Marine Club
- c/o Marinetech North West

ERA Technology Ltd
Cleeve Road
Leatherhead
Surrey KT22 7SA
UK   ¶
Tel:(0372) 374151
Tlx:264045
Fax:(0372) 374496

ERIC Document Reproduction Service
Box 190
Arlington VA 22210
USA
Tel:(703) 823 0500

Erico Petroleum Information Ltd
Erico House
92-99 Upper Richmond Road
London SW15 2TE
UK   §
Tel: (081) 788 1072
Tel: (081 789 1812

ESA/IRS
　IRS-Dialtech (UK)
　Dept of Trade & Industry
　Room 392, Ashdown House
　123 Victoria Street
　London SW1E 6RB　　§
　and
　Science Reference
　Information Service
　25 Southampton Buildings,
　Chancery Lane
　London WC2 1AW
　UK　　　　　　　§
　Tel: (071) 323 7951
　Fax: (071) 323 7930
　and
　ESRIN (Europe)
　Via Galileo Galilei
　I-00044 Frascati
　Italy
　Tel:+39 94 1801
　Fax:+39 94 180361

Esbjerg Industrial
　Development Service
　33 Skolegade
　6700 Esbjerg
　Denmark
　Tel:+45 5 1237 44

ESC Publishing Ltd
　Mill Street
　Oxford OX2 0JU
　UK
　Tel:(0865) 249248
　Tlx:83147

ESDU International Ltd
　27 Corsham Street
　London N1 6UA
　UK
　Fax:(071) 490 2701

ESL Information Services Inc
　345 East 47th Street
　New York NY 10017
　USA　　　　　　¶
　Tel:(212) 705 7527
　Fad:(212) 753 9568

Esso Exploration &
　Production Norway Inc
　PO Box 560
　4001 Stavanger
　Norway
　Tel:+47 46 97555

Esso Exploration &
　Production UK Ltd
　Ermyn Way
　Leatherhead
　Surrey KT22 8UX
　UK
　Fax:(0372) 222678

Esso Petroleum Co Ltd
　Esso House
　Victoria Street
　London SW1E 5JW
　UK
　Tel:(071) 834 6677
　Fax:(071) 245 2078

Estuarial & Coastal Marine
　Science
　- see Institute of
　Oceanographic Sciences

European Diving Technology
　Committee
　- Commission of the
　European Communities

Euromoney Publications plc
　Nestor House
　Playhouse Yard
　London EC4V 5EX
　UK
　Fax:(071) 329 4349

Euromonitor Publications Ltd
　87/88 Turnmill Street
　London EC1M 5QU
　UK
　Tel:(071) 251 8024
　Tlx:21120
　Fax:(071) 608 3149

Europa Publications Ltd
　18 Bedford Square
　London WC1B 3JN
　UK
　Tel:(071) 580 8136
　Fax:(071) 636 1664

European Association for
　Grey Literature
　Exploitation
　(Rob H A Wessels, EAGLE
　Secretary)
　Schuttersveld 2
　NL 2611 WE Delft
　The Netherlands

European Association of
　Exploration Geophysicists
　PO Box 298
　3700 AG Zeist
　The Netherlands　　¶
　Tel:+31 3404 569 97
　Fax:+31 3404 626 40

European Association of
　Petroleum Geologists
　Utrecht Seweg No 62
　P O Box 298
　3700 AG Zeist
　The Netherlands

European Biotechnology
　Information Project
　- see British Library

European Chemical News
　Quadrant House
　The Quadrant
　Sutton, Surrey SM2 5AS
　UK　　　　　　§
　Tel: (081) 661 8125
　Fax: (081) 661 3375

European Diving Technology
　Committee
　c/o Belgische Geologische
　Dienst
　Jennerstraat 13
　1040 Bruxelles
　Belgium

European Fuel Information
　Centre
　c/o Branch Nationale des
　Negociants en Produits
　Petroliers
　18 rue Leonard de Vinci
　F 75116 Paris
　France
　Tel:+33 1 4500 8205

European Liquefied
　Petroleum Gas Assocn
　6 rue Galilee
　75782 Paris Cedex 16
　France
　Tel:+33 1 4723 5274
　Fax:+33 1 4723 5279

European Petrochemical
　Association
　Avenue Louise 250, Bte 74
　1050 Bruxelles
　Belgium
　Tel:+32 02 640 23 10
　Tlx:22186

European Underwater
　Biomedical Society
　North Sea Hyperbaric
　Centre
　Howe Moss Drive, Dyce
　Aberdeen AB2 0GL
　Scotland

European Study Conferences
　Douglas House, Queens Sq
　Corby NN17 1PL
　UK
　Fax:(0536) 304218

Export Council of Norway
　20 Pall Mall
　London SW1Y 5NC
　UK

Expro Science Publications
Brouwersgracht 236
1013 HE Amsterdam
The Netherlands

Exxon Production Research
33119 Mercer Street
Houston TX 77027-6019
USA

F

Fabritius A S
Brobekkveien 80
Alnabru
PO Box 1156 Sentrum
N-0107 Oslo 1
Norway   §
Tel:+47 2 640888
Tlx:18137
Fax:+47 2 650200

Fairbase Database Ltd
PO Box 91 04 46
3000 Hannover 91
Germany   §
Tel: (511) 44 33 30
Fax: (511) 44 27 70

Fairchild Microfilms
Fairchild Publications Inc
7E.12th Street
New York NY 10003
USA
Tel:(212) 741 4000

Faxon Europe
PO Box 197
1000 AD Amsterdam
The Netherlands   §

**Federal Institute for
Geosciences & Natural
Resources**
Marine Information &
Documentation System
Stilleweg 2
D 3000 Hannover 51
Germany   §
Tel: (511) 643 2819
Fax: (511) 643 2304

**Federal Interagency Cttee
on the Health & Safety
Effects of Energy
Technology**
- US Dept of Commerce,
NTIS

**Federal Ministry of
Petroleum & Energy**
PMB 12574
Jfaakubu Gowon Street
Lagos
Nigeria
Tlx: 21126

**Federation Francaise des
Petroliers Independants**
10 Rue Laborde
75008 Paris
France
Tel:+33 1 43 870001
Tlx:281997

**Federation of Danish
Industries**
H c Andersen Boulevard 18
1596 Copenhagen V
Denmark
Tel:+45 1 152233
Tlx:22993
Fax:+45 1 323281

**Federation of European
Petroleum & Gas
Equipment Manufacturers**
PO Box 190
2700 AD Zoetermeer
The Netherlands
Tel:+31 079 531100

Federation Petroliere Belge
rue de la Science 4
1040 Bruxelles
Belgium
Tel:+32 2 51230 03
Fax:+32 51105 91

Fein-Marquart Associates Inc
7215 York Road
Baltimore MD 21212
USA   §
Tel: (301) 821 5980
Fax: (301) 296 0712

Filtration Society
c/o Dept of Chemical
Engineering
University of Technology
Loughborough
Leicestershire LE11 3TU
UK
Tel:0509 232663
Tlx:34319

Financial Times
Southwark Bridge
London SE1 9HL
UK
Fax:(071) 407 5700

**Financial Times Business
Information Ltd**
Greystoke Place, Fetter Lane
London EC4A 1ND
UK
Tel:01 405 6969
Tlx:883694
and
PO Box 12
Sunbury-on-Thames
TW16 7UD
UK   §
Tel: (0932) 761444
Fax: (0932) 781425
- *FT Profile*   §

**Finnish Institute of Marine
Research**
PO Box 33
SF-00931 Helsinki
Finland
Tel:+358 0 331 044

**Finnish Petroleum
Federation**
Fabianinkatu 8
PO Box 188
Helsinki 13
Finland
Tel:358 65 58 31
Fax:+358 17 13 14

Finsbury Data Services
68-74 Carter Lane
London EC4V 5GA
UK
Tel:(071) 248 9828
Tlx:892520

Fire Research Station
Dept of the Environment
Borehamwood
Hertfordshire WD6 2BL
UK
Tel:(081) 953 6177
Tlx:8951648
Fax:(081) 207 5299

Fisheries Inspectorate
- see Ministry of
Agriculture, Fisheries &
Food Fisheries & Food

Fiz Chemie GmbH
Postfach 12 60 50
Steinplatz 2
1000 Berlin 12
Germany   §
Tel:(30) 31 90 03
Fax:(30) 31 32 03-7

FIZ-Karlsruhe
7514 Eggenstein-
Leopoldshafen 2
Germany   §
Tel:+49 7247 808555
Fax:+49 7247 808666
- *also STN International*

FIZ-Technik
Postfach 60 05 47
Ostbahnhofstrasse 13
6000 Frankfurt/Main 1
Germany   §
Tel:+49 69 4308-2-25
Tlx:4 189 459
Fax:+49 69 4308-2-00

**Florida Institute of
Technology**
Underwater Technology
Department
1707 NE Indian River Drive
Jensen Beach FL 33457
USA
Tel:(305) 334 4200

Fluid Engineering Centre
- see BHR Group Ltd

**Food & Agriculture
Organisation**
Via delle Terme di Caracalla
Rome 00100
Italy   ¶
Tel:+39 6 5797
Tlx:6181 Foodagvi
Fax:+39 6 579 73152
-*Fisheries Information, Data
& Statistics Service*

Forecast Plus
- see DRI/McGraw Hill

**Fort Bovisand Underwater
Centre**
Plymouth
Devon PL9 0AB
UK
Tel:(0752) 42570
Tlx:45639 (P27)

Foster Associates Inc
1015 15th Street NW
Suite 1100
Washington DC 10005
USA   §
Tel:(202) 408 7710
Fax:(202) 408 7723

Fraser Williams
London House
London Road South
Poynton
Cheshire SK12 1YP
UK   §
Tel:(0625) 871126
Fax:(0625) 871128

Fred B Rothman & Co
10368 W Centennial Road
Littleton CO 80127
USA
Tel:(303) 979 5657

**Freeman (W H) & Company
Ltd**
660 Market Street
San Francisco CA 94104
USA
Tel:(415) 391 5870
Tlx:340542
and
20 Beaumont Street
Oxford OX1 2NQ
UK
Tel:(0865) 726975
Tlx:83677
Fax:(0865) 790391

**French Assn of Independent
Petroleum Companies**
- see Federation Francaise
des Petroliers Independants

**Fridtjof Nansen Inst at
Polhogda**
Fr Nansenvei 47
0162 Oslo 1
Norway

**Fridtjof Nansen Institute
Library**
Fr Nansensvei 17
1324 Lysaker
Norway

FT Electronic Publishing
Southwark Bridge
London SE1 9HL
UK   §
Tel: (071) 873 3000
Fax: (071) 925 2125

FT North Sea Letter
10 Cannon Street
London EC4Y 4BY
UK

**Fulmer Research Institute
Ltd**
Hollybush Hall
Stoke Poges
Slough, SL2 4QD
UK
Tel:(0281) 62181
Tlx:849374
Fax:(0281) 63178

G

Gale Research Company
Book Tower
Detroit MI 48226
USA
Tel:(313) 961-2242

Gas Consumers Concil
6th Floor
Abford House
15 Wilton Road
London SW1V 1LT
UK
Tel: (071) 730 4696

Gas Research Institute
8600 W Bryn Mawr Avenue
Chicago, IL 60631
USA
Tel:(312) 399 8100

Gastech Ltd
2 Station Road
Rickmansworth
Herts WD3 1QP
UK
Tel:(0923) 776363
Fax:(0923) 777206

Gaz de France
Direction Generale
23 rue Philibert Delorme
75840 Paris Cedex 17
France
Tel:+33 1 47 542020
Tlx:650483
and
Direction des Etudes et
Techniques Nouvelles
361 Avenue de President
Wilson
93210 La Plaine Saint-Denis
France   §
Tel:+33 1 49 225800
Fax:+33 1 49 225760

G-Cam-Serveur
Europeenne de Donnees
1 rue de Boccador
75008 Paris
France   §
Tel:+33 1 47 208834
Fax:+33 1 47 201143
and
Tour Maine-Montparnasse
33 avenue du Maine
75755 Paris Cedex 15
France
Tel:+33 1 4538 7072
Tlx:203933

GCPES
Rua das Granjas 13
Granja Vianna
Cotia
Sao Paulo 06700
Brazil
Tel:+55 11 813 4792

Gebruder Borntraeger
Johannesstrasse 3A
D 7000 Stuttgart 1
Germany
Tel:+49 711 62354143

General Electric Information Services - GEIS
401 North Washington
Street
Rockville MD 20850
USA   §
Tel:(301) 294 5405
Fax:(301) 294 5501

GEISCO
5050 Quorum Drive
Suite 500
Dallas TX 75240
USA
Tel:(214) 788 8214
and
Geisco House
25/29 High Street
Kingston-upon-Thames
Surrey KT1 1LN
UK
Tel:(01) 546 8821
895705

General Directorate of Environmental Protection
- Department of the
Environment

General Microfilm Co
70 Coolidge Hill Road
Watertown MA 02171
USA
Tel:(617) 926 5557

General Videotex Corporation - Delphi
3 Blackstone Street
Cambridge MA 02139
USA   §
Tel:(617) 491 3393
Fax:(617) 491 6642

Geo Abstracts / Elsevier
Regency House
34 Duke Street
Norwich NR3 3AP
UK   §
Tel: (0603) 626327
Fax: (0603) 667934

GeoBased Systems Inc
PO Box 13545
Research Triangle Part
NC 27709
USA
Tel:(919) 361 5717
Tlx:294116

Geochemical Society
c/o Department of Geology
McMaster University
Hamilton
Ontario L8S 4M1
Canada
Tel:(416) 525 9140
and
c/o Dept of Geological
Sciences
Wright State University
Dayton OH 45435
USA
Tel:(513) 873 3455

Geocomp Corp
342 Sudbury Road
Concord MA 01742
USA
Tlx:466193

GeoFiz
Bundesanstalt fur
Geowissenschaften und
Rohstoffe
Postfach 51 01 53
Stilleweg 2
D-3000 Hannover 51
Germany   §
Tel:(511) 643 2819
Fax:(511) 643 2304

Geolab Nor AS
Hornebergveien 5
PO Box 1581
Nidarroll
N7002 Trondheim
Norway
Tel:+47 75 964000

Geological Association of Canada
c/o Dept of Earth Sciences
Memorial University of
Newfoundland
St Johns
Newfoundland A1B 3XS
Canada
Tel:(709) 737 8000

Geological Society of America
PO Box 9140
3300 Penrose Place
Boulder CO 80301-9140
USA   ¶
Tel:(303) 447 2020
Fax:(303) 447 1133

Geological Society of London
Burlington House
Piccadilly
London W1V 9HG
UK
Tel:(071) 434 9944
Fax:(071) 439 8975

Geological Survey of Canada
Dept of Energy, Mines &
Resources
601 Booth Street
Ottawa
Ontario K1A 0EB
Canada

Geological Survey of Denmark
Thoravej 31
2400 Copenhagen NV
Denmark
Tel:+45 1 106600
Fax:+45 1 196868

Geological Survey of Ireland
14 Hume Street
Dublin 2
Eire
Tel:+353 1 760855

Geological Survey of Norway
Leiv Erikssons Vei 39
Trondheim
Norway

Geological Survey of the Netherlands
PO Box 157
2000 AD Haarlem
The Netherlands
Tel:+31 23 30 03 00
Tlx: 71105
Fax:+31 23 25 16 14

Geologische Bundesanstalt
Rasumofskygasse 23
Postfach 154
1031 Vienna
Austria
+43 222 72 5674

Geologists Association
Burlington House
Piccadilly
London W1V 9AG
UK
Tel:(01) 734 2356

Geophysical Directory Inc
PO Box 130508
Houston TX 77219
USA
Tel:(713) 529 8789
Fax:(713) 529 3646

Geophysics Directorate
- see British Geological
Survey

**Geoscience Information
Society**
c/o American Geological
Institute

Geosystems
PO Box 40
Didcot, OX11 9BX
UK   §
Tel: (0235) 813913

GeoVision
1600 Carling Avenue
Suite 350
Ottawa
Ontario K1Z 8R7
Canada
Tel:(613) 722 9418
Tlx:0534517
Fax:(613) 7225285
and
5680 Peachtree Parkway
Norcross GA 30092
USA   §
Tel:(404) 448 8224
Fax:(404) 447 4524

GEP-ASTEO
15 rue Beaujon
75008 Paris
France
Tel:+33 1 46 227738
Tlx:280900 F fedem
Fax:+33 1 45 635986

**German Business
Information**
Bahnhofstrasse 27a
Postfach 1323
8043 Munich 80
Germany   §
Tel:(89) 9 50 60 95
Fax:(89) 9 50 39 78

**German Shipbuilding &
Ocean Industries Assn**
(Verband fur Schiffbau und
Meerestechnik e v)
An Den Alter 1
D-2000 Hamburg 1
Germany
Tel:+49 40 246205
Tlx:2162496
Fax:+49 40 246287

Germanischer Lloyd
PO Box 11 1606
Vorsetzen 32
2000 Hamburg 11
Germany
Tel:+49 40 361491
Tlx:212838
and
4th Floor
Riverdale House
68 Molesworth Street
London SE13 7EY
UK
Tel:(081) 318 1319
Tlx:28744
Fax:(081) 852 6115

GERMINAL
- c/o Bureau de Recherches
Geologiques et Minieres

Getty Research Library
PO Box 770070
Houston TX 77215-0070
USA

Gilbert Eliott & Company
381 Salisbury House
London Wall
London EC2M 5SB
UK
Tel:(071) 628 6782
Tlx:888886
Fax:(071) 628 3500

**GKSS Gesellschaft fur
Kernenergieverwertung
inSchiffbau und Schiffahrt
mbH**
Postfach 160
2054 Geesthacht
Germany
Tel:+49 04152 121

**Glasgow College of
Technology**
- see Institution of
Environmental Science

**Global Corrosion
Consultants Ltd**
Hazledine House
Telford, TF3 4JL
UK
Tel:(0952) 502502

Goodfellow Associates
71 Eccleston Square
London SW1V 1PJ
UK
Tel:(071) 821 1377

**Gordon & Breach Science
Publishers**
Reading Bridge House, 5th
Floor
Reading Bridge Approach
Reading RG1 8PP
UK
Tel:(0734) 560080
Fax:(0734) 568211
and
PO Box 786
Cooper Station
New York, NY 10276
USA
Tel:(212) 206 8900
Fax:(212) 645 2459
and
58 rue Lhomond
75005 Paris
France
Tel:+33 1 43 36 2404
Fax:+33 1 47 07 1526

**Government of
Newfoundland & Labrador**
Petroleum Directorate
PO Box 4750
St John's
Newfoundland A1C 5T7
Canada
Tel:(709) 737 2323
Tlx:016-4034

**Gower Publishing Company
Ltd**
Gower House
Croft Road
Aldershot
Hants GU11 3HR
UK
Tel:(0252) 331551
Tlx:858001
Fax:(0252) 344405

Grafton Books Ltd
- see HarperCollins

**Graham & Trotman Ltd
(Kluwer)**
Sterling House
66 Wilton Road
London SW1V 1DE
UK
Tel:(071) 821 1123
Tlx:298878 GramcoG
Fax:(071) 630 5229

Granada Publishing
-see HarperCollins
Publishers

**Greek Ministry of Industry,
Energy & Technology**
General Directorate for
Petroleum & Energy
80 Michalacopoulou Street
Gr 10192 Athens
Greece
Tel:+30 1 770 8615-1
Fax:+30 1 778 3711

Greenwell Montagu
Bow Bells House
Bread Street
London EC4M 9EL
UK
Tel:(071) 236 2040
Tlx:883006
Fax:(071) 248 0702

Greig, Middleton & Co
78 Old Broad Street
London EC2M 1JE
UK
Tel:(071) 920 0481
Tlx:887296
Fax:(071) 377 0353

**Group of Experts on the
Scientific Aspects of
Marine Pollution**
- see Food & Agriculture
Organisation

Gulf Canada Resources Ltd
401 9th Avenue SW
PO Box 130
Calgary
Alberta T2P 2H7
Canada   ¶
Tel:(403) 233 4000
Tlx: 038-24551
Fax:(403) 233 5143

Gulf Publishing Company
Book & Software Div:
Energy Bibliography & Index
Texas A&M University
Library
PO Box 2608
Houston TX 77252-2608
USA   §
Tel:(713) 529 4301
Fax:(713) 520 4438
and
PO Box 2608
3301 Allen Parkway
Houston TX 77001
USA
Tel:(713) 529 4301
Tlx:762908

H

Hale, Robert
- see Robert Hale Ltd

**Hamburg Messe Und
Congress**
Postfach 30 23 60
2000 Hamburg 36
Germany

Harcourt Brace Jovanovich
757 Third Avenue
New York NY 10017
USA
Tel:(212) 888 4444
Tlx:127891
and
24-28 Oval Road
London NW1 7DX
UK
Tel:(071) 267 4466
Fax:(071) 482 2293

HarperCollins Publishers
77-85 Fulham Palace Road
Hammersmith
London W6 8JB
UK
Tel:(081) 741 7070
Fax:(081) 307 4440

Hart Publications Inc
PO Box 1917
Denver CO 80201
USA
Tel:(713) 520 4444

Harvester Press
- Simon & Schuster
International Group

Health & Safety Executive
Broad Lane
Sheffield S3 7HQ
UK   ¶§
Tel:(0742) 78141
Tlx: 54556
Fax:(0742) 755802+720006
and
Baynards House
1 Chepstow Place
London W2 4TF
UK
Tel:(071) 221 0870
Tlx: 25683 SafetyG
Fax:(071) 727 2254
and
Explosions & Flame
Laboratory
Harpur Hill
Buxton
Derbyshire SK17 7AW
UK
Tel:(0198) 6211
Tlx:668113
and
Occupation Medicine &
Hygiene Labs
St Hugh's House
Trinity Road
Bootle
Merseyside L20 3OY
UK
Tel:(051) 951 4000

**Heat Transfer & Fluid Flow
Service**
Harwell Laboratories
Didcot OX11 0RA
UK   §
Tel: (0235) 24141)

Heinemann
- see Reed International
Books

**Hemisphere Publishing
Corporation**
1025 Vermont Avenue NW
Washington DC 20005
USA
Tel:(202) 783 3958

**Her Majesty's Stationery
Office**
- see HMSO

Heriot Watt University
Riccarton
Edinburgh EH14 4AS
Scotland
Tel:(031) 449 5111
Fax:(031) 451 3164
- Offshore Engineering
Information Service   ¶
-*Dept of Petroleum Eng'g*
- *Institute of Acoustics*
and
Dept of Electrical &
Electronic Eng
Mountbatten Building
Grassmarket
Edinburgh EH1 2HT
Scotland
Tel:(031) 225 8432

Highlands & Islands
Enterprise
27 Bridge Street
Inverness IV1 1QR
Scotland   ¶
Tel:(0463) 234171
Tlx:75267
Fax:(0463) 244469

Highline Community College
Marine Diving Technical
Programme
721 Cliff Drive
Santa Barbara CA 93105
USA

HMSO
Publications Division
St Crispin's
Duke Street
Norwich NR3 1PD
UK
Tel:(0603) 22211
and
Publications Centre
PO Box 276
London SW8 5DT
UK
Tel:(071) 873 9090
Tlx:297138
Fax:(071) 873 8463
and
Agency Publications
(official international
publications -eg UN, EEC)
PO Box 276
London SW8 5DT
UK
Tel:(071) 622 2216

HMSO Bookshops
80 Chichester Street
Belfast BT1 4JY
N Ireland
Tel:(0232) 238451
and
258 Broad Street
Birmingham B1 2HE
UK
Tel:(021) 643 3740
and
Southey House
Wine Street
Bristol BS1 1BQ
UK
Tel:(0272) 264306
and
71 Lothian Road
Edinburgh EH3 9AZ
Scotland
Tel:(031) 228 4181
and
49 High Holborn
London WCIV 6HB
UK
Tel:(071) 873 0011
and
9-21 Princess Street
Albert Square
Manchester M60 8AS
UK
Tel:(061) 834 7201

HMW Enterprises Inc
604 Salem Road
Etters, PA 17319
USA
Tel:(717) 938 4691

Hoare Govett Investment
Research
4 Broadgate
London EC2M 7LE
UK
Tel:(071) 601 0101
Tlx:297801
Fax:(071) 256 8500

Holleman (A J) Engineering
Ltd
Box 5317
St John's
Newfoundland A1C 5W1
Canada
Tel:(709) 753 0040
Tlx: 016 4714

Hollobone (T A) & Co Ltd
177a High Street
Beckenham, Kent BR3 1AH
UK
Tel:(081) 663 3859
Fax:(081 633 3860
- *AODC*

Hoover Institution Press
Stanford University
Stanford CA 94305
USA
Tel:(415) 497 3373

Hopewell Agency
154 Second Street
Trenton NJ 08611
USA   §
Tel: (609) 394 0370

Hotline Energy Reports
70 West 6th Street, # 415
Denver CO 80205
USA
Tel:(303)623 7130

Howell Publications
PO Box 27561
Houston TX 77227
USA
Tel:(713) 961 4432

HR Wallingford
Wallingford
Oxfordshire OX10 8BA
UK   ¶
Tel:(0491) 35381
Tlx:848552
Fax:(0491) 32233

Hughes Tool Company
PO Box 2539
Houston TX 77252-2539
USA   §
Tel:(713) 924 2222
Tlx:1620075
Fax:(713) 924 2009

Humbolt State University
Library
Documents Department
Arcata CA 95521
USA
Tel:(707) 826 3419

Hungarian Hydrocarbon
Institute
PO Box 32
22443 Szazhalombatta
Hungary
Tel:+36 26 528 10
Fax: +36 26 5 46 15

Hunter Publishing Ltd
950 Lee Street
Des Plaines IL 60016
USA
Tel:(312) 296 0770

Hydraulics Research Ltd
- see HR Wallingford

Hydrographer of the Navy
Hydrographic Office
Ministry of Defence (Navy)
Taunton
Somerset TA1 2DN
UK   ¶
Tel:(0823) 337900
Tlx:46274
Fax:(0823) 284077

Hyman Unwin Ltd
37/39 Queen Elizabeth
Street
London SE1 2QB
UK
Tel:(071) 407 0709
Tlx:886245
Fax:(071) 831 9489

Hydrographic Society
- c/o NE London Polytechnic

I

IAWPRC, UK Committee
- c/o Marine Biological Assn

IBC Technical Services Ltd
Bath House
56 Holborn Viaduct
London EC1A 2EX
UK
Tel:(071) 236 4080
Tlx:888870 OYEZ SE
Fax:(071) 589 0849

**ICI Industrial Software
Products**
PO Box 7 Brunner House
Winnington
Northwich
Cheshire CW8 4DJ
UK
Tel:(0606) 705575
Tlx:629655
Fax:(0606) 75885

ICIS-LOR Group Ltd
18 Upper Grosvenor Street
London W1X 9PD
UK   §
Tel:(071) 493 3040
Tlx:296557
Fax:(071) 499 8200

ICL
Computer Graphics Division
322 Euston Road
London NW1 3BD
UK
Tel:(071) 387 7030
Tlx:22971

IFI/Plenum Data Company
302 Swann Avenue
Alexandria VA 22301
USA   §
Tel:(703) 683 1085
Fax:(703) 683 0246

IFREMER
66 Avenue d'Iena
75116 Paris
France
Tel:+33 1 47 235528
Tlx:610775
and
Technoplolis 40
155 rue Jean-Jacques
Rousseau
92138 Issy les Moulineaux
Tel:+33 1 46 482100
Tlx: 610775
Fax:+33 1 46 482296
and
Centre de Brest
Boite Postale 70
29280 Plouzane
France   ¶
Tel:+33 98 224013
Tlx:940627
Fax:+33 98 224545

IIR Ltd
28th Floor, Centre Point
103 New Oxford Street
London WC1A 1DD
UK
Tel:(071) 412 0141
Fax:(071) 412 0145

**IKU - Continental Shelf &
Petroleum Research
Institute**
S P Andersens Veg 15b
N 7034 Trondheim
Norway   ¶
Tel:+47 7 541100
Tlx:55434
Fax:+47 7 591102

Image Graphics Inc
917 Bridgeport Avenue
Shelton CT 06484
USA
Tel:(203) 926 0100
Fax:(203) 9269705

IMCO Services
2400 West Loop South
PO Box 22605
Houston TX 77027
USA
Tel:(713) 561 1393

**Imperial College of Science
& Technology**
Exhibition Road
London SW7 2BU
UK
Tel:(071) 589 5111
Tlx:261503 IMPCOL LD
Fax:(071) 584 7596
*- London Centre for Marine
Technology
and*
Royal School of Mines
London SW7 2BP
UK
Tel:(071) 225 8478
Fax:(071) 589 6806
*-Dept of Mineral Resources
Engineering
- Rock Mechanics
Information Service*

**Independent Chemical
Information Services ICIS**
Latour Grand House
28 avenue Alfred Belmontet
92210 St Cloud/Paris
France
Tel:+33 1 771 0308

**Independent Petroleum
Association of America**
1101 16th Street NW
Washington DC 20036
USA   §
Tel: (202) 857 4722

**Indian Institute of
Technology**
Madras 600 036
India
and
Powai
Bombay 400 076
India
Tel:+91 22 011 71385

**Indonesian Petroleum
Association**
J L Menteng Raya 3
Jakarta
Indonesia
Tel:+62 21 350235

Industrial Promotions
PO Box 53
Rickmansworth
Herts WD3 2AG
UK
Tel:(0923) 778311

Industrial & Marine Publications
2 Queensway
Redhill
Surrey RH1 1QS
UK
Tel:(0737) 768611

Industrial & Trade Fairs International
Blenheim Court
Solihull
Birmingham B91 1BG
UK
Tel:(021) 705 6707

Industriele Raad voor de On- en Offshore
- see Netherlands Industrial Council for Oceanology

Industry Department for Scotland
Alhambra House
45 Waterloo Street
Flasgow G2 6AT
Scotland
Fax:(041) 248 2855

Infield Systems Ltd
15 ARtillery Passage
Bishopsgate
London E1 7LJ
UK   §
Tel:(071) 377 0102
Fax:(071) 247 5035

INFOCEAN
BP 332
F-29273 Brest
France

INFOIL II
- see MTD/Norwegian Petroleum Directorate

Infomart Online
1450 Don Mills Road
Don Mills
Ontario M3B 2X7
Canada   §
Tel:(416) 445 6641
Fax:(416) 445 3508

Infonorme - London Information
Index House
Ascot, Berkshire SL5 7EU
UK
(0344) 23377
849426 Loninf G
(0344) 292294

Information Access Company
362 Lakeside Drive
Foster City CA 94404
USA   §
Tel:(415) 387 5000
Fax:(415) 378 5499

Information Inc
1725 K Street
Suite 1414
Washington DC 20006
USA   §
Tel:(202) 833 1174
Fax:(202) 331 3842

Info-One International
7th Floor
77 Pacific Highway
North Sydney NSW 2060
Australia   §
Tel:(2) 959 5075
Fax:(2) 929 5127

INIST (Diffusion)
see Centre National de la Recherece Scientifique   §

INKADAT
Fachinformationszentrum
Karlsruhe
7514
Eggenstein-Leopoldshafen 2
Germany   §
Tel:+49 7247 824568
Tlx:7836487

Inmarsat Maritime Satellite Communications
40 Melton Street
London NW1
UK
Fax:(071) 387 2115

Inspectorate of Explosives & Flammables
PO Box 355
3100 Tonsborg
Norway

Institut Francais de l'Energie
3 rue Henri-Heine
Paris 75016
France
Tel:+33 1 45 244614
Tlx:615867

Institut Francais du Petrole
Direction de l'Economie et de la Documentation
1-4 Avenue de Bois-Preau
BP 311
92506 Rueil-Malmaison
France   §
Tel:+33 1 47 490214
Tlx:IFP A 203050
Fax:+33 1 47 526429

Institut Francaise de Recherche pour l'Exploitation de la Mer
- see IFREMER

Institut Maurice-Lamontagne Peches et Oceans
CP 1000, 850 Route de la Mer
Mont-Joli
Quebec G5H 3Z4
Canada
Tel:(418) 775 6500
Tlx:051-8-8303

Institut National de Plongee Professionnelle et d'Intervention en Milieu Aquatique
Port de la Pointe Rouge
Marseille 13008
France
Tel:+33 91 733 462
Tlx:430315

Institute de l'Information Scientifique et Technique
- see Centre National de la Recherche Scientifique

Institute for Energy Technology
Box 173
1751 Halden
Norway
Tel:+47 3 181100
Tlx:76335
Fax:+47 3 181120
and
Box 40
2007 Kjeller
Norway
Tel:+47 6 812560
Tlx:74573 energ n
Fax:+47 6 816356

Institute for Geological Research
02150 Otaniemi
Finland
Tel:+385 461011

**Institute for Marine
Environmental Research**
Prospect Place
The Hoe
Plymouth PL1 3DH
UK
Tel:(0752) 21371

**Institute for Mechanical
Construction**
Leeghwaterstraat 5
PO Box 29
Delft
The Netherlands

**Institute for Scientific
Information**
3501 Market Street
Philadelphia PA 19104
USA
Tel:(215) 386 0100
and
132 High Street
Uxbridge
Middlesex UB8 1DP
UK
Tel:(0895) 70016
Tlx:933693

**Institute of Corrosion
Science & Technology**
Exeter House (1 Corr ST)
48 Holloway Head
Birmingham B1 1NQ
UK
Tel:(021) 622 1912

Institute of Diving
PO Box 876
Panama City
Panama
Tel:+507 769 7544

Institute of Energy
18 Devonshire Street
London W1N 2AU
UK   §
Tel:(071) 580 7124
Fax:(071) 580 4420

**Institute of Energy
Development**
PO Box 19243
Oklahoma City OK 73144
USA
Tel:(405) 691 4449

**Institute of Energy
Economics**
Energy Data and Modelling
Centre (EDMC)
10 Mori Building
1-18-1 Toranomon
Minato-Ku
Tokyo 105
Japan
Tel:+81 3 501 2901
Fax:+81 3 504 2412

Institute of Gas Technology
2424 South State Street
Chicago IL 60616-3896
USA   ¶§
Tel:(312) 567 3650
Tlx:256189
Fax:(312) 567 5209
- *IGNET*

Institute of Hydrography
Avenida do Brasil
Lisbon 5
Portugal

**Institute of Industrial
Research & Standards**
Ballymun Road
Dublin 9
Eire
Tel:+353 1 371101
Tlx:0027 5449
Fax:+353 1 379620

**Institute of Information
Scientists**
44-45 Museum Street
London WC1Y 1LY
UK
Tel:(071) 831 l8003

Institute of Marine Engineers
76 Mark Lane
London EC3R 7JN
UK   ¶
Tel:(071) 481 8493
Tlx:886841
Fax:(071) 488 1854

Institute of Materials
PO Box 471
1 Carlton House Terrace
London SW1Y 5BE
UK
Tel:(071) 839 4071

Institute of Metals
- see Institute of Materials

Institute of Naval Medicine
Alverstoke
Gosport
Portsmouth PO12 2DL
UK
Tel:(0705) 822351

**Institute of Oceanographic
Sciences**
Deacon Laboratory
Brook Road, Wormley
Godalming
Surrey GU8 5UB
UK   ¶
Tel:(0428) 79 4141
Tlx:858833
Fax:(0428) 793066
and
The Bidston Observatory
Bidston
Birkenhead
Merseyside L43 7RA
UK
Tel:(051) 653 8633

Institute of Petroleum
61 New Cavendish Street
London W1M 8AR
UK   ¶
Tel:(071) 636 1004
Tlx:264380
Fax:(071) 255 1472

**Institute of Sound &
Vibration Research**
-see University of
Southampton

**Institute of Water &
Environmental Managemnt**
15 John Street
London WC1N 2EB
UK
Tel:(071) 831 3110

**Institution of Chemical
Engineers**
165-171 Railway Terrace
Rugby CV21 3HQ
UK   ¶§
Tel:(0788) 78214
Tlx:311780
Fax:(0788) 60833

Institution of Civil Engineers
1-7 Great George Street
London SW1 3AA
UK   ¶
Tel:(071) 222 7722
Tlx:946185
Fax:(071)222 7500

Institution of Electrical Engineers
Savoy Place
London WC2R 0BL
UK ¶
Tel:(071) 240 1871
Tlx:261176
Fax:(071)497 3557

Institution of Electrical & Electronic Engineers
345 E 47th Street
New York, NY 10017
USA
Tel:(212) 644 7900
Tlx:236411
and
445 Hoe Lane
Piscataway NJ 08854
USA

Institution of Environmental Science
Glasgow College of Technology
7 Cowcaddens Road
Glasgow G4
Scotland
Tel:(041) 332 7090

Institution of Gas Engineers
17 Grosvenor Crescent
London SW1X 7ES
UK
Tel:(071) 245 9811

Institution of Geologists
2nd Floor
Geological Society Apts
Burlington House
London W1V 9HG
UK
Tel:(071) 734 0751

Institution of Mechanical Engineers
1 Birdcage Walk
London SW1H 9JJ
UK
Tel:(071) 222 7899
Tlx:917944
Fax:(071) 222 9881

Institution of Metallurgists
- now see Institute of Materials

Institution of Mining Engineers
Danum House
6a South Parade
Doncaster DN1 2DY
UK
Fax:(0302) 340554

Institution of Mining & Metallurgy
44 Portland Place
London W1N 4BR
UK
Tel:(071) 580 3802
Tlx:261410
Fax:(071) 436 5388

Institution of Structural Engineers
11 Upper Belgrave Street
London SW1X 8BH
UK
Tel:(071) 235 4535
Fax:(071) 235 4294

Instituto Argentino del Petroleo
Maipu 645, 3rd Floor
1006 Buenos Aires
Argentina

Instituto Brasiliero de Petroles
S 1034a 1038
Caixa Postal 343
CEP 29 073
Estado do Rio de Janeiro
Brazil

Instituto Geologico y Minero de Espana
Rios Rosas 23
28003 Madrid
Spain
Tel:+34 91 441 65 00

Instituto Mexicano del Petroleo
Eje Central Lazaro
Cardenas 152
Mexico 14 DF
Mexico
Tel:+52 5 67 66 00

Instituto Nacional de Hidrocarburos
Pl de la Castellana 89
28046 Madrid
Spain
Tel:+34 1 348 8100
Fax:+34 1 455 7671

Inter Documentation Co AG
Poststrasse 14
6300 Zug
Switzerland

Intergovernmental Oceanographic Commission
c/o UNESCO
7 Place de Fontenoy
75700 Paris
France
Tel:+33 1 45 681000
Tlx:204461

International Assn for Bridge & Structural Engin'g
ETH Honnggerberg
8093 Zurich
Switzerland
and c/o Instn of Structural Engineers

International Assn of Drilling Contractors
PO Box 4287
Houston TX 77210
USA ¶
Tel:(713) 578 7171
Tlx: 516985 INTDRILL
Fax:(713) 578 0589

International Assn of Geophysical Contractors
Suite 400
5335 West 48th Svenue
Denver CO 80212
USA
Tel03) 458 8404

International Association for Hydaulic Research
Rotterdamseweg 185
POB 177
2600 MH Delft
The Netherlands

International Association of Underwater Engineering Contractors
177a High Street
Beckenham
Kent BR3 1AH
UK
Tel:(081) 663 3859
Fax:(081) 663 3860

International Atomic Energy Agency
Vienna International Centre
PO Box 100
Wagramenstrasse 5
1400 Wien
Austria

International Center for Marine Resource Development
- see University of Rhode Island ¶

International Centre for Scientific & Technical Information (ICSTI)
Ulitsa Kuusinena 21B
125252 Moscow
Russia   §
Tel:+7 095 198 7460
Fax:+7 095 943 0089

International Chamber of Shipping
30-32 St Mary Axe
London EC3A 8ET
UK
Tel:(071) 283 2922
Fax:(071) 626 8135

International Council for the Exploration of the Sea
Palaegade 2-4
1261 Copenhagen K
Denmark
Tel:+45 1 154225
Tlx:2036 22498

International Energy Agency
2 rue Andre-Pascal
75775 Paris Cedex 16
France   §
Tel:+33 1 45 249887
Tlx:630190
Fax:+33 1 45 249988

International Federation for Information Processing
rue Du Marche 3
Ch 1204 Geneva
Switzerland
Tel:+41 22 282649

International Human Resources Development Corpn
137 Newbury Street
Boston MA 02116
USA
Tel:(617) 536 0202

International Information Centre for Natural Gas & Hydrocarbon Gases
BP 311
1/4 Avenue de Bois-Preau
92506 Rueil Malmaison
France
Tel:+33 47 52 60 12
Tlx: IFP A 203050 F
Fax:+33 1 47 526429

International Labour Office
1211 Geneva 22
Switzerland
Tel:+41 22 99611
Tlx:22271
and
Vincent House
Vincent Square
London SW1P 2NB
UK
Tel:(071) 828 6401
Tlx:886836 INTALAB

International Maritime Organization
4 Albert Embankment
London SE1 7SR
UK
Tel:(071) 735 7611
Tlx:23588
Fax:(071) 587 3210

International Ocean Institute
Royal University of Malta
Msida
Malta

International Oceanographic Foundation
4600 Rickenbacker
Causeway
Miami FL 33149
USA   ¶
Tel:(305) 361 4888
Fax:(105) 361 4711

International Palaeonto-logical Association
US Geological Survey
E-305 Natural History Bldg
Smithsonian Institution
Washington DC 20560
USA
Tel:(202) 343 3523

International Petroleum Exchange of London
International House
1 St Katherine's Way
London E1 9UN
UK
Tel:(071) 481 0643
Tlx:927479
Fax:(071) 481 8485

International Petroleum Industry Environmental Conservation Assn
1 College Hill (first floor)
London EC4R 2RA
UK
Tel:(071) 248 3447
Fax:(071) 489 9067

International Pipe Line Contractors Assocn
95 Boulevard Berthier
75017 Paris
France
Tel:+33 1 43 806153

International Publications Service
114 East 52nd Street
New York NY 10016
USA
Tel:(212) 685 9351

International Research & Development Co Ltd
Fossway
Newcastle upon Tyne
NE6 2YD
UK
Tel:(091) 265 0451
Tlx:537086

International Science Services
Conventionen 62-49
3703 HN Zeist
The Netherlands
Tel:+31 0 3405 5109

International Society of Offshore & Polar Engineers
PO Box 1107
Golden CO 80402-1107
USA   ¶
Tel:(303) 273 3673
Fax:(303) 420 3760

International Tanker Owners Pollution Federation
Staple Hall, Stonehouse Ct
87-90 Houndsditch
London EC3A 7AX
UK   §
Tel:(071) 621 1255
Tlx:887514 tovlop
Fax:(071) 626 5913

International Thomson Publishing Ltd
38 Hans Crescent
London SW1X 0LZ
Fax:(071) 225 2761

International Thomson Publishing Services Ltd
Cheriton House
North Way
Andover
Hampshire SP10 5BE
UK
Tel: (0264) 332424
Tlx: 47214
Fax:(0264) 364418

International Union of Geological Sciences
Ottawa Secretariat
601 Booth Street
Room 177 Ottawa
Ontario K1A 0E8
Canada

Iraq National Oil Company
PO Box 476
Baghdad
Iraq
Tel:+964 887 11 15
Tlx:2204 inoh IK

Irish Gas Association
24a d'Olier Street
Dublin 2
Eire
Tel:+353 1 787111

Irish Offshore Services Association
Confederation House
Kildare Street
Dublin 2
Eire
Tel:353 1 779801
Fax:353 1 777823

IRL Press
1911 Jefferson Davis, # 907 Highway
Arlington VA 22202
USA
Tel:(703) 998 2980
Tlx:899144
and
PO Box 1
Eynsham
Oxford OX8 1JJ
UK
Tel:(0865) 882283
Tlx:83147

Istituto Technico Industriale
Statale 'Alessandro Rossi'
52 Via Legione Gallienco
36100 Vicenzia
Italy
Tel:+39 444 506040

J

JAI Press Inc
36 Sherwood Place
POB 1678
Greenwich CT 06836
USA
Tel:(203) 661 7602

James Capel & Company
Petroleum Services Dept
James Capel House
POBox 551
6 Bevis Marks
London EC3A 7JQ
UK
Tel:(071) 621 0011
Tlx:888866
Fax:(071) 621 0496

James Lorimer & Co
35 Buthin Street
Toronto
M5A 3V8
Canada

Japan Hydrographic Department
Marine Safety Agency
5 Chome,7 Sukiji
Chuo-Ku
Tokyo
Japan

Japan Information Centre of Science & Technology
5-2 Nagatacho, 2 Chome
Chiyoda-ku
CPO Box 1478
Tokyo 100
Japan   §
Tel:+81 3 581 6411
Fax:+81 3 581 6446

Japan Marine Science & Technology Centre
2-15 Natsushima-Cho
Yokasuka-Shi 237
Japan
Tel:+81 468 65 2865

Japan Petroleum Development Association
Keidanren Kaikan, 9-4
1-chome
Ohtemachi,
Chiyoda-Ku
Tokyo 100
Japan
Tel:+81 3 279 5841

Japan Petroleum Institute
17 F Shin-Aoyama East Building
1-1 1-chome,
Minami-Aoyama
Minato-Ku
Tokyo 107
Japan
Tel:+81 3 475 1235

John Brown Constructors & Engineers Ltd
20 Eastbourne Terrace
London W2 6LE
UK
Tel:(071) 262 8080
Tlx:263521
Fax:(071) 262 0387

John M Campbell Publisher
1215 Crossroads Boulevard
Norman OK 73072
USA
Tel:(405) 321 1388

John Wiley and Sons
605 Third Avenue
New York NY 10158
USA
Tel:(212) 850 6000
Tlx:12-7063
and
Baffins Lane
Chichester
West Sussex PO19 1UD
UK   §
Tel:(0243) 770215
Tlx:86290
Fax:(0243)f 775878

Johns Hopkins University
Baltimore MD 21218
USA
Tel:(301) 338 7875
and
Chesapeake Bay Institute
Tel:(301) 366 3300
- *Johns Hopkins University Press*

Johnson Reprint Microeditions
111 Fifth Avenue
New York NY 10003
USA
Tel:(212) 614 3200

K

Kaihatsu Computing Centre
1-8-2 Marunouchi
Chiyoda-ku
Tokyo 100
Japan   §
Tel:+81 3 213 0921

Kartelias School of Diving
3 Karageorgi Servias
Kastella
Piraeus
Greece
Tel:+30 1 412 2047

Kenny (J P) & Partners Ltd
Thames Plaza
5 Pinetrees, Chertsey Lane
Staines, Middx TW18 3DT
UK
Tel:(0784) 6222
Fax:(0784) 462273

Kent Marketing Services Ltd
227 Colborne Street
London
Ontario N6B 2S4
Canada

Keynote Publications Ltd
28-42 Barren Street
London EC1P 8QE
UK
Tel:(071) 253 3006

KIVI
PO Box 30424
2500 GK The Hague
The Netherlands
Tlx:33641

Kluwer Academic Publishers
Group
PO Box 989
3300 AZ Dordrecht
The Netherlands
Tel:+31 78 178222
Tlx:20083
and
Distribution Center
PO Box 322
3300 AH Dordrecht
The Netherlands
Tel:+31 78 172811
Tlx:20083 KADC
and
PO Box 358
Accord Station
Higham MA 02018 0358
USA
Tel:(617) 871 6600
Tlx:200190 KLUER U

Knighton Enterprises Ltd
2 Marlborough Street
Farringdon, Oxon SN7 7JP
UK
Tel:(0367) 242525
Fax:(0367) 241125

Knowledge Systems Inc
4124 Walney Road
Chantilly Va 22021
USA   §
Tel:(703) 631 8622

Kogan Page Limited
120 Pentonville Road
London N1 9JN
UK
Tel:(071) 278 0433
Tlx:263088 KOGAN G
Fax:(071) 837 6348

Korea Ocean Research &
Development Institute
PO Box 29 An San
Seoul 425-600
Korea
Tel:+82 2 863 4770
Tlx:Kordi K27675
Fax:+82 345 6698

Kraus Reprints & Periodicals
Div of Kraus-Thomson
Organisation
One Water Street
White Plains NY 10601
USA
Tel:(914) 761 9600

Krieger Publishing Company
PO Box 9542
Melbourne FL 32901
USA
Tel:(407) 724 9542
Fax:(407) 951 3671

Kuwait Petroleum
Corporation
Salhia Complex
Faahed Al Salem Street
PO Box 26565
Safat
Kuwait
Tel:+965 245 5455
Tlx:44874/8

Kyodo News International Inc
50 Rockefeller Plaza, Rm806
New York NY 10020
USA   §
Tel:(212) 586 0152
Fax:(212) 307 1532

Kyukuto Boeki Kaisha Ltd
CPO Box 330, 7th Floor
New Otemachi Building
2-1, 2-Chome,Otemachi
Chiyoda-Ku
Tokyo 100-90
Japan
Tel:+81 3 244-3511

L

Langham Oil Conferences Ltd
376 Mail Street
Queniborough
Leicester LE7 8DB
UK
Tel:(0664) 424776
Fax:(0664) 424832

Law of the Sea Institute
-see Univ of Rhode Island

Law Society: Solicitors
European Group
113 Chancery Lane
London WC2A 1PL
UK
Tel:(071) 424 1222
Tlx:261203

Learned Information Inc
143 Old Marlton Pike
Medford NJ 08055-8750
USA
Tel:(609) 654 6266
Fax:(609) a654 4309
and
Learned Information
(Europe) Ltd
Woodside
Hinksey Hill
Oxford OX1 5AU
Fax:(0865 736354

Lehigh University
Maine Geotechnical
Laboratory
Building 17
Bethlehame, PA 18015
USA

Leith International
Conferences
c/o Jewel & Esk Valley
College
24 Milton Road East
Edinburgh EH15 2PP
Scotland
Tel:(031) 669 8461

Lexington Books
125 Spring Street
Lexington, MA 02173
USA
Tel:(617) 862 6650
Tlx:923455

Libraries Unlimited Inc
PO Box 263
Littleton CO 80160-0263
USA
Tel:(303) 770 1220

Library Association
7 Ridgemount Street
London WC1E 7AE
UK
Tel:(071) 636 7543
Fax:(071) 436 7218

Library Microfilms
737 Loma Verde Avenue
Palo Alto CA 94303
USA
Tel:(415) 494 1812

Library of Congress
Washington DC 20540
USA
Tel:(202) 426 5093
- *Photoduplication Service*
Tel:(202) 287 5650

Lincoln Electric Company
22801 St Clair Avenue
Cleveland OH 44117
USA
Tel:1 216 481 8100
Tl x:4332077

**Liquiefied Petroleum Gas
Industry Technical Assn**
17 Grosvenor Crescent
London SW1X 7ES
UK
Tel:(071) 245 9511

**Lloyd's Maritime
Information Services Ltd**
Dunster House
17-21 Mark Lane
London EC3R 7AP
UK
Tel:(071) 490 17201
Tlx:987321
Fax:(071) 929 3133
and
1 Singer Street
London EC2A 4LQ
UK   §
Tel:(071) 490 1720
Fax:(071) 250 3142

Lloyd's of London Press Ltd
Sheepen Place
Colchester
Essex CO3 3LP
UK
Tel:(0206) 772277
Tlx:9873231 Lloyds
Fax:(0206) 46273
and

1 Singer Street
London EC2A 4LQ
UK
Tel:(071) 250 1500
Tlx:987321

Lloyd's Register of Shipping
71 Fenchurch Street
London, EC3M 4BS
UK   ¶
Tel:(071) 709 9166
Tlx:888379
Fax:(071) 488 4796
and
Kronprinsessegade 26
1306 Copenhagen K
Denmark
Tel:+45 1 154015
Tlx:19851
and
Printing House, Manor Royal
Crawley, RH10 2QN
UK
Tel:(0444) 236060

Lomond Publications
POB 56
Mount Airy MD 21771
USA
Tel:(301)829 1496

**London Marine Technology
Centre**
- see Imperial College

London Oil Reports
51 Harrowby Street
London W1
UK
Tel:(071) 723 7997
Tl x:296557 LOR G

Long Beach Public Library
Ocean and Pacific Avenue
Long Beach, CA 90802
USA
Tel:(213) 437 2949

Longman Cartermill Ltd
Technology Centre
St Andrew's, Fife KY16 9EA
Scotland   §
Tel:(0334) 77660
Fax:(0334) 77180

Longman Group Ltd
Burnt Mill
Harlow, CM20 2JE
UK
Tel:(0279) 26721
Tlx:81259
Fax:(0279) 31067

Lorne Maclean Marine
34 Apsley End Road
Shillington
Hitchin, SG5 3LX
UK
Tel:(0462) 712761

Louisiana State University
Sea Grant College Program
134 Wetland Resources Bldg
Baton Rouge LA 70803
USA   ¶
Tel:(504) 388 6445
Fax:(504) 388 6331

Lundberg Surveys Inc
PO Box 3996
North Hollywood CA 91609
USA
Tel:(818) 768 5111
Tlx:(510)_ 1003197

M

Macdonald & Janes
74 Worship Street
London EC2A 2EN
UK
Tel:(071) 377 4600
Tlx:885233

Mackay Consultants
Balgownie Technology Cent
Balgownie Drive
Aberdeen AB22 8GW
Scotland
Tel:(0224) 822547
Fax:(0224) 822587

Maclean Hunter
30 Old Burlington Street
London W1X 2AE
UK
Tel:(071) 437 0644
Tlx:24555

Macmillan Distribution Ltd
Houndmills
Basingstoke
Hants RG21 2XS
UK
Tel:(0256) 29242
Tlx:85493

Macmillan Publishers Ltd
4 Little Essex Street
London WC2R 3LF
UK
Tel:(071) 836 6633
Tlx:262024
Fax:(071) 439 1440

Marathon International Petroleum (GB) Ltd
174 Marylebone Road
London NW1 5AT
UK
Fax:(071) 486 0222

Marcel Dekker
Hutgasse 4
Postfach 812
CH-4001 Basel
Switzerland
Tel:+41 61 258482
and
270 Madison Avenue
New York NY 10016
USA
Tel:(212) 696-9000
Tlx:421419

MARIN
- see Maritime Research
Institute Netherlands

Marine Biological Association of the UK
The Laboratory
Citadel Hill
Plymouth, PL1 2PB
UK
Tel:(0752) 21761
Tlx: 903241

Marine Board
- see National Research
Council   ¶

Marine Clearance Divers Grp
Navclearmin
Naval Base
Slijkenssteenweg 1
8400 Oostende
Belgium
Tel:+32 59 801402

Marine Conservation Society
4 Gloucester Road
Ross on Wye
Herefordshire HR9 5BN
UK
Tel:(0989) 66017

Marine Directorate
- see Department of
Transport

Marine Environmental Research
Taviton Mill House
Tavistock
Devon
UK

Marine Information Advisory Service
- see Inst of Oceanographic
Sciences   §

Marine Management (Holdings) Ltd
c/o 76 Mark Lane
London EC3R 7JN
UK
Tel:(071) 481 8493
Tlx:886841

Marine Policy
PO Box 63
Guildford
Surrey GU1 5BH
UK

Marine Publications International
42-43 Lower Marsh
London SE1 7RQ
UK
Tel:(071) 928 4491

Marine Science Communications
270 Madison Avenue
New York, NY 10016
USA

Marine Sciences & Information Directorate
- Canadian Dept of
Fisheries & Environment

Marine Survey Office
27 Eden Quay
Dublin 1
Eire
Tel:+353 1 744900

Marine Surveys Directorate
- see British Geological
Survey

Marine Technology Centre
NTNF
Hakon Hakonsonset 34
7000 Trondheim
Norway

Marine Technology Centre, Newcastle
- see Univ of Newcastle
upon Tyne

Marine Technology Directorate
- see MTD Ltd   ¶

Marine Technology Society
Suite 906
1828 L Street NW
Washington DC 20006
USA   ¶
Tel:(202) 775 5966
Fax:(202) 429 9417

Marine Technology Support Unit
Department of Energy
Harwell Laboratory B.10.28
Didcot
Oxfordshire OX11 0RA
UK   §
Tel:(0235) 435985
Fax:(0235) 432753

Marine & Estuarine Management Division
NOS/NOAA/OCRM
1825 Connecticut
Avenue,NW
Suite 714
Washington
DC 20235
USA
Tel:(202) 673 5122

Marinetech North West
Coupland III Bldg
University of Manchester
Manchester M13 9PL
UK
Tel:(061) 273 3278
Fax:(061) 273 8835

Marintek
- see Sintef Group   ¶

Maritime Research Information Service
Maritime Transportation
Research Board
National Academy of Science
2101 Constitution Ave NW
Washington DC 20418
USA

Maritime Research Institute Netherlands
Blaak 16
PO Box 21873
301 TA Rotterdam
The Netherlands
Tel:+31 10 130960
Tlx:26585

Martin Dunitz
154 Camden High Street
London NW1 0NE
UK
Tel:(071) 482 2202
Tlx:296307 NUNBKS

Martinus Nijhoff Publishers
- see Kluwer Academic
Publishers Group

**Massachusetts Coastal Zone
Management Office**
20th Floor
100 Cambridge Street
Boston MA 02202
USA ¶
Tel:(617) 727 9530
Fax:(617) 727 2754

**Massachusetts Institute of
Technology**
Microproduction Laboratory
Room 14-0551
Cambridge MA 02139
USA
and
Sea Grant Program
Room E38-356
Cambridge MA 02139
USA ¶
Tel:(617) 253 5944
Fax:(617) 258 8125

**Mathematical Geologists of
the United States**
c/o John C Davis
Geological Research Section
Kansas Geological Survey
Lawrence KS 66045
USA

Matthew Bender & Co Inc
PO Box 658
Albany NY 12201
USA
Tel:(518) 487 3000
Fax:(518) 487 3584
and
17 Clifton Road
London N3 2AS
UK
Tel:(071) 794 1617

Maus (J M J) Ltd
6A-10 Tudor Road
Hampton
Middlesex TW12 2NQ
UK
Tel:(081) 979 0201
Tlx:929805
Fax:(081) 979 0208

**MBC Information Services
Ltd**
Paulton House
8 Shepherdess Walk
London N1 7LB
UK §
Tel:(071) 490 0049
Fax:(071) 490 2979

McGill University
Marine Science Centre
Box 6070 Station A
Montreal
Quebec H3C 8G1
Canada
Tel:(514) 392 5723

McGraw-Hill Book Company
1221 Avenue of the
Americas
New York NY 10020
USA
Tel:(212) 512 2000
Tlx:232365
Fax:(212) 512 4871
and
McGraw-Hill House
Shoppenhangers Road
Maidenhead
Berkshire SL6 2QL
UK
Tel:(0628) 23432
Tlx:84884
Fax:(0628) 35895

McGraw-Hill Inc
Princeton-Hightstown Road
North-1
Hightstown NJ 08520
USA §
Tel:(609) 426 5523
Fax:(609) 426 7352

MCI International Inc
1 International Drive
Rye Brook, NY 10573-6830
USA §
Tel:(914) 934 6166
Fax:(071) 480 7228

McLaren Micropublishing
PO Box 972, Station F
Toronto
Ontario M4Y 2N9
Canada

Mead Data Central
9393 Springboro Pike
PO Box 933
Dayton OH 45401
USA §
Tel:(513) 865 6800
Fax:(513) 865 6909
and
International House
1 St Katherine;s Way
London E1 9UN
UK §
Tel:(071) 488 9187
Fax:(071) 480 7228

Meadowfield Press Ltd
ISA Building
Dale Road Industrial Estate
Shildon
Co Durham DL4 1QZ
UK
Tel:(0670) 55860

**Mechanical Engineering
Publications Ltd**
- see MEP Ltd

Medical Research Council
Decompression Sickness
Panel
20 Park Crescent
London WIN 4AL
UK
Tel:(071) 636 5422

**Mekaniske Verksteders
Landsforening**
c/o Beama
8 Leicester Street
London WC2H 7BN
UK
Tel:(071) 437 0678

**Memorial University of
Newfoundland**
Queen Elizabeth II Library
St John's
Newfoundland A1B 3Y1
Canada ¶
Tel:(709) 737 7425
Fax:(709) 737 3118
- see Ocean Engineering
Information Center

MEP Ltd
24 Northgate Avenue
Bury St Edmunds
Suffolk IP32 6BW
UK
Tel:(0284) 63277
Tlx:8167376

Metals Society
- see Institute of Materials

Meteorological Office
Ministry of Defence
London Road
Bracknell
Berkshire RG12 2SZ
UK
Tel:(0344) 420242
Tlx:848160 WeaklG
and
231 Corstorphine Road
Edinburgh EH12 7BB
Scotland
Tel:(031) 334 9721

Metes (J S) & Company (PTE) Ltd
Tanglin P O Box 158
Singapore 9124
Republic of Singapore
Tel:+65 737 3617

Metocean Consultancy Ltd
Hamilton House
Kings Road
Haslemere
Surrey GU27 2QA
UK
Tel:(0428) 56925
Tlx:9312110367
Fax:(0428) 61530

Microfilms International Marketing Company
Pergamon Press Inc
Maxwell House, Fairview Park
Elmsford NY 10523
USA
Tel:(914) 592 7700

MicroInfo Ltd
PO Box 3
Omega Park
Alton
Hants GU34 2PG
UK   §
Tel:(0420) 68648
Tlx:858431
Fax:(0420) 69889

Micromedia Limited
158 Pearl Street
Toronto
Ontario, M5H 1L3
Canada
Tel:(416) 593 5211

Middlesex Polytechnic
- now Middlesex University

Middlesex University
Bounds Green
New Southgate
London N11 2NQ
UK

Midwest Oil Register
PO Box 700597
Tulsa OK 74170
USA
Tel:(918) 742 9925

Mid-Continent Oil & Gas Association
711 Adams Office Building
Tulsa OK 74103
USA   §
Tel:(918) 582 5166

Mineralogical Society of America
Suite 330
1130 17th Street NW
Washington DC 20036
USA   ¶
Tel:(202) 775 4344
Fax:(202) 775 0018

Mineralogical Society of Canada
c/o Dept of Geological Sciences
McGill University
3450 University Street
Montreal H3A 2A7
Canada
Tel:(514) 392 5835

Mineralolwirtschaftsverband ev
Steindamm 71
2000 Hamburg 1
Germany

Minerals Management Service
Office of Offshore Information Services
Room 2070,Main Building
Washington DC 20240
USA   ¶
Tel:(202) 343 3421
and
Office of Offshore Information Serv
1951 Kidwell Drive
Suite 601 MS642
Vienna VA 22180
USA
Tel:(703) 285 2604
and
Alaska OCS Region
949 E 36th Avenue, Room 110
Anchorage AK 99508-4302
USA   ¶
Tel:(907) 271 6000
Fax:(907) 271 6805
and
OCS Information Program, Offshore Info & Pub'ns
381 Elden Street
Heradon VA 22070-4716
USA
Tel:(703) 787 1080
and
Technology Assessment & Research Branch, MS-647
12203 Sunrise Valley Drive
Reston VA 22091
USA
Tel:(703) 648 7752
and

Pacific OCS Region
Public Enquiries Office
1340 West Sixth Street
Los Angeles CA 90017
USA
Tel:(213) 894 2062

Mining and Metallurgical Society of America
275 Madison Avenue
Suite 2301
New York NY 10016
USA
Tel12) 684 4150

Ministere des Mines et des Hydrocarbures
BP 874
Libreville
Gabon
Tlx:5352 go minergie

Ministerio de Enercia y Minas
Sentrum Simon Bolivar
Caracas
Venezuela
Tlx:22592

Ministry of Agriculture Fisheries & Food
Great Westminster House
Horseferry Road
London SWlP 2AE
UK
Tel:(071) 216 7467
Tlx:21271
- *Fisheries Inspectorate*

Ministry of Defence
(HQ)Main Building
Whitehall
London SWlA 2HB
UK
Tel:(071) 218 9000
Tlx:825911
-*Navy Department Bank Block,Old Admiralty Bldg*
-*Directorate of Research & Technology*
- see Meterological Office
- see Hydrographer of the Navy

Ministry of Energy & Infrastructure
National Center of Scientific & Technological Inform'n
Atidim Industrial Park
PO Box 43074
Tel Aviv 61430
Israel
Tel:+972 3 492037/8
Fax:+972 3 492033

Ministry of Industry
Rue Ahmed Bey
Immeuble de Collsee
Algiers
Algeria
Tlx:52748

**Ministry of Industry &
Energy**
Rua da Horta Seca 15
1294 Lisbon Codex
Portugal
Tel:+351 346 30 91
Fax:+351 347 s59 01

**Ministry of Industry &
Energy**
Paseo de la Castellana 160
Madrid 16
Spain
Tel:+34 1 4588010
Tlx:42112

Ministry of Oil
PO Box 5077
Kuwait
Tlx: 22363

Ministry of Oil
PO Box 256
Tripoli
Libya
Tlx:61508

Ministry of Petroleum
Taleghani Street
Teheran
Iran
Tel:+98 61 51

**Ministry of Petroleum &
Mineral Resources**
PO Box 247
Riyadh
Saudi Arabia
Tlx:20058

**Ministry of Petroleum &
Mineral Resources**
PO Box 9
Abu Dhabi
United Arab Emirates
Tlx: 2273

**Mission Interministerielle de
la Mer**
3 place de Fontenoy
75700 Paris
France
Tel:+33 1 42 735505
Tlx:MISMER 201052

MIT Press
28 Carleton Street
Cambridge MA 02142
USA
Tel:(617) 253 5234

Mobil Exploration Norway Inc
PO Box 510
4001 Stavanger
Norway

**Mobil Research &
Development Corp**
PO Box 1026
Princeton NJ 08540
USA
Tel:(609) 737 3000

**Monks Wood Experimental
Station**
Abbots Ripton
Huntingdon
Cambridgeshire PE17 2LS
UK

**Morgan Grampian Book
Publishing Co Ltd**
Morgan Grampian House
30 Calderwood Street
London SE18 6QH
UK
Tel:(081) 855 7777
Tlx:896238

MTD Ltd
19 Buckingham Street
London WC2N 6EF
UK   ¶
Tel:(071) 321 0674
Fax:(071) 930 4323

Muir-Carby Bottkjaer NV
56-60 Gresham Street
London EC2V 7BB
UK
Tel:(071) 600 4503
Tlx:916136
Fax:(071 606 2316

N

**Nanhai East Oil Corporation
of China**
Floors 4/5,Guangzhou Hotel
Guangzhou City
China
Tel:+86 2061556
Tlx:44168 NHEOC CN

**Nanhai West Oil Corporation
of China**
Potou
Zhanjiang
Guangdong Province
China
Tel:+86 24 3111
Tlx:44086 COSSB CN

Nansen Institute
- see Fridtjof Nansen
Institute

**Nantes Chambre de
Commerce**
PO Box 18X
Palais de la Bourse
44040 Nantes
France
Tel:+33 40 733214
Tlx:700693

**National Academy of
Engineering Science**
2101 Constitution Avenue
NW
Washington DC 20418
USA
Tel:(202) 393 8100

National Academy Press
2101 Constitution Ave NW
Washington DC 20418
USA   ¶
Tel:(202) 334 3313
Fax:(202) 334 2793

**National Assn of Corrosion
Engineers**
1440 South Creek Drive
Houston TX 77084
USA
Tel:(713) 492 0535

**National Center of Scientific
& Technological
Information (COSTI)**
- see Ministry of Energy &
Infrastructure   §

**National Defense Research
Institute**
Box 1165
581 11 Linkoping
Sweden

**National Drilling Contractors
Association**
3008 Millwood Avenue
PO Box 11187
Columbia SC 29211
USA
Tel:(803) 252 5646

National Economic Development Office
Millbank Tower
Millbank
London SW1P 4QX
UK
Tel:(071) 217 4095

National Energy Authority
Grensasvegur 9
1208 Reykjavik
Iceland
Tel:+354 1 83600
Tlx: 2339 or-kust is
Fax:+354 12 688896

National Energy Board, Canada
Environment Directorate
311 - 6tgh Avenue SW
Calgary, Alberta T2P 3H2
Canada   ¶
Tel:(403) 294 2675
Fax:(403) 292 5503

National Energy Information Centre
FEDEX
US Department of Energy
Forrestal Building
Washington DC 20585
USA
Tel:(202) 252 8800

National Engineering Laboratory
East Kilbride
Glasgow G75 2QU
Scotland   ¶
Tel:(0355) 220222
Tlx:777888
Fax:(0355) 236930

National Geophysical Data Center
- see US Natl Oceanic &
Atmospheric Admin   §

National Hydrocarbons Authority
(Ente Nazionale Idrocarburi)
Piazza le Enrico Mattei 1
00144 Rome
Italy
Tel:+39 6 659001
Tlx:610636

National Hyperbaric Centre
123 Ashgrove Road
Aberdeen AB2 5FA
Scotland
Fax:(0224) 684378

National Information Services Corporation
Wyman Towers
Suite 6
3100 St Paul Street
Baltimore MD 21218
USA   §
Tel:(301) 243 0797
Fax:(301)454 8061

National Institute for Occupational Safety & Health
4676 Columbia Parkway
Cincinnati OH 45226
USA   §
Tel:(513) 533 8317
and
5600 Fishes Lane
Rockville MD 20852
USA
and
PO Box 630
Colton CA 92324
USA

National Institute for Petroleum & Energy Research
BETC Crude Oil Analysis
Data Bank
PO Box 2128
Bartlesville OK 74005
USA
Tel:(918) 336 2400

National Institute of Standards & Technology
A 323 Physics Building
Gaithersburg MD 20899
USA   §
Tel:(301) 975 2208

National Iranian Petrochemical Company
Karim Khan Zand Boulevard
PO Box 7484
7739 Teheran
Iran
Tel:+98 82 20 81-7

National Library of Medicine
8600 Rockville Pike
Bethesda MD 20894
USA   §
Tel:(301) 496 6193
Fax:(301) 496 0822

National Library of Scotland
George IV Bridge
Edinburgh EH1 1EW
Scotland
Fax:(031) 225 9944

National Maritime Institute
- now British Maritime
Technology

National Meteorological Library
- see Meteorological Office

National Mining Office for Hydrocarbons
Via Molise 2
Rome
Italy
Tel:+39 6 4705.2364

National Ocean Industries Association
Suite 410
1100 NW 17th Street
Washington DC 20036
USA

National Oceanic & Atmospheric Administration
- see under 'US'

National Oceanographic Data Center
- see US National Oceanic &
Atmospheric Administration

National Paint & Coatings Association
1500 Rhode Island Avenue
NW
Washington DC 20005
USA

National Petroleum Council
1625 NW K Street
Washington DC 20006
USA
Tel:(202) 393 6100
Tlx:892770

National Physical Laboratory
Queens Road
Teddington
Middlesex TW11 0LW
UK   ¶
Tel:(081) 977 3222
Tlx:262344
Fax:(081) 943 2155

National Power
Oakdale
Oakdale Avenue
off Ripon Road
Harrogate HG1 2YX
UK
Tel:(0423) 704409
Fax:(0423) 520883

National Radiological Protection Board
Chilton
Didcot
Oxfordshire OX11 0RQ
UK   ¶
Tel:(0235) 831600
Fax:(0235) 833891)

National Research Council
Transportation Research
Board
2101 Constitution Ave NW
Washington DC 20418
USA   ¶
Tel:(202) 334 2934
Fax:(202) 334 2033
and
Marine Board
2101 Constitution Avenue
Washington DC 20418
USA   ¶
Tel:(202) 334 3119
Fax:(202) 334 3789

National Research Council of Canada
Montreal Road
Ottawa
Ontario K1A 0R6
Canada
- *Assc Committee on Geotechnical Research*
and
Atlantic Regional
Laboratory Library
1411 Oxford Street
Halifax
Nova Scotia, B3H 3Z1
Canada
- and see Canada Institute
for Scientific & Technical
Information (CISTI)

National Science Foundation
1800 G Street NW
Washington DC 20550
USA   §
Tel:(202) 357 5000

National Sea Grant Depository
University of Rhode Island
Pell Library Building
Narraganset Bay Campus
Narraganset RI 02282
USA   ¶
Tel:(401) 792 6114
Fax:(401) 792 6160

National Technology Information Service (NTIS)
- see US Dept of Commerce
§

Natural Environmental Research Council
Polaris House
North Star Avenue
Swindon
Wiltshire SN2 1EU
UK
Tel:(0793) 40101
Tlx:444293
- *Research Vessel Service*

Naval Sea Systems Command
NAVSEA
- see US Dept of the Navy

Nederlands Normalisatie Instituut
- see Netherlands
Standards Institute

Nederlandse Aardolie Maatschappij bv
Schepersmaat 2
9405 TA Assen
The Netherlands
Tel:+31 5920 69111
Tlx:32113

Nederlandse Assn en Gas Exploratie en Produktie Associatie
Bezuidenhoutweg 29
2594 AC The Hague
The Netherlands
Tel:+31 70 478871

NEL
- see National Engineering
Laboratory

Netherlands Government Publishing Office
Staatsuitgeverij
The Hague
The Netherlands

Netherlands Industrial Council for Oceanology
Martinus Nijhofflaan 2
2624 ES delft
The Netherlands
Tel:+31 15 568258
Fax:+31 15 568100

Netherlands Marine Research Foundation
Laan van Nieuw Oost Indie
131
2593 BM The Hague
The Netherlands
Tel:+31 70 3440700
Fax:+31 70 3832173

Netherlands Maritime Information Centre
Postbus 21873
3001 AW Rotterdam
The Netherlands   §
Tel:+31 10 413 0960
Fax:+31 10 411 2857

Netherlands Organisation for Applied Scientific Research
Juliana van Stolberglaan
148
2595 CL The Hague
The Netherlands
Tel:+31 70 814481
Tlx:31660

Netherlands Standards Institute
Kalfj Eslaan 2
2623 AA Delft
The Netherlands
Tel:+31 15 611061
Tlx:312123

New England River Basins Commission
NERBC/RALI Project
55 Court Street
Boston MA 02108
USA

New South Wales Dept of Minerals & Energy
Minerals & Energy House
50-55 Christie Street
Saint Leonards NSW 2065
Australia   §
Tel:+61 2 901 8888

New York Times Information Bank
229 W 43rd Street
New York NY 10036
USA
Tel:(212) 556 1234

New Zealand Oceanographic Institute
177 Thordon Quay
PO Box 12346
Wellington
North Island
New Zealand

New Zealand Oil & Gas Ltd
PO Box 3149
Wellington
New Zealand
Tel:+64 4 725 408
Tlx:3011809 NZOG
Fax:+64 4 725 558

**Newcastle upon Tyne
Polytechnic**
- now University of
Northumbria at Newcastle

**Newfoundland Ocean
Industries Association**
Box 44 Atlantic Place
Suite 805 215 Water Street
St John's
Newfoundland A1C 6C9
Canada   ¶
Tel:(709) 753 8123
Fax:(709) 753 6010

**Newfoundland & Labrador
Dept of Mines & Energy**
Energy Resource Centre
6th Floor, Atlantic Place
PO Box 8700
St John's
Newfoundland A1B 4J6
Canada   ¶
Tel:(709) 729 2416
Fax:(709) 729 2508

**Newfoundland & Labrador
Government**
PO Box 4750
St John's
Newfoundland A1C 5T7
Canada
Tel:(709) 576 2781

Newport Associates
7400 East Orchard Road
Suite 320
Englewood CO 80111
USA   §
Tel:(303) 779 5515
Fax:(303) 779 0908

Newsbank Inc
Readex Microprint Corp
58 Pine Street
New Canaan CT 06840
USA
Tel:(203) 966 1100
Fax:(203) 966 6254

NewsNet Inc
945 Haverford Road
Bryn Mawr PA 19010
USA   §
Tel:(215) 527 8030
Fax:(215) 527 0338

News-A-Tron Corp
1 Peabody Street
Salem MA 01970
USA   §
Tel:(508) 744 4744
Fax:(508) 744 9738

NewsSource
1450 Don Mills Road
Don Mills
Ontario M3B 2X7
Canada   §
Tel:(416) 445 6641
Fax:(416) 445 3508

Nichols Publishing Company
PO Box 96
New York NY 10024
USA
Tel:(212) 580 8079

**Nigerian National Petroleum
Corporation**
Falomo Office Complex
PMB 12701
Ikoyi
Lagos
Nigeria
Tel:+234 1 60 31 00
Tlx:21609 NG

Nihon University
1-8 Surugadai
Kanda
Chiyoda-Ku
Tokyo 101
Japan

**Noble, Denton, Woodcock &
Associates Pty**
Murdoch House
5 The Esplanade
Mount Pleasant
Western Australia 6155
Australia

**Noraton Information
Corporation**
401-6080 Young Street
Halifax
Nova Scotia B3K 5L2
Canada
Tel:(902)453 4620
Tlx:(019)22771

Nordisk Institut for Sjorett
Biblioteket for
Petroleumsvett
Karl Johansgt 47
0162 Oslo 1
Norway

**Norges Geologiske
Undersokelse**
- see Geological Survey of
Norway

Norges Geotekniske Institutt
- see Norwegian
Geotechnical Inst

**Norges Skippsforsknings-
institutt**
- see Norwegian Ship
Research Inst

**Norges Standardiserings-
forbund**
- see Norwegian Standards
Assn

**Norges Teknisk Naturviten-
skapelige Forskningsgrad**
- see Royal Norwegian
Council for Scientific &
Industrial Res

Noroil Publishing House Ltd
Torvbeen 10
Postboks 480
4001 Stavanger
Norway
Tel:+47 4 57 06 44
Tlx:33327
Fax:+47 4 570461

Noroil Publishing House Ltd
Kingsgate Business Centre
12-50 Kingsgate Road
Kingston Upon Thames
Surrey KT2 5AA
UK

Norsk Hydro AS
PO Box 200
1321 Stabekk
Norway
and
Hovedkontoret
Bygdoy Alle 2
02157 Oslo 2
Norway
and
PO Box 4313
5013 Bergen
Norway
and
Kjorbokollen
PO Box 490
1301 Sandvika
Norway
and
Lars Hillesgt 30
5013 Nygardstangen
Norway
Tel:+47 5 216000
Tlx:40920 hydro n
Fax:+47 5 216196

Norsk Petroleumsforening
- see Norwegian Petroleum
Society

Norsk Senter for Informatik
Postboks 350 80
Blindern
0314 Oslo 3
Norway
Tel:+47 2 45 25 08
Tlx:72042 NSI N

Norske Hydro Oil & Gas
Hydro House
69 London Road
Twickenham
Middlesex TW1 1EE
UK
Fax:(081)892 1686

Norske Olje Revy
- see Norwegian Information
Publishers AS

Norske Tekniske Hogskole
- see Norwegian Institute of
Technology

**North East London
Polytechnic**
Maryland House
Manbey Park Road
London E15 1EY
UK
and
Underwater Technology
Group
Romford Road
London E15 4LZ
UK

**North East Scotland
Development Authority**
8 Albyn Place
Aberdeen AB1 1YH
Scotland
Tel:(0224) 643322
Tlx:739277
Fax:(0224) 631187

North Sea Books
Greenfield
Broughton
Biggar ML12 6HQ
Scotland
Tel:(0899) 4316

North Sea Monitor
Vossiusstraat 20
PO Box 5614
1007 AP Amsterdam
The Netherlands

North Sea Observer
PO Box 235
Skoyen
N 0212 Oslo 2
Norway

**North West University
Consortium**
Coupland III Building
The University
Manchester M13 9PL
UK

**North-Holland Publishing
Company (Elsevier)**
335 Jan van Galenstraat
PO Box 211
Amsterdam
The Netherlands
Tel:+31 20 5803911
Tlx:18582 ESPA NL
and
PO Box 1991
1000 BZ Amsterdam
The Netherlands
Tel:+31 20 5862 911
Tlx:10704 espom nl

**Norwegian Centre for
Information**
Forskningsveien 1
0314 Oslo 3
Norway
Tel:+47 2 169 5880
Tlx:11536 CIIR N

**Norwegian Chamber of
Commerce**
Cockspur Street
London SW1
UK
Tel:(071) 930 0181

**Norwegian Geotechnical
Institute**
Sogasveien 72
PO Box 40
Taasen
0801 Oslo 8
Norway   ¶
Tel:+47 2 23 03 88
Tlx:19787 ngi n
Fax:+47 2 23 04 48

**Norwegian Hydrodynamic
Laboratories**
River & Harbour Laboratory
PO Box 4118
Volentinlyse
7001 Trondheim
Norway

**Norwegian Hydrotechnical
Laboratories**
Kablbuveien 153
7034 Trondheim NTH
Norway
Tel:+47 7 593400
Tlx:520520
Fax:+47 7 943345

**Norwegian Information
Publishers AS**
Kongensgt 6
PO Box 873 Sentrum
0104 Oslo 1
Norway
Tel:+47 2 41 72 00
Tlx:162276

**Norwegian Institute for
Water Research**
Grooseveien 36
4890 Grimstad
Norway
Tel:+47 4 14 39 33
and
PO Box 333
Blindern
0314 Oslo 3
Norway
Tel:+47 2 25 52 80
Tlx:74190 niva n
and
Route 866
2312 Otterstad
Norway
Tel:+47 65 76 752
and
Breiviken 2
5035 Bergen-Sandviken
Norway
Tel:+47 5 25 97 00

**Norwegian Institute of
Technology**
7034 Trondheim-NH
Norway
Tel:+47 7 593000
Tlx: 0025 55186
- *Scientific & Industrial Res
Fndn*
- *Group Oceanographic
Centre*
and
Marine Technology Centre
Tel:+47 7 59 40 00

**Norwegian Marine
Technology Research
Institute**
Haakon Haakonsonsgt 34
PO Box 4125 Valentinlyst
N-7001 Trondheim
Norway
Tel:+47 7 595500
Tlx:55146 nsfit n
Fax:+47 7 595776
- Sintef / Marintek

**Norwegian Maritime
Directorate**
THV Meyerset 7
Oslo
Norway

Norwegian Ministry of Industry & Crafts
Akersgaten 42
Oslo 1
Norway
Tel:+47 2 419010

Norwegian Ministry of Local Government & Labour
Pilestredet 33
Oslo 1
Norway
Tel:+47 2 202270

Norwegian Oil Review
- see Norwegian Information Publishers AS

Norwegian Petroleum Consultants AS
PO Box 23
1371 Asker
Norway

Norwegian Petroleum Directorate
Professor Olav Hanssens vei 10
PO Box 600
4001 Stavanger
Norway   §
Tel:+47 4 876056
Tlx:42863 NOPED
Fax:+47 4 551571

Norwegian Petroleum Directorate
The Infoil Secretariat
PO Box 600
4000 Stavanger
Norway
Tel:+47 4 3 875057
Tlx:42853 noped

Norwegian Petroleum Directorate
Lagardsveien 80
PO Box 600
4001 Stavanger
Norway
Tel:+47 4 53 31 60
Tlx:33100

Norwegian Petroleum Society
Petro-Forum
PO Box 1897 Vika
0124 Oslo 1
Norway
Tel:+47 2 20 70 25
Tlx:77 322 nopet

Norwegian Ship Research Insitute
Marine Technology Centre
POB 4125 Valentinlyst
7001 Trondheim
Norway

Norwegian Society of Automatic Control
Kronprinsengst 17
Oslo 2
Norway
Tel:+47 2 18213

Norwegian Society of Chartered Engineers
Kronprinsengst 17
Oslo 2
Norway
Tel:+47 2 74363

Norwegian Standards Association
PO Box 7020
Homansbyen
0306 Oslo 3
Norway
Tel:+47 2 46 60 94
Tlx:19050 nsf n
Fax:+47 2 46 44 57

Norwegian State Diving School
P O Box 6
5034 Ytre Lakesevag
Norway

Norwegian Underwater Technology Centre (NUTEC)
Gravdalsveien 255
5034 Ytre Laksevag
Bergen
Norway   ¶
Tel:+47 5 261601
Tlx: 0025 42892
Fax:+47 5 344720

NOWEA
Postfach 32 02 03
400 Dusseldorf 30
Germany
Tel:+49 211 848835

Noyes Data Corporation Inc
Millroad at Grand Avenue
Park Ridge NJ 07656
USA
Tel:(201) 391 8484

Nuclear Energy Agency
OECD
38 Boulevard Suchet
75016 Paris
France
Tel:+33 1 45 24 82 0
Tlx:630668

O

Occidental College
Mary Norton/Clapp Library
1600 Campus Road
Los Angeles
California CA 90041
USA
Tel:(213) 259 2640

Ocean Engineering Information Centre
Bartlett Building
Memorial University of Newfoundland
St John's
Newfoundland A1B 3X5
Canada   ¶§
Tel:(709) 737 8377
Tlx: 016 4101
Fax:(709) 737 4706
- Centre for Cold Ocean Resource Engineering

Ocean Engineering Information Service
PO Box 989
La Jolla CA 92037
USA
Tel:(714) 454 1922

Ocean Engineering
- see Pergamon Press

Ocean Oil Information
3050 Post Oak (Suite 200)
Houston TX 77056
USA   ¶
Tel:(713) 621 9720
Tlx:244-66-077 Pubru
Fax:(713) 963 6285

Oceana Publications Inc
75 Main Street
Dobbs Ferry
New York NY 10522
USA
Tel:(914) 693 1320

Oceandril Inc
777 North Eldridge
Suite 740
Houston TX 77079
USA   §
Tel:(713) 558 0099

Oceanic Engineering Society Newsletter, IEEE
- see Sabbagh Assoc Inc

Oceanic Society
Executive Office
Magee Avenue
Stamford CT 06902
USA
Tel:(203) 327 9786

Oceanroutes Inc
680 W Maude Avenue
Sunnyvale CA 94086
USA   §
Tel:(408) 245 3600
Fax:(408) 245 5301

Oceans Institute of Canada
1236 Henry Street, 5th Floor
Halifax
Nova Scotia B3H 3J5
Canada   ¶
Tel:(902) 494 6552
Fax:(901) 494 1334

**OCLC Online Computer
Library Center Inc**
6565 Frantz Road
Dublin OH 43017-0702
USA   §
Tel:(614) 764 6000
Fax:(614) 764 6096

**OCS Dept of Commerce &
Economic Development**
PO Box D
Juneau AK 99811
USA
Tel:(907) 465 2500

**Office of Minerals
Management Information**
- see US Department of the
Interior

**Office of Ocean & Coastal
Resource Management**
Suite 706
1875 Connecticut Ave NW
Washington DC 20235
USA   ¶
Tel:(202) 606 4111
Fax:(202) 606 4329

Office Of Oil Imports
- see US Department of
Energy

Offshore
1200 Post Oak Boulevard
Houston TX 77056
USA

Offshore Certification Bureau
Greencoat House, 5th Floor
Francis Street
London SW1P 1DB
UK
Tel:(071) 602 7282
Tlx:265300

Offshore Conf's & Exhib's
55 Fife Road
Kingston u Thames KT1 1TA
UK
Tel:(081) 459 5831

Offshore Data Services Inc
PO Box 19909
Houston TX 77224
USA   ¶§
Tel:(713) 781 2713
Fax:(713) 781 9594

**Offshore Energy Technology
Board**
- see Offshore Supplies Off

Offshore Engineer
1 Heron Quay
London E14 9XF

**Offshore Engineering
Information Centre**
-see Heriot-Watt Univ Libr ¶

Offshore Fire Training Centre
Forties Road
Montrose DD10 9ET
Scotland
Tel:(0674) 72230

Offshore Information Lit'ture
127 Eaton Manor
Hove, Sussex BN3 3QD
UK
Tel:(0273) 733241

Offshore Minerals Group
- see Warren Spring
Laboratory

Offshore Oil & Gas Directory
247-249 Collins St, 3rd Flr
Melbourne, Victoria 3000
Australia
Tel:+61 3 654 6150
Tlx:echo aa 32708

**Offshore Petroleum Board
Library**
Suite 5, TD Place
140 Water Street
St John's
Newfoundland A1C 6H6
Canada
Tel:(709) 778 1450

**Offshore Petroleum Industry
Training Board**
Forties Road
Montrose
Angus DD10 9ET
Scotland
Tel:(0674) 2230
Tlx:76104 pitmbt
Fax:(0674) 77335

Offshore Supplies Office
Dept of Energy
1 Palace Street
London SW1E 5HE
UK
Tel:(071) 238 359
Tlx:918777
Fax:(071) 834 3771
and
Alhambra House
45 Waterloo Street
Glasgow G2 6AS
Scotland   ¶
Tel:(041) 221 8777
Tlx:779379
Fax:(041) 221 1718
and
Greyfriars House
Gallowgate
Aberdeen AB9 2ZU
Scotland
Tel:(0224) 641242
Tlx:739283
Fax:(0224) 642104

Offshore Supply Association
Faulenstrasse 23
2800 Bremen 1
Germany
Tel:+49 421 3045 296
Tlx:246534
Fax:+49 421 3245235

**Offshore Technology
Conference**
PO Box 833868
Richardson TX 75083-3868
USA   ¶
Tel:(214) 669 0072
Fax:(214) 669 0135

OGM Publishing Co Inc
One Poydras Plaza
639 Loyola Ave, Suite 1100
New Orleans LA 70113
USA
Tel:(504) 522 1100

Ohio State University Press
Room 316
2070 Neil Avenue
Columbus OH 43210
USA
Tel:(614) 422 6930

Oil Business Information Service
27 High Street
Inverkeithing
Fife KY11 1NL
Scotland

Oil Companies International Marine Forum
6th Floor, Portland House
Stag Place
London SW1E 5BM
UK
Tel:(071) 828 7696
Fax:(071) 245 2556/2

Oil Daily
1401 New York Avenue,
Suite 500
Washington DC 20005
USA
Tel:(202) 662 0700

Oil Division
- see Department of Energy

Oil Industry International Exploration & Production Forum (E+P Forum)
25-28 Old Burlington Street
London W1X 1LB
UK
Tel:(071) 437 6291
Fax:(071) 434 3721

Oil News Service
Springfield House
Dollar
Clackmannanshire
Scotland

Oil Spill Intelligence Report
- see Centre for Short-Lived Phenomena

Oil & Colour Chemists Association
Priory House
967 Harrow Road
Wembley
Middlesex HA0 2SF
UK
Tel:(081) 908 1086
Tlx:922670

Oil & Gas Commission
Institute of Petroleum
Exploration
Kanlagarth Road
Dehra-Dun
India

Oil & Gas Consultants Inc
c/o Society of Petroleum
Engineers, Tulsa
USA

Oil & Gas Directory
PO Box 130508
Houston TX 77219
USA
Tel:(713) 529 8789
Fax:(713) 529 3646

Oil & Gas European Magazine
Newmann Reichardt Str 34
Hamburg 70
Germany

Oil & Gas Journal
1421 S Sheridan Road
PO Box 1260
Tulsa OK 74101
USA   §

Oil & Pipelines Agency
35-38 Portman Square
London W1H OEY
UK
Fax:(071) 935 3510

Oildom Publishing Company of Texas
3314 Mercer Street
Houston TX 77027
USA
Tel:(713) 622 0676
Fax:(713) 623 4768

Oilfield Publications Ltd
PO Box 11
Ledbury, HR8 1BN
UK
Tel:(0531) 4563
Tlx:35566
Fax:(0531) 4239

Oilman
76 Oxford Street
London W1N 9FD
UK

Oklahoma Petroleum Directory
PO Box 700684
Tulsa OK 74170
USA
Tel:(918) 299 0194

Olje og Energidepartementet
- see Royal Ministry of Oil & Energy, Norway

Oljedirektoratet
- see Norwegian Petroleum Directorate

ONS Foundation
PO Box 410
4001 Stavanger
Norway
Tel:+47 4 33250

Open University
Dept of Earth Sciences
Walton Hall
Milton Keynes
MK7 6AA

Optech Ltd
East Street
Farnham
Surrey GU9 7XX
UK   §
Tel:(02s52) 714340
Fax:(0252) 711121

ORBIT Search Service
Achilles House
Western Avenue
London W3 0UA
UK   §
Tel:(081) 992 3456
Tlx:8814514
Fax:(081) 933 7335

Oregon State Library
State Library Building
Salem
OR 97310
USA
Tel:(503) 378 4368

Oregon State University
Engineering Experiment Station
Corvallis OR 97331
USA
Tel:(503) 754 0123
and
Library, Documents Division
Tel:(503) 734 3411

Organisation for European Economic Development
2 rue Andre Pascal
F-75775 Paris Cedex 16
France   §
Tel:+33 1 4524 9887
Fax:+33 1 4524 9988

Organisation of Arab Petroleum Exporting Countries
Box 20501
Safat 13066
Kuwait
Tel:+965 244 82 00
Tlx:221 66
Fax:+965 242 68 85

Organisation of Petroleum Exporting Countries
Obere Donaustrasse 93
1020 Vienna 2
Austria
Tel:+43 222 21 11 20
Tlx:134474
Fax:+43 222 26 43 20

Oslo Commission
48 Carey Street
London WC2A 2JE
UK
Tel:(071) 242 9927
Fax:(071) 831 7427

Oslo Port Authority
Prinsengate 2
0153 Oslo 1
Norway
Fax:+47 2 41 68 60
Tlx:771910 port n

Osterreichische Gesellschaft fur Erdolwissenschaften
Erdbergstr 72
1031 Vienna
Austria
Tel:+43 713 23 48
Tlx:132138

Otto Veith Verlag
Postfach 701606
Neumann Reichardtstr 34
2000 Hamburg 70
Germany

Outer Continental Shelf Oil & Gas Information Prog
- see US National Oceanic & Atmospheric Administration

Oxford Economic Research Associates
Blue Boar Ct, Alfred Street
Oxford OX1 4EH
UK
Tel:(0865) 251142
Fax:(0865) 251172

Oxford Institute for Energy Studies
57 Woodstock Road
Oxford PX2 6FA
UK
Fax: (0865) 310527

Oxford University Press
Walton Street
Oxford OX2 6DP
UK
Fax:(0865) 56767
Tlx:837330 Oxpres G
Fax:(0865) 56646

P

Paint Research Association
Waldegrave Road
Teddington
Middlesex TW1 8LD
UK   ¶
Tel:(081) 977 4427
Tlx: 918720
Fax:(081) 943 4705

Paleontological Research Institution
1259 Trumansburg Road
Ithaca NY 14850
USA
Tel:(607) 273 6623

Paleontological Society
Dr Donald L Wolberg, Secretary
c/o New Mexico Bureau of Mines & Mineral Resources
Socorro New Mexico 87801
USA   ¶
Tel:(505) 835 5140
Fax:(505)835 6333

Pallister Resource Management Ltd
640, 300 5th Avenue SW
Calgary
Alberta T2P 3C4
Canada   ¶
Tel:(403) 233 8480
Fax:(403) 233 8486

Parent Information Enterprises
2500 Tanglewilde
Suite 265
Houston TX 77063
USA   §
Tel:(713) 781 7690
Fax:(713) 783 8552

Paris Commission
48 Carey Street
London WC2A 2JE
UK
Tel:(071) 242 9927
Tlx:21185
Fax:(071) 831 7427

Pasha Publications Inc
1401 Wilson Boulevard
Suite 900
Arlington VA 22209
USA   §
Tel:(703) 528 1244
Fax:(703) 528 1243

Peches et Oceans
- see Institut Maurice-Lamontagne

Pennsylvania State University
College of Earth & Mineral Sciences
215 Wagner Building
University Park PA 16802
USA
Tel:(814) 865 1327
- *Pennsylvania State University Press*

PennWell Publishing Company
PO Box 1260
Tulsa OK 74101
USA   §
Tel:(918) 835 3161
Tlx:211012
Fax:(918) 831 9497
and
PO Box 682
London SW9 0EA
UK
Tel:01 675 7464
Tlx:8951165

Pentech Press Ltd
- see John Wiley

Pergamon Financial Data Services (PFDS)
Paulton House
8 Shepherdess Walk
London N1 7LB
UK   §
Tel:(071) 490 0049
Fax:(071) 490 2979

Pergamon Press
Headington Hill Hall
Oxford OX3 0BW
UK
Tel:(0865) 64881
Tlx:83177
Fax:(0865) 60285
and
Maxwell House
Fairview Park
Elmsford NY 10523
USA
Tel:(914) 592 7700
Tlx:137328

Peter Peregrinus Ltd
Station House
Nightingale Road
Hitchin
Herts SG5 1SA
UK
Tel:(0462) 53331
Tlx:825962

Petrocompanies
25-31 Ironmonger Row
London EC1V 3PN
UK

Petroconsultants Group SA
PO Box 228
1211 Geneve 6
Switzerland
Tel:+41 22 368811
Tlx:27763 PETR CH

Petroconsultants Inc
2 Houston Center
909 Fannin, P-330
Houston TX 77010
USA
Tel:(713) 654 1368

Petroconsultants (CES) Ltd
Burleigh House
18 Newmarket Road
Cambridge CB5 8EG
UK
Tel:(0223) 315933
Tlx:81507
Fax:(0223) 60899
and
36 Upper Brook Street
London W1Y 1PE
UK
Tel:(071) 499 5021
Fax:(071) 629 1257

Petroflash! Inc
PO Box 798
Lakewood NJ 08701
USA   §
Tel:(201) 367 1600
Tlx:701295

Petroleum Abstracts
- see University of Tulsa ¶§

Petroleum Argus Ltd
83-93 Shepperton Road
London N1 3DF
UK   §
Tel:(081) 359 8792
Fax:(081) 226 0695

**Petroleum Association of
Japan**
No. 9-4, 1-Chome,
Ohtemachi
Chiyoda-ku
Tokyo
Japan
Tel:+81 3 279 38

**Petroleum Communication
Foundation**
Suite 1250
633 6th Avenue SW
Calgary
Alberta T2P 2Y5
Canada   ¶
Tel:(403) 264 6064
Fax:(403) 237 6286

Petroleum Directorate
- see Canada Newfoundland
Offshore Petroleum Board

Petroleum Economist
25-31 Ironmonger Row
London EC1V 3PN
UK
Tel:01 251 3501
Tlx:27161

Petroleum Engineer
PO Box 1589
Dallas TX 75221
USA

Petroleum Extension Service
- see University of Texas at
Austin   ¶

**Petroleum Industry
Research Foundation**
122 East 42nd Street
New York NY 10168
USA
Tel:(212) 867 0052

**Petroleum Industry Training
Service**
13 2115 27th Avenue NE
Calgary
Alberta T2E 7E4
Canada   ¶
Tel:(403) 250 9606
Tlx:03 822 806
Fax:(403) 291 9408

Petroleum Information
PO Box 158
Claremont, WA 6010
Australia
Tel:+61 9 383 3477

**Petroleum Information
Bureau**
4 Brook Street
London W1
UK

**Petroleum Information
Corporation**
4100 East Dry Creek Road
Littleton CO 80122
USA   §
Tel:(303) 740 7100
Tlx:450244
Fax:(303) 694 1754
and
Land Data Services Division
PO Box 2612
Denver CO 80201-2612
USA   §
Tel:(303) 740 7100
and
Exploration Systems
Division
4150 Westheimer Road
PO Box 1702
Houston TX 77251
USA   §
Tel:(713) 961 5660
Fax:(713) 961 7917

Petroleum Information Ltd
Norman House
105-109 Strand
London WC2R 0BY
UK
Tel:(071) 240 8371
Tlx:946309

Petroleum Institute
Environmental
Conservation Executive
PO Box 740
North Sydney
NSW 2060
Australia

**Petroleum Institute of
Pakistan**
4th Floor
PIDC House
Dr Ziauffin Ahmed Road
Karachi
Pakistan

**Petroleum Intelligence
Weekly**
575 Broadway, 4th Floor
New York NY 10012
USA   §
Tel:(212) 941 5500

Petroleum Management
78878 San Filipe, # 100
Houston TX 77063
USA

Petroleum News Southeast Asia Ltd
6th Floor
146 Prince Edward Road
West Kowloon
Hong Kong

Petroleum Publishers Incorporated
PO Box 129
Brea CA 92621
USA
Tel:(213) 691 1419

Petroleum Review
61 New Cavendish Street
London W1M 8AR
UK

Petroleum Science & Technology Institute
Dunedin House
25 Ravelston Terrace
Edinburgh EH4 3EX
Scotland   ¶
Tel:(031) 451 5231
Fax:(0231) 451 5232

Petroleum Society of CIM
320, 101 6th Avenue SW
Calgary
Alberta T2P 3P4
Canada   ¶
Tel:(403) 237 5112
Fax:(403) 262 4792

Petroleum Training Association, North Sea
Flint House
80 High Street
Lowestoft
Suffolk NR32 1XN
UK

Petroleum Training Validation
Forties Road
Montrose DD10 9ET
Scotland
Fax:(0674) 76494

Petroscan
United Communications Grp
4550 Montgomery Avenue
Suite 700 N
Bethesda MD 20814
USA   §
Tel:(301) 961 8700
Fax:(301) 961 8666

Phillips Petroleum Co
16th Floor
Phillips Building
Bartlesville OK 74004
USA
Tel:(918) 661 4269
and
The Adelphi
1-11 John Adam Street
London WC2N 0BW
UK

Phillips & Drew
120 Moorgate
London EC2M 6XP
UK
Tel:(071) 628 4444
Tlx:291163
Fax:(071) 588 0252

Pipeline Industries Guild
17 Grosvenor Crescent
London SW1X 7ES
UK
Tel:(071) 235 7938

Pipeline & Gas Journal
PO Box 22267
Houston Tx 77227
USA
Tel:(713) 622 0676
Fax:(712) 623 4768

Pipelines Inspectorate
- see Dept of Energy

Pipes & Pipelines International
PO Box 21
Beaconsfield
Bucks HP4 1NS
UK
Tel:(04946) 5139

Pitman Books Ltd
128 Long Acre
London WC2E 9AN
UK
Tel:(071) 379 7383
Tlx:261367 pitman

Plastics & Rubber Institute
11 Hobart Place
London SW1W 0HL
UK
Tel:(071) 245 9555
- *from January 1993 see Institute of Materials*

Plenum Publishing Co
233 Spring Street
New York NY 10013
USA
Tel:(215) 255 0713
Tlx:2341139
and
88/90 Middlesex Street
London E1 7EZ
UK
Tel:(01) 377 0686

Plymouth Marine Laboratory
Citadel Hill
Plymouth PL1 2PB
UK   ¶
Tel:(0752) 221761
Fax:(0752) 226865

Plymouth Polytechnic
- now University of Plymouth

PMG Chamber of Mines & Petroleum
PO Box 1032
Port Moresbny
Papua New Guinea
Fax:+675 21 2988

Polar Continental Shelf Project
- see Canadian Dept of Energy, Mines & Resources

Portland State University Library
Regional Depository
PO Box 1151
Portland OR 97207
USA
Tel:(503) 229 3673

POWER
- see US Dept of Energy

Praeger Publishers Inc
383 Madison Avenue
New York NY 10017
USA
Tel:(212) 688 9100
Tlx:125166

Precision Microfilming Services
3770 Kempt Road
Halifax
Nova Scotia B3K 4X8
Canada   ¶
Tel: (902) 455 5451
Fax:(902) 454 5570

Predicasts (USA)
1101 Cedar Avenue
Cleveland OH 44106
USA   §
Tel:(216) 795 3000
Fax:(216) 229 9944
and
8-10 Denman Street
London W1V 7RF
UK
Tel:(071) 734 3817
Fax:(071)734 5934

Prentice-Hall International
301 Sylvan Avenue
Englewood Cliffs NJ 07632
USA
Tel:(201) 592 2000
Tlx:135423
- see also Simon & Schuster

**Princeton Microfilm
Corporation**
PO Box 2073
Princeton NJ 08543
USA
Tel:(609) 452 2066

Princeton University
School of Engineering &
Applied Science
41 William Street
Princeton NJ 08540
USA
Tel:(609) 452 4900

**Prodive Centre for
Underwater Technology**
Falmouth Oil Exploration
Base
Falmouth Docks
Cornwall TR11 4NR
UK
Tel:(0326) 315691
Tlx:45362
Fax:(0326) 315716

**Professional Assn of Diving
Instructors**
2064 North Bush Street
Santa Anna CA 92706
USA

Professional Data Services
Congressional Information
Services
4520 East-West Highway
Suite 800
Bethesda MD 20814
USA
Tel:(301) 365 1550

**Professional Divers'
Association of Australasia**
90 Queen Street
Melbourne
Victoria
Australia
Tel:+61 3 67 4636

Professional Publications Inc
1609 Northwest Boulevard
Columbus OH 43212
USA
Tel:(614) 488 8236

Profile Information
Sunbury House
79 Staines Road
Sunbury-on-Thames
Middlesex TW16 7AH
UK
Tel:(0932) 761444
Tlx:8811720
Fax:(0932) 761444

PSI Energy Software Inc
800 Gessner Road
Suite 1220
Houston TX 77024-4257
USA   §
Tel:(713) 496 4850

**Public Petroleum
Corporation (DEP SA)**
Mesogion 35754 Akademias
Street
15231 GR-Hallandri
Athens
Greece
Tel:+30 650 13 40

Q

QL Systems Ltd
901 St Andrew's Tower
275 Sparks Street
Ottawa
Ontario K1R 7X9
Canada   §
Tel:(613) 238 3499

Queen Mary College
- see University of London

Queen's University of Belfast
Belfast BT7 1NN
Northern Ireland
Tel:(0323) 45133

Questel
83-85 Bld Vincent Auriol
75646 Paris Cedex 13
France   §
Tel:+3 1 4423 6464
Fax:+33 1 4423 6465
and
UK User Support
- see Fraser Williams   §

Quick Source Inc
1010 Wayne Ave, Suite 510
Silver Spring MD 20910
USA   §
Tel:(301) 650 8865
Fax:(301) 565 9412

R

RAPRA Technology Ltd
Shawbury
Shrewsbury
Shropshire SY4 4NR
UK   ¶
Tel:(0939) 250383
Tlx:35134
Fax:(0939) 250383

RAI Exhibition Centre
Europaplein
1078 GZ Amsterdam
The Netherlands

Rauma-Repola Oy
PO Box 176
26107 Rauma
Finland
and
Rauma-Repola Offshore
Engineering
PO Box 306
33101 Tampere
Finland

**Ray Piper Company Inc
(Alaska Update)**
PO Box 270718
Houston TX 77277
USA

**Reed Information Services
Ltd**
Windsor Court
East Grinstead House
East Grinstead
W Sussex RH19 1XA
UK
Tel:(0342) 326972
Tlx:95127
Fax:(0342) 315130

Reed International Books
Michelin House
81 Fulham Road
London SW3 6RB
Tel:(071) 581 9393
Tlx:920191
Fax:(071) 589 8456

Reed, Thomas, Publications Ltd
- see Thomas Reed
Publications Ltd

Reidel (D) Publishers
PO Box 989
3300 AZ Dordrecht
The Netherlands
Tel:+31 78 178222
Tlx:20083 kadc

Research Vessel Service
No 1 Dock
Barry
Glamorgan CF6 6UZ
Wales
Tel:0416 737451
Tlx:497101

Resource Planning Consultants Inc
1800 West Loop Street
Suite 890
Houston TX 77027
USA
Tel:(713) 840 0041

Resource Publications Inc
3210 Marquart
Houston TX 77027
USA
Tel:(713) 961 4191

Reuters Ltd
85 Fleet Street
London EC4P 4AJ
UK   §
Tel:(071) 250 1122
Tlx:23222
Fax:(071) 324 8144

Rhein-Main-Donau AG
Postfach 40 15 69
8000 Munchen 40
Germany
Tel:+49 89 38071

Rice University
PO Box 1892
Houston TX 77001
USA
Tel:(713) 527 4858

River & Harbour Laboratory
Klaebuveien 153
7000 Trondheim
Norway

Robert Gordon University
352 King Street
Aberdeen
AB9 2TQ
Scotland
Tel:(0224) 636611
and
St Andrew's Street
Aberdeen AB1 1HG
Scotland
Tel:0224 636611
and
Offshore Technology Centre
School Hill
Aberdeen AB9 1FR
Scotland
Tel:(0224) 633611

Robert Gordon's Institute of Technology
- now the Robert Gordon
University

Robert Hale Ltd
Clerkenwell House
45-47 Clerkenwell Green
London EC1R 0HT
UK
Tel:(071) 251 2661
Fax:(071) 490 4958

Robert Stanger & Company
1129 Broad Street
Shrewsbury NJ 07701
USA
Tel:(201) 389 3600

Rock Mechanics Information Service
- see Imperial College   §

Rocky Mountain Mineral Law Foundation
Porter Administration
Building
7039 East 18th Avenue
Denver CO 80220
USA   §
Tel:(303) 321 8100

Rogalandsforskning Rogaland Research Institute
PO Box 2503
Ullandhaug
N-4004 Stavanger
Norway
Tel:+47 4 87 50 00
Tlx:405013 rogfo n

Routledge
- see International Thomson

Royal Aeronautical Society
4 Hamilton Place
London WIV 0BQ
UK
Tel:(071) 449 3515
Fax:(071) 499 6230

Royal Bank of Scotland
42 St Andrew Square
Edinburgh EH2 2YE
Scotland
Fax:(031) 557 6565

Royal Commission on the *Ocean Ranger*
- see Canada Newfoundland
Offshore Petroleum Board

Royal Danish Navy School
Dykkerkursus Nyholm
Homnen
1433 Copenhagen K
Denmark

Royal Institute of Navigation
1 Kensington Gore
London SW6 2AT
UK
Tel:((071) 589 5201

Royal Institution of Chartered Surveyors
12 Great George Street
London SW1P 3AD
UK
Tel:(071) 222 7000
Tlx:915443

Royal Institution of Great Britain
21 Albemarle Street
London WIX 4BS
UK
Tel:(071) 409 2992

Royal Institution of Naval Architects
10 Upper Belgrave Street
London SW1X 8BQ
UK   ¶
Tel:(071) 235 4622
Tlx:265844
Fax:(071) 245 6959

Royal Ministry of Oil & Energy
Tollburgt 31
PO Box 8148 DEP
0033 Oslo 1
Norway
Tel:+47 2 419010

Royal Norwegian Council for Scientific & Industrial Res
Sognsveien 72
Oslo 8
Norway
Tel:+47 2 237685

Royal School of Mines
- see Imperial College

Royal Society
6 Carlton House Terrace
London SW1Y 5AG
UK
Tel:(071) 839 5561
Tlx:917876
Fax:(071) 930 2170

Royal Society for the Prevention of Accidents
Occupational Safety Division
Canon House, The Priory
Queensway
Birmingham B4 6BS
UK   ¶
Tel:(021) 200 2461
Tlx:336546
Fax:(021) 200 1254

Royal Society of Chemistry
Thomas Graham House
Science Park
Milton Road
Cambridge CB4 4WFD
UK   §
Tel:(0223) 420066
Fax:(0223) 423623
and
Burlington House
London W1V 0BN
UK
Tel:(071) 734 9864
Tlx:268001

Royal Society of Edinburgh
22-24 George Street
Edinburgh EH2 2PQ
Scotland
Tel:(031) 225 6057
Fax:(031) 220 6889

Royal Swedish Navy Diving Schools
Marinens Dykericentrum
S130 61 Horsfjarden
Sweden
Tel:+46 750 63000

Rubber & Plastics Research Association
- now RAPRA Technology

Rutgers University
Center for Coastal & Environmental
Studies
Doolittle Hall
New Brunswick NJ 08903
USA

S

Sabbagh Associates Inc
2634 Round Hill Lane
Bloomington IN 47401
USA

Saga Petroleum AS
IPO Box 9
1322 Hovik
Norway

Sage Data Inc
104 Carnegie Center
Princeton NJ 08540
USA
Tel:(609) 924 3000

Salvage Association
107-112 Leadenhall Street
London EC3A 4AP
UK
Tel:(071) 623 1299
Fax:(071) 626 4963

SAMSOM Data Systemen bv
Wilhelminalaan 1
Postbus 180
2504 EB Alphen a/d Rijn
The Netherlands
Tel:+31 1720 6219
Tlx:39704

San Diego State University
Government Publications
Dept, Library
820 E Street
San Diego CA 92101
USA
Tel:(619) 236 5800
and
Government Publications
Dept, Library
5300 Campanile Drive
San Diego CA 92182
USA
Tel:(619) 265 5832

San Francisco State University
J Paul Leonar Library
1630 Holloway Avenue
San Francisco CA 94132
USA
Tel:(415) 469 1557

Savant Research Studies
2 New Street
Carnforth LA5 9BX
UK
Tel:(0524) 734505

Scarecrow Press
52 Liberty Street
PO Box 656
Metuchen NJ 08840
USA
Tel:(201) 548 8600

SCICON
Brick Close
Kiln Farm
Milton Keynes MK11 3EJ
UK
Tel:(0908) 565656
Tlx:826693
Fax:(0908) 561427

Science & Engineering Research Council
Polaris House
North Star Avenue
Swindon SN2 1ET
UK
Tel:(0793) 411292
Tlx:411658
Fax:(0793) 411501

Scientific Press
540 University Avenue
Palo Alto CA 94301
USA
Tel:(415) 322 5221

Scientific Surveys Ltd
PO Box 21
Beaconsfield HP9 1HW
UK
Fax:(0494) 670155

Scientific Symposia
33 Bowling Green Lane
London EC1R 0DA
UK
Tel:01 887 1241

Scottish Enterprise
10 Queen's Road
Aberdeen
Scotland
Tel:(0224) 641791
Tlx:73586
Fax:(0224) 644325
and
120 Bothwell Street
Glasgow G2 7JP
Scotland
Tel:(041) 248 2700
Tlx:777600
Fax:(041) 221 1897

Scottish Office
Industry Dept for Scotland
-Energy
New St Andrew's House
St James's Centre
Edinburgh EH1 3TA
Scotland
Tel:(031) 556 8400
Fax:(031) 244 4785
and
Sea Fisheries Inspectorate
56 High Street
Lossiemouth IV31 6AA
Scotland
Fax:(0343 814635

Scottish Offshore Partnership
c/o Technology Reports
Centre, British Library

**Scottish Offshore Training
Assn Limited**
Blackness Road
Altens
Aberdeen
Scotland
Tel:(0224) 873983
Fax:(0224) 873221

**Scripps Institution of
Oceanography**
University of California
A007 La Jolla
La Jolla CA 92093
USA

**Scuola Professionale
Subacquea e Iperbarica**
'Marco Polo'
Via Salaria n 1075
Roma 00138
Italy
Tel:+39 6 840 1650

Sea Fish Industry Authority
Sea Fisheries House
10 Young Street
Edinburgh EH2 4JG
Scotland
Tel:(031) 225 2515
Tlx:727225
Fax:(031) 220 0445
and
Industrial Development Unit
St Andrew's Dock
Hull HU3 4QE
UK
Tel:(0482) 27837
Tlx:527261
Fax:(0482) 223310
and

Marine Farming Unit
Ardtoe, Acharocle
Argyll PH36 4
Scotland   ¶
Tel:(096) 785213
Fax:(096) 785343

Sea Fisheries Inspectorate -
see Ministry of Agriculture,
Fisheries & Food

Sea Technology
Suite 1000
1117 North 10th Street
Arlington VA 22209
USA

Sea Technology Magazine
- see Compass Publications

Seatrade Conferences
Fairfax House
Colchester
Essex CO1 1RJ
UK
Tel:(0206) 45121
Fax:(0206) 45190

Secretariat d'Etat a la Mer
3 Place de Fontenoy
75700 Paris
France
Tel:+33 1 42 735505
Tlx:MISMER 201052

**Secretariat for the Nordic
Energy Libraries**
PO Box 49
DK-4000 Roskilde
Denmark   §
Tel:+45 42 371212
Tlx:43116
Fax:+45 46 755627

**Seismological Society of
America**
2620 Telegraph Avenue
Berkeley CA 94704
USA
Tel:(415) 848 0954

**Selection & Industrial
Training Administration**
334-350 Royal Exchange
Manchester M2 7FB
UK

Sell's Publications Ltd
55 High Street
Epsom
Surrey
UK
Tel:(037) 27 26376
Tlx:265619

Selvigs Publishing AS
PO Box 9070
Vaterland
0134 Oslo 1
Norway
Tel:+47 2 425867
Tlx:170620

Sharnell Ltd
Deemouth Centre
South Esplanade East
Aberdeen AB 3BB
Scotland

Sharp (I P) Associates
Exchange Tower, Suite 1900
2 First Canadian Place
Toronto
Ontario M5X 1E3
Canada   §
Tel:(416) 364 5361
Tlx: 0622259
Fax:(416) 364 2910
and
Heron House
10 Dean Farrar Street
London SW1H 0DX
UK
Tel:(071) 222 7033
Tlx: 8954178

**Shell Exploration &
Production (Norway) Inc**
PO Box 560
4001 Stavanger
Norway

**Shell International
Petroleum Co Ltd**
Shell Centre
London SE1 7NA
UK
Tel:(071) 934 1234
Fax:(071) 934 8060

**Shell UK Exploration &
Production**
Shell Mex House
Strand
London WC2R 0DX
UK
Fax:(071) 257 4053

Sheppards & Chase
Clements Hse, Gresham St
London EC2V 7AU
UK
Fax:(071) 606 5830

Shipbuilding & Electrical
Engineering Division
- see Dept of Trade &
Industry

SIA Ltd
Computer Systems House
Ebury Gate
23 Lower Belgrave Street
London SW1W 0NW
UK
Tel:(071) 730 4544
Fax:(071) 730 6762

Sijthoff & Noordhoff Intl
Publishers (KAPG)
- see Kluwer Academic
Publishers Group

Simon & Schuster
International Group
Campus 400
Maylands Avenue
Hemel Hempstead Herts
HP2 7EZ
UK
Tel:(0442) 881900
Fax:(0442) 882099

Single Buoy Moorings Inc
Engineering Dept
Po Box 157
Monaco City
Monaco
and
2 King Street
Twickenham TW1 3SN
UK
Fax:(081) 891 3225

SINTEF Group
Scientific & Industrial
Research Foundation
Norwegian Institute of
Technology
7034 Trondheim-NTH
Norway   ¶
Tel:+47 7 593000
Tlx: 0025 55186
Fax:+47 7 591480
and
SINTEF IKU
Hakon Magnussonsgt 1B
PO Box 1883 Jarlesletta
7002 Trondheim
Norway
Tel:+47 7 92 06 11
Tlx:55434 IKU N

and
SINTEF MARINTEK
Haakon Haakonsonsgt 34
PO Box 4125 Valentinlyst
7001 Trondheim
Norway
Tel:+47 7 59 55 00
Tlx: 55146 NSFIT N
Fax:+47 7 595776
and
PO Box 173 Asnes
3201 Sandefjord
Norway
Tel:+47 34 73033
Fax:+47 34 73516

SIRA Ltd
South Hill
Chislehurst, Kent BR7 5EH
UK
Tel:(081) 467 2636
Tlx:896649
Fax:(081) 467 6515

Skibsteknisk Laboratorium
- see Danish Ship Research
Labs

SLAMARK International SpA
88 via G Trevis
00147 Roma
Italy
Tel:+39 6 514 0176
Tlx:61-3408

Smith Rea Energy Analysts
Ltd
3 Beer Cart Lane
Canterbury
Kent CT1 2NJ
UK
Tel:(0227) 763456
Tlx:96118
Fax:(0227) 454385

Smithsonian Science
Information Exchange Inc
1730 M Street NW, Rm 300
Washington DC 20036
USA
Tlx:4992963

SOCIDOC
142 rue Montmartre
75002 Paris
France
Tel:+33 1 40 268321
Tlx:220528

Societe National
Elf-Aquitaine
- see Elf-Aquitaine

Society for Mining,
Metallurgy & Exploration
(SME) Inc
PO Box 625002
Littleton CO 80162-5002
USA   ¶
Tel:(303) 973 9550
Fax:(303) 973 3845

Society for Underwater
Technology
75 Mark Lane
London EC3R 7JN
UK   ¶
Tel:(071) 481 0750
Tlx:8868414
Fax:(071) 481 4001

Society of Consulting Marine
Engineers & Ship
Surveyors
6 Lloyds Avenue
London EC3N 2AX
UK
Tel:(071) 488 3010

Society of Economic
Geologists
PO Box 571
Golden CO 80402
USA
Tel:(303) 236 5538

Society of Exploration
Geophysicists
PO Box 702740
Tulsa OK 74170-2740
USA   §
Tel:(918) 493 3516
Tlx:796392 SEG TUL

Society of Naval Architects
& Marine Engineers
601 Pavonia Ave, # 400
Jersey City NJ 07306
USA   ¶
Tel:(201) 798 4800
Fax:(201) 798 4975

Society of Petroleum
Engineers
PO Box 833836
222 Palisades Creek Drive
Richardson TX 75083 3836
USA   §
Tel:(214) 669 3377
Tlx:730989 SPEDAL
Fax:(214) 669 0135

Society of Photo-Optical Instrument Engineers
405 Fieldston Road
PO Box 10
Bellingham WA 98225
USA
Tel:(206) 676 3290

Society of Piping Engineers & Designers
One Main Street
Houston TX 77023
USA
Tel:(713) 221 8090

Society of Professional Well Log Analysts
6001 Gulf Freeway, #C129
Houston TX 77023
USA
Tel:(713) 928 8925

Softsearch Inc
1560 Broadway, Suite 900
Denver CO 80202
USA   §
Tel:(303) 831 3400
Fax:(303) 831 3466

Solent Exhibitions
20 Burleigh Crescent
Inverkeithing
Fife KY11 1DQ
Scotland

Sonarmarine ltd
Orpington
Kent BR5 3RN
UK
Tel:(0689) 32111

South Australian Dept of Mines & Energy (SADME)
191 Greenhill Road
Parkside, SAustralia 5063
Australia   §
Tel:+61 8 274 7500
Fax:+61 8 272 7597

South Huanghai Oil Corporation
2451 Hongqiad Road
Shanghai
China
Tel:+86 21 1329388
Tlx:33167

South Pacific Underwater Medicine Society
43 Canadian Bay Road
Mount Eliza
Victoria 3930
Australia

Southam Business Information & Communications Group
1450 Don Mills Road
Don Mills
Ontario M3B 2X7
Canada   §
Tel:(416) 445 6641
Fax:(416) 445 3508

Spearhead Exhibitions
55 Fife Road
Kingston upon Thames
KT1 1TA
UK
Tel:(081) 549 5831

Spill Control Association of America
17117 W Nine Mile Road
Suite 1515
Southfield MI 48075
USA
Tel:(313) 552 0500

Spill Control Newsletter
- Environment Canada

Spokane Public Library
Reference Department
West 906 Main Avenue
Spokane WA 99201
USA
Tel:(509) 838 4226

Spon (E & F N)
2-6 Boundary Row
London SE1 8HN
UK
Tel:(071) 865 0066
Tlx: 290164 Chapmag
Fax:(071) 522 9623

Springer Verlag
Heidelberger Platz 3
1000 Berlin 33
Germany
Tel:+49 30 8207 1
Tlx:183319 spbln d
Fax:+49 30 8214091
and
8 Alexandra Road
London SW19 7JZ
UK
Tel:(081) 947 1280
Fax:(081) 947 1274
and
175 Fifth Avenue
New York NY 10010
USA
Tel:(212) 460 1500
Tlx:12-5994 BOOKSP
Fax:(212) 473 6272

SRI International
World Petrochemical Program
333 Ravenswood Avenue
Menlo Park CA 94025
USA
Tel:(415) 859 6308
Tlx:334486

Standard Oil Company
4440 Warrensville Center Road
Cleveland OH 44128
USA

Standards Association of Australia
80 Arthur Street
North Sydney
New South Wales
Australia
Tel:+61 2 9296022
Tlx:26514

Standards Council of Canada
45 O'Connor Street, # 1200
Ottawa
Ontario K1P 6N7
Canada   ¶
Tel:(613) 238 3222
Tlx: 053-4403
Fax:(613) 995 4564

Stanford Maritime Press
Stanford CA 94305
USA
Tel:(415) 323 9471

Stanford University Libraries
Government Documents Department
Stanford CA 94305
USA
Tel:(415) 723 9108

Stanley Thornes (Publishers)
Educa House, Old StationDr
Leckhampton, Cheltenham
Glos GL53 0DN
UK
Tel:(0242) 228888
Tlx:43592
Fax:(0242) 221914

State Pollution Control Authority (STF)
POP Box 8100
Dep
0032 Oslo 1
Norway

Statens Dykkerskole
- see Norwegian State Diving School

Statoil
3960 Starhelle
Norway
and
PO Box 74
1360 Nesbru
Norway
and
PO Box 1508, Nidarholl
7001 Trondheim
Norway
and
PO Box 300
4001 Stavanger
Norway
and
PO Box 308
5501 Haugesund
Norway
and
Gullfaks Produksjon
Souheimset 23
PO Box 2474
5037 Stolheimsvik
Norway

Stevens Institute of
Technology
Dept of Ocean Engineering
Hoboken NJ 07030
USA

STI National Institute of
Technology
Akersveien 24C
POB 8116
Dep 0032
Oslo 1
Norway
Tel:+47 2 20 45 50
Tlx:72494 sti n
Fax:+47 2 204550

Stichting Coordinate
Maritiem Onderzoek
PO Box 21873
3001 AW Rotterdam
The Netherlands
Tel:+31 10 130960
Tlx:26585

STN International
- c/o Fiz Karlsruhe
- c/o Royal Society of
Chemistry   §

Stock Beech
The Bristol and West
Building
Broad Quay
Bristol BS1 4DD
UK
Tel:(0272) 260051
Tlx:44739

STOP
9c Droningens Tvaergade
1302 Copenhagen K
Denmark
Tlx:19798

Straatliche
Prufungskommission dur
das Tauchergewe
in Schleswig-Holstein
Spohienblatt 50
2300 Keil
Germany
Tel:+49 431 61014

Strathclyde Regional
Council
Industrial Development Unit
Strathclyde House
20 India Street
Glasgow G2 4PF
Scotland
Tel:(041) 227 3866
Fax:(041) 227 2870

Submex Limited
19-21 Roland Way
London SW7 3RF
UK
Tel:(071) 373 7340
Tlx:8814824 Submex
Fax:(071) 373 7340

Subsea Engineering News
P O Box 213
Swindon SN6 8UA
UK

Sunderland Polytechnic
- now University of
Sunderland

SUT
- see Society for Underwater
Technology

Swedish Corrosion Institute
PO Box 5607
S4 86 Stockholm
Sweden

Swedish Dept for Geological
Research
PO Box 670
75128 Uppsala
Sweden
Tel:+46 18 155280

Swedish Geotechnical
Institute
Fack
58101 Linkoping
Sweden

Swedish National Maritime
Resources Commission
Box 295
401 24 Gothenburg
Sweden
Tel:+46 6 31156070

Swedish Petroleum Institute
Sveavagen 21
1134 Stockholm
Sweden
Tel:+46 8 23 58 00
Fax:+46 8 21 03 25

Swedish Trade Fair Fndation
Box 5222
402 24 Gothenburg
Sweden

Swedocean
Box 5506
114 85 Stockholm
Sweden
Tel:+46 8 783 8000
Fax:+46 8 660 3378

Sweet & Maxwell Ltd
South Quay Plaza
183 Marsh Wall
London E14 9FT
UK
Tel:(071) 538 8686
Fax:(071) 538 9508

Systech Consultancy
Services
c/o Chaganlal Vishram &Co
107 - 109 Sherif Devji Street
Bombay 400 003
India
Tel:+91 22 343158
Tlx:1173516

T

Tapir
- see Technical University of
Norway

Taylor & Francis Ltd
4 John Street
London WC1N 2ET
UK
Tel:(071) 405 2237
Tlx:858540
Fax:(071) 831 2035
and
Rankine Road
Basingstoke
Hants RG24 0PR
UK
Tel:(0256) 840366
Fax:(0256) 479438

Technical Research Centre of Finland (VTT)
Espoo
Finland

Technical University of Denmark
Acoustics Laboratory
Building 352
Lundtoftevej 100
DK-2800 Lyngby
Denmark
Tel:+45 2 88 16 22

Technical University of Norway
7034 Trondheim
Norway
Tel:+47 7 20354

Technip
- see Editions Technip

Technische Hogeschool Eindhoven
- see Einhoven University of Technology

Technische Hogeschool te Delft
- see Delft University of Technology

Technische Hogeschool Twente
- see Twente University of Technology

Technology Reports Centre
- British Library Document Supply Centre

Tecnon (UK) Ltd
12 Calico House
Plantation Wharf
York Place
London SW11 3TN
UK   §
Tel:(071) 924 3955
Fax:(071) 978 5307

Telerate Systems Inc
One World Trade Center
New York NY 10048
USA   §
Tel:(212) 938 5400

Telerate (Europe/Gulf) Ltd
Winchmore House
12-15 Fetter Lane
London EC4A 1BR
UK   §
Tel:(071) 583 0044
Fax:(071) 583 1837

Telesystemes Questel
83-85 Bld Vincent Auriol
75013 Paris
France
Tel:+33 1 45 826464
Tlx:204594

Tennessee Microfilms
PO Box 1096
Nashville TN 37202
USA

Texaco Inc
PO Box 245
Bellaire TX 77401
USA

Texas A & M University
Sterling C Evans Library
College Station TX
77843-5000
USA   ¶
Tel:(409) 845 8111
Fax:(409) 845 6238
and
Sea Grant Program
Dept of Marine Resources
Tel:(713) 845 4515
and
- see Gulf Publishing

Themedia Ltd
PO Box 2
Chipping Norton OX7 5QX
UK
Tel:(0608) 84700

Thomas Reed Publications
184 High Street West
Sunderland
Tyne & Wear SR1 1UQ
UK
Tel:(091) 567 5211/5
Fax:(091) 514 4104

Thomas Telford Limited
1 Heron Quay
London E14 9XF
UK
Tel:(01) 987 6999
Tlx:298105 Civils

Thompson Wright Associates
P O Box 892
Golden CO 80402
USA

Thomson Communications (Scandinavia) AS
Struenseegade 7-9
2200 Copenhagen N
Denmark
Tel:+45 1 378055
Fax:+45 1 373639

TNO Technical Scientific Services
Schoemakerstraat 97
Delft
The Netherlands   §
Tel:+31 15 697283
Fax:+31 15 564800

Tolley Publishing Company Ltd
2 Addiscombe Road
Croydon CR9 5A
UK
Tel:(081) 688 4163
Fax:(081) 686 3155

Transportation Research Board
-see National Research Council   ¶§

Transportation Research Information Services
- see US National Academy of Sciences   §

Tsintichimneftemash
c/o International Centre for Scientific & Technical Information   §

Turkieye Petrolieri Anonim Ortakligi - Turkish Petroleum Corporation
Mudafaa Caddesi 22 Pk 209
Bakanliklar
Ankara
Turkey

TWD/TNO
PO Box 214
2600 Delft AE
The Netherlands
Tel:31 (15) 569330
Tlx:38071

Twente University of Technology
POB 217
Enschede
The Netherlands
Tel:+31 53 899111

TWI
(The Welding Institute)
Abington Hall
Abington
Cambridge CB1 6AL
UK   ¶
Tel:(0223) 891162
Tlx:81183
Fax:(0223) 892588

U

UEG (Underwater Engineering Group)
- now MTD LTd

Ufficio Nazionale Minerario pergl
- see Natl Mining Office for Hydrocarbons

Uichida Rokakuho
2-37-4 Kitazawa
Setaqaya-ku
Tokyko
Japan

UK Marine & Freshwater Sciences Library Group
c/o Marine Biological Assn

UK Module Constructors Assn Ltd
c/o Pannell Kerr Foster
2 Rothsay Terrace
Edinburgh
Scotland
Tel:(031) 831 7393
Tlx:2l95928
Fax:(031) 225 6017

UK Offshore Operators Association
3 Hans Crescent
London DE1X 0LN
UK
Tel:(071) 690 6255
Tlx:938291
Fax:(071) 589 8961

UK Online User Group
- see Institute of Information Scientists

UN Economic Commission for Europe
Palais des Nations
1211 Geneve 10
Switzerland
Tel:+41 22 346011
Tlx:389696

UN Ocean Economics & Technology Branch
- publications: Kluwer / Graham & Trotman

Undersea & Hyperbaric Medical Society
6950 Rockville Pike
Bethesda MD 20014
USA
Tel:(301) 571 1818

Underwater Association
The Marine Laboratory
Ferry Road
Hayling Island
Hants PO11 0DG
UK
Tel:(0705) 463231

Underwater Engineering Group
- now part of MTD Ltd

Underwater Systems Design
2 Montserrat Road
Lee on Solent
Hampshire PO13 9LT
UK

Underwater Technology Conference
c/o Den Norske Creditbank
POB 4040
5001 Bergen
Norway

Underwater Technology
- see Society for Underwater Technology

Underwater Training Centre
40 Kingsway
Cronulla
New South Wales 2230
Australia
Tel:+61 527 1744
Tlx:AA 23976

Underwater Training Centre
46 Karaka Park Place
St Heliers
Auckland 5
New Zealand
Tel:+64 9 542 460

Underwater Training Centre
Inverlochy
Fort William
Inverness-shire PH33 6LZ
Scotland
Tel:(0397) 3786/2821
Tlx:779703

UNESCO Press
7 Place de Fontenoy
75700 Paris
France
Tel:+33 1 45 681000
Tlx:204461-204379

Union des Chambres Syndicales de l'Industrie du Petrole
16 avenue Kleber
75116 Paris
France
Tel:+33 1 45 021120
Tlx:630545

Union Francaise des Geologues
77-79 rue Claude-Bernard
75005 Paris
France
Tel:+33 1 47 079195

Unione Petrolifera
Viale Civilta del Lavoro 38
000144 Rome
Italy
Tel:+39 6 59 14841
Tlx:611455

United Communications Group
4550 Montgomery Avenue
Suite 700 N
Bethesda MD 20814
USA   §
Tel:(301) 961 8700
Fax:(301) 961 8666

United Engineering Center
345 E 47th Street
New York NY 10017
USA
Tel:(212) 752 6800

United Nations Publications
Palais des Nations
1211 Geneva 10
Switzerland
and
Room LX2300
United Nations Building
New York NY 10017
USA
Tel:022 346011
Tlx:289696
- available through HMSO in UK

UNITI Bundsverband mittelstandischer Mineralol-unternehmen ev
Buchstrasse 10
2000 Hamburg 76
Germany

Universite des Sciences Sociales de Grenoble
Reseau d'information sur l'economie de l'energie
BP 47X
38040 Grenoble Cedex
France   §
Tel:+33 76 548178

Universite Laval
Centre de Recherches sur l'Eau
Cite Universitaire
Ste Foy
Quebec G1K 7P4
Canada
Tel:(418) 656 2131
Tlx: 51 31621

Universitetsforlaget
P O Box 2977
Toyen
N 0608 Oslo 6
Norway

University College of Cape Breton
POB 5300
Sydney
Nova Scotia B1P 6L2
Canada
Tel:(902) 539 5300

University College of Swansea
Singleton Park
Swansea SA2 8PP
Wales
Tel:(0792) 25678

University College of Wales
Old College
King Street
Aberystwyth
Dyfed SV23 3AX
Wales
Tel:(0970) 3177

University College London
- see University of London

University Microfilms International
White Swan House
Godstone
Surrey RH9 8LW
UK
Tel:(0883) 844123
Tlx:95212 IPI G
and
300 N Zeeb Road
Ann Arbor MI 48106
USA
Tel:(313) 761 4700

University of Alaska Anchorage
3211 Providence Drive
Anchorage AK 99508
USA
Tel:(907) 786 1871
and
Environment & Natural Resources Institute
707 A Street
Anchorage AK 99501
USA
Tel:(907) 257 2733
Fax:(907) 276 6847
- Arctic Environmental Information & Data Center
and
Juneau
PO Box 4G
Juneau AK 99811 0571
USA
Tel:(907) 465 2920
and
Sea Grants Program
Elmer E. Rasmusen Library
Fairbanks AK 99775 1007
USA
Tel:(907) 474 7624
and
Southeast
11120 Glacier Highway
Juneau, AK 99801
USA   ¶
Tel:(907) 789 4472
Fax:(907) 789 4595

University of Alberta
Computing Services
General Services Building
Room 352
Edmonton
Alberta T6G 2HI
Canada   §
Tel:(403) 492 5212

University of Aston in Birmingham
Gosta Green
Birmingham B4 7ET
UK
Tel:(021) 359 3611

University of Bath Press
University of Bath
Claverton Down
Bath BA2 7AY
UK
Tel:(0225) 61244

University of Birmingham
PO Box 363
Birmingham B15 2TT
UK
Tel:(021) 472 1301

University of Bradford
Bradford
West Yorkshire BD7 1DP
UK
Tel:(0274) 33466

University of Bristol
9 Old Park Hill
Bristol BS2 8BB
UK
Tel:(0272) 24161

University of British Columbia
Vancouver
British Columbia, V6T 1W5
Canada
Tel:(604) 228 2211

University of Calgary
- see Arctic Institute of North America

University of California Santa Barbara
Santa Barbara CA 93106
USA
Tel:(805) 961 2311
and
Government Publications Department, Library
Tel:(805) 961 4109
and
Davis
Government Documents Dept, Sheilda Library
Davis CA 95616
USA
Tel:(916) 752 1642
and
Irvine
Government Publications Department, Library
PO Box 19557
Irvine CA 92713
USA
Tel:(714) 856 7234
and
Los Angeles
Library Public Affairs Service
US Document Dept
405 Hilgard Avenue
Los Angeles CA 90024
USA
Tel:(213) 825 1201
and
Berkeley
Documents Department
Room 350, General Library
Berkeley CA 94720
USA
Tel:(415) 642 6657
and

Earthquake Engineering
Research Centre
University Extension
2223 Fulton Street
Berkeley CA 94720
USA
Tel:(415) 642 2331
and
Continuing Education in
Engineering
2223 Fulton Street
Berkeley CA 94720
USA   ¶
Tel:(510) 642 4151
Fax:(510) 643 8683
and
Government Publications
Department, Library
PO Box 5900
Riverside CA 92507
USA
Tel:(714) 787 3220
and
San Diego
Documents Department
Central University Library
Mail Stop C-075-P
La Jolla CA 92093
USA
Tel:(619) 534 3338
and
-see Scripps Institution of
Oceanography

University of Cambridge
Cambridge CB2 1TN
UK
Tel:(0223) 358933
- *School of Physical Sciences*
- *Department of Engineering*

**University of Central
England in Birmingham**
Department of Librarianship
Birmingham B42 2SU
UK
Tel:(021) 356 6911

University of Chicago Press
5801 Ellis Avenue
Chicago, IL 60637
USA
Tel:(312) 753 3344
Tlx:245603

University of Delaware
College of Marine Studies
Newark DL 19711
USA
Tel:738 2000

University of Dundee
- see Centre for Petroleum
& Mineral Law Studies

University of Edinburgh
Library
George Square
Edinburgh EH8 9YL
Scotland
Tel:(031) 667 0100

University of Florida
Coastal Engineering
Archives
433 West Hall
Gainesville, FL 32611
USA
Tel:(904) 392 3261

University of Glasgow
Glasgow, G12 8QQ
Scotland
Tel:(041) 339 8855
- *Marine Technology Centre*
- *Dept of Architecture &
Ocean Engineering*

University of Hawaii
2444 Dole Street
Honolulu, HI 96822
USA
and
Dept of Ocean Engineering
811 Olomehaii Street
Honolulu, HI 96813
USA
and
Sea Grant Program
2540 Maile Way
Honolulu, HI 96822
USA

University of Liverpool
PO Box 147
Liverpool L69 3BX
UK
Tel:(051) 709 6022

University of London
Senate House
London WC1E 7HU
UK
Tel:(071) 636 8000
and
Queen Mary College
Mile End Road
London E1 4NS
UK
Tel:(081) 980 4811
and
University College London
Dept of Mechanical Engin'g
Gower Street
London W1E 6BT
UK
Tel:(071) 387 7050

University of Manchester
Pollution Research Unit
Oxford Road
Manchester, M13 9PL
UK
Tel:(061) 273 3333
and
**UMIST-Institute of Science
& Technology**
Sackville Street, PO Box 88
Manchester M60 1QD
UK
-see Corrosion & Protection
Centre (CAPCIS)

University of Michigan
Dept of Naval Arch &
Marine Eng'g
North Campus
Ann Arbor MI 48109
USA
Tel:(313) 764 1817
and
Michigan Information
Transfer Source (MITS)
106 Harlan Hatcher
Graduate Library
Ann Arbor MI 48104-1205
USA   ¶
Tel:(313) 763 5060
Fax:(313) 936 3630
and
University of Michigan Press
Ann Arbor MI 48106
USA
Tel:(313) 764 4394

University of New Brunswick
Dept of Surveying &
Engineering
Fredericton
New Brunswick
Canada

University of New Hampshire
Durham NH 03824
USA
Tel:(603) 862 1234

**University of Newcastle upon
Tyne**
Armstrong Building
Queen Victoria Street
Newcastle uTyneNE1 7RU
UK
Tel:0632 328511
Tlx:53654
- *Marine Technology Centre*
-- *Dept of Naval Architecture
& Shipbuilding*
- *School of Marine Technology*
- see Decompression
Sickness Central Register

University of North Carolina
Chapel Hill NC 27514
USA
Tel:933 6981
and
Sea Grant College Program
1235 Burlington
Laboratories
Raleigh NC 27650
USA
Tel:737 2011

**University of Northumbria at
Newcastle**
Ellison Place
Newcastle upon Tyne
NE1 8ST
UK
Tel:091 232 6002

University of Oklahoma
Energy Resources Center
Informations Systems Prog
PO Box 3030
Norman OK 73070
USA
Tel:(405) 360 1600
Tlx:910/830 6521
and
University of Oklahoma
Press
1005 Asp Avenue
Norman OK 73019
USA
Tel:(405) 325 5111

University of Oregon Library
Documents Section
Eugene OR 97403
USA
Tel:(503) 686 3053

University of Plymouth
School of Maritime Studies
Drake Circus
Plymouth PL4 8AA
UK
Tel:(0752) 264651
Fax:(0752) 222792

University of Rhode Island
Kingston RI 02881
USA
Tel:(401) 792 2147
- *Law of the Sea Institute*
and
Graduate School of
Oceanography
Tel:(401) 792 6184
and

International Center for
Marine Resource
Development
126 Woodward Hall
Kingston RI 02881
USA
Tel:(401) 792 2479
Fax:(401) 789 3342
and
Department of Ocean
Engineering
202 Lippitt Hall
Kingston RI 02881
USA
Tel:(401) 792 2273
Fax:(401) 782 1066
and
Marine Advisory Service
Narragansett RI 02882
USA
and
National Sea Grant
Depository

University of Salford
Centre for Underwater
Science & Technology
Salford M5 4WT
UK
Tel:(061) 736 5843

University of Southampton
Southampton SO9 5NH
UK
Tel:(0703) 559122
- *Department of
Oceanography*
- *Institute of Sound &
Vibration Research*
- *Faculty of Engineering &
Applied Science*

**University of Southern
California**
Doheny Memorial Library
Government Documents
Department
University Park MC 0182
Los Angeles CA 90089-0182
USA
Tel:(213) 740 2339
Fax:(213) 749 1221
and
Inst for Marine & Coastal
Studies
Sea Grant Program, SSW 38
Los Angeles CA 90007
USA
Tel:(213) 741 2311

University of Strathclyde
George Street
Glasgow G1 1XW
Scotland
Tel:(041) 552 4400
and
Project Mass (MTD)
100 Montrose Street
Glasgow G4 0LZ
Scotland
Tel:(041) 552 4400
and
Dept of Shipbuilding &
Naval Arch
Livingstone Tower
26 Richmond Street
Glasgow G1 7HX
Scotland
Tel:(041) 552 4460
Tlx:77472

University of Sunderland
Langham Tower
Ryhope Road
Sunderland
Tyne & Wear SR2 7EE
UK

University of Texas at Austin
University Station
Austin TX 78713-7508
USA
Tel:(512) 471 1232
and
Petroleum Extension
Service - PETEX
Balcones Research Center
Austin TX 78712
USA
Tel:(512) 471 3154
Tlx:767161
Fax:(512) 471 9410
- see Bureau of Economic
Geology

University of Tulsa
Information Services:
Petroleum Abstracts
600 South College
101 Harwell
Tulsa OK 74104-3189
USA
Tel:(918) 631 2296/7
Tlx:49-7543
Fax:(918) 599 9361

**University of Wales Inst of
Science & Technology**
Dept of Marine Studies
King Edward VII Avenue
Cardiff CF1 3NU
Wales
Tel:(0222) 22656

University of Washington Libraries
Government Documents
Seattle WA 98195
USA
Tel:(206) 543 4664

University of Wisconsin Library
Interlibrary Loan Dept
728 State Street
Madison
Madison WI 53706
USA

Urban Verlag GmbH
Neumann-Reichardt
Strasse D34
2000 Hamburg 70
Germany
Tel:+49 40 6 567071
Fax:+49 40 6 567075

US Army Coastal Engineeriug Research Center
Kingham Building
Telegraph & Leaf Roads
VA 22060
USA

US Army Corps of Engineers
Cold Regions Research & Engineering Laboratory
72 Lyme Road
Hanover, NH 03755-1290
USA   ¶§
Tel: (603) 646 4738
Fax:(603) 646 4278
(library 646 4712)

US Bureau of Mines
2401 E Street NW
Washington DC 20390
USA

US Coast Guard
Dept of Transportation
G-MTH-1
2100 2nd Street SW
Washington DC 20593
USA   §
Tel:(202) 267 1217
Tlx:172343
Fax:(202) 267 0025

US Dept of Commerce
National Technical Information Service
5285 Port Royal Road
Springfield VA 22161
USA   §
Tel:(703) 487 4812
Fax:(703) 321 8547

and
Suite 620
425 NW 13th Street
Washington DC 20004
USA
Tel:(202) 724 3383
- Federal Interagency Cttee on the Health & Safety Effects of Energy Technology

US Dept of Energy
1000 Independence Avenue SW
Washington DC 20588
USA   §
Tel:(202) 252 2363
and
POWER
Energy Library
Washington DC 20545
USA
Tel:(301) 353 4166
and
Energy Information Administration
1000 Independence Avenue SW
Mail Stop E1421
Washington DC 20585
USA   §
Tel:(202) 2252 1155
and
Office of Scientific & Technical Information (OSTI)
PO Box 62
Oak Ridge TN 37831
USA   §
Tel:(615) 576 1188
and
Office of Oil Imports
PO Box 19267
Washington DC 20036
USA   §
Tel:(202) 653 3445
and
Bartlesville Project Office
PO Box 1398
Bartlesville OK 74005
USA   ¶§
Tel:(918) 336 2400
Fax:(918) 337 4418

US Dept of Health & Human Services
National Institute for Occupational Safety & Health
4676 Columbia Parkway
Cincinnati OH 45226
USA   §
Tel:(513) 533 8317

US Dept of the Interior
Office of the Secretary
Room 6151
18th & C Streets NW
Washington DC 20240
USA
Tel:(202) 343 7351
- see US Geological Survey
- see Minerals Management Service   ¶

US Dept of the Navy
Naval Sea Systems Command, NAVSEA
Director of Ocean Engineering
Code 00C
Washington DC 20362-5101
USA   ¶
Tel:(703) 697 7403
Fax:(703) 692 5822
and
Naval Construction Battalion Center
Civil Engineering Laboratory
Port Hueneme CA 93043
USA
and
Office of Naval Research
Arlington, VA
USA
and
Naval Ocean Systems Center
San Diego, CA
USA
and
Navy Oceanographic Office
Library (Code 1600)
Washington DC 20373
USA
Tel:(202) 763 1435
and
Naval Underwater Systems Center
Newport RI 02840
USA
Tel:(401) 841 4338
and
New London CT
USA
and
Naval Research Laboratory
Underwater Sound Reference Division
PO Box 8337
Orlando FL 32806
USA

US Dept of Transportation
Office of Aviation
Information Management &
Office of Pipeline Safety
400 7th Street SW
Washington DC 20590
USA   §
Tel:(202) 366 4387
- see US Coast Guard

**US Environmental
Protection Agency**
401 M Street SW
Washington DC 20460
USA
Tel:(202) 382 2190
Fax:(202) 755 2155
and
- Office of Solid Waste &
Emergency Response   §
- Office of Pesticides & Toxic
Substances Public Data
Branch   §
Tel:(202) 382 3524

US Geological Survey
Sun Rise Valley Drive
Reston VA 22092
USA
and
National Earthquake
Information Center (NEIC)
Earthquake Data Base
System (EDBS)
Mailstop 967, PO Box 25046
Denver Federal Center
Denver CO 80225-0046
USA   §
Tel:(303) 236 1500
Fax:(303) 236 1519
and
Books and Open File
Reports
Federal Center, Building 41
Box 25425
Denver CO 80225
USA
and
Map Distribution
Federal Center, Building 41
Box 25286
Denver CO 80225
USA
and
Alaska Distribution Section
New Federal Building
Box 12 101 Twelfth Avenue
Fairbanks AK 99701
USA

**US Government Printing
Office**
Superintendent of
Documents
Washington DC 20402
USA
Tel:(202) 783 3238

US Marine Safety Council
c/o US Coast Guard HQ
Washington DC 20590
USA

**US National Academy of
Sciences**
Transportation Research
Board
National Research Council
2101 Constitution Avenue
NW
Washington DC 20418
USA
Tel:(202) 334 3250
Fax:(202) 334 2854
- Transportation Research
Information Services

**US National Ocean Survey
Map Library**
6000 Executive Boulevard
Rockville MD 20810
USA
Tel:(301) 443 8031

**US National Oceanic &
Atmospheric
Administration (NOAA)**
Herbert C Hoover Building
14th & Constitution Avenue
NW
Room 5128
Washington DC 20230
USA
Tel:(202) 377 3436
- Environmental Research
Laboratories
and
National Oceanographic
Data Center
Washington DC 20235
USA
and
Sanctuaries & Reserves Div
1825 Connecticut Ave NW
Washington DC 20235
USA   ¶
Tel:(202) 606 4125
Fax:(202) 606 4329
and
Committee on Atmospheres
& Oceans
6009 Executive Boulevard
Rockville MD 20852
USA

and
Information Services
Building
6001 Executive Boulevard
Rockville, MD 20852
USA
- Library & Information
Services Div, Rockville
Tel:(301)443 8330
and
Environment Conservation
Division
Aquatic Toxicology
Fisheries Center
2725 Montlake Boulevard
East
Seattle WA 98112
USA
and
National Geophysical Data
Center, Code E.GCI
Department 720
325 Broadway
Boulder CO 80308-3328
USA   §
Tel:(303) 497 6120
Fax:(303) 497 6513
and
Alaska Regional Office
PO Box 021668
Juneau AK 99802-1668
USA
Tel:(907) 586 7221
and
Outer Continental Shelf
Environmental Assessment
Program of NOAA
701 C Street - Box 56
Anchorage, AK 99513
USA
Tel:(907) 271 3652
and
Outer Continental Shelf Oil
& Gas Information Prog
Fax:(703) 285 2285

**US Office of the
Oceanographer of the Navy**
Hoffman Building II
200 Stovall Street
Alexandria VA 22332
USA

UTC
PO Box 95
Sandsli
5049 Bergen
Norway
Tel:+47 5 40915

V

Van Nostrand Reinhold Company
135 W 50th Street
New York NY 10020
USA
Tel:(212) 265 8700
Tlx:214253

Van Nostrand Reinhold (UK) Ltd
- see International Thomson Publishing

VCH Publishers (UK) Ltd
8 Wellington Court
Wellington Street
Cambridge CB1 1HW
UK
Tel:(0223) 321111

VEG Gasinstitut Nv
PO Box 137
Wilmersdorf 50
7300 AC Apeldoorn
The Netherlands
Tel:+31 55 422922

Verband Fur Schiffbau & Meerestechnik Ev
- see German Shipbuilding & Ocean Industries Association

Verein Deutscher Ingenieure
Graf-Recke Strasse 84
Postfach 1139
4000 Dusseldorf 1
Germany
Tel:+49 211 62141
Tlx:8586525 VDX

Vereniging Technische Commissie Vloeibaar Gas
Waalseweg 1
5711 BM Someren
The Netherlands
Tel:+31 4937 4707

Vereniging van de Nederlandse Aardolie Industrie
see Assn of the Dutch Petroleum Ind

Veritec
- see Det norske Veritas

Vestlash Group Inc
627 E Street NW
Suite 300
Washington DC 20004
USA   §
Tel:(202) 347 6973
Fax:(202) 628 1133

Vikoma Ltd
88 Place Road
Cowes
Isle of Wight
UK
Tel:(0983) 296021

Vinn Information Centre
PO Box 250
8501 Narvik
Norway

Voight Industries Inc
PO Box 200
Lubec ME 04652
USA   §
Tel;(207) 733 5593

Vortex
2 Chute Lodge
Chute Forest
Andover
Hampshire SP11 9DG
UK
Tel:(0264) 70777

Vrijhof Ankers BV
Allegro 114
2825 BG Krimpen Ad Ijssel
The Netherlands

W

Warburg Securities
1 Finsbury Avenue
London EC2M 2PA
UK
Tel:(071) 280 2222

Warren Publishing Inc
2115 Ward Court NW
Washington DC 20037
USA   §
Tel:(202) 872 9200
Fax:(202) 293 3435

Warren Spring Laboratory
Oil Pollution Group
Gunnels Wood Road
Stevenage
Herts SG1 2BX
UK   ¶
Tel:(0438) 741122
Tlx: 82250 WSL
Fax:(0438) 360858

Warsash Nautical Bookshop
31 Newtwon Road
Warsash
Southampton SO3 6FY
UK
Tel:(0489) 52384

Washington Area Council
Engineering Laboratories
8811 Colesville Road, Suite 225
Silver Spring MD 20910
USA
Tel:(301) 588 8668

Washington State Library
Regional Depository
Documents Section
Olympia WA 98501
USA
Tel:(206) 753 4027

Water Research Centre
WRC Environment
PO Box 16
Henley Road, Medmenham
Marlow
Bucks SL7 2HD
UK   ¶
Tel:(0491) 571531
Tlx:848632
Fax:(0491) 579094

Welding Institute
- now TWI

Welding Research Council
c/o United Engineering Centre

Werkgroep Noordzee
Vossiusstraat 20-III
1071 AD Amsterdam
The Netherlands

Wessex Institute of Technology
High Street
Southampton SO1 0AA
UK
Tel:(0703) 221397

West European Graduate Education in Marine Technology
c/o MTD Ltd

West Publishing Company
50 West Kellogg Boulevard
PO Box 64526
St Paul KN 55164-0526
USA
Tel:(612) 228 2500

Western Union Telegraph Company
1 Lake Street
U Saddle River NJ 07458
USA   §
Tel:(201) 825 5000

Western Washington University
Mabel Zoe Wilson Library
516 High Street
Bellingham WA 98225
USA
Tel:(206) 676 3075

Wharton Econometric Forecasting Associates Group (WEFA)
Ebury Gate
23 Lower Belgrave Street
London SW1W 0NW
UK   §
Tel:(071) 730 8171
Fax:(071) 730 1400
and
150 Monument Road
Bala Cynwyd PA 19004
USA
Tel:(215) 667 6000
Fax:(215) 668 9524

Whico Atlas Company
PO Box 25143
Houston TX 77005
USA
Tel:(713) 523 6673

Whitney Communication Corporation
1401 New York Avenue
Suite 500
Washington DC 20005
USA
Tel:(202) 662 0700

Whittles Publishing Services
Roseleigh House
Harbour Road
Latheronwheel
Caithness KW5 6DW
Scotland
Tel/Fax:(05934) 240

Whole World Publishing Inc
400 Lake Cook Road
Suite 207
Deerfield IL 60015
USA
Tel:(800) 323 4305

Wiley
- see John Wiley & Sons

William Heinemann Ltd
-see Reed International Books

William S Hein & Co Inc
1285 Main Street
Buffalo NY 14209
USA
Tel:(716) 882 2600

Wilson (H W) Company
via Thompson Henry Ltd
London Road
Sunningdale
Berkshire SL5 0EP
UK
Tel: (0344) 22009
Fax:(0344) 26120
and
950 University Avenue
Bronx, NY 10452
USA   §
Tel:(212) 588 8400
Fax:(212) 590 1617

Witherby & Co Ltd
32-36 Aylesbury Street
London EC1R 0ET
UK
Tel:(071) 235 5413
Fax:(071) 251 1296

Wood Group
Greenbank Crescent
East Tullos
Aberdeen AB1 4BG
Scotland
Tel:(0224) 874904

Woods Hole Oceanographic Institution
Office of Research Libraries
Woods Hole MA 02543
USA   ¶
Tel:(617) 548 1400
Email:OMNET-WHOI.LIB
LIBRA

Woodside Offshore Petroleum Pty Ltd
1 Adelaide Terrace
Perth, W Australia 6000
Australia
Tel:+61 47 224 4111

Woodside Research
PO Box 6359
Station D
Calgary
Alberta T2P 2C9
Canada   §
Tel:(403) 269 6003

World Federation of Pipeline Contractors Assn
4100 First City Center
1700 Pacific Avenue
Dallas TX 75201 4618
USA
Tel:(214) 969 2700

World Health Organization
1211 Geneva 27
Switzerland

World Information Systems
PO Box 535
Harvard Square Station
Cambridge MA 02238
USA   §
Tel:(627) 491 5100
Fax:(627) 492 3312

World Microfilm Publications Ltd
Microworld House
2-6 Foscote Mews
London W9 2HH
UK
Tel:(071) 266 2202
Fax:(071) 266 2314

World Oil
3301 Allen Parkway
Houston TX 77019
USA

Worldwide Videotex
PO Box 138
Babson Park Branch
Boston MA 02157
USA   §
Tel:(617) 449 1603

Yard Ltd
Avonbridge House
Bath Road
Chippenham SN15 2BB
UK
Fax:(0249) 655723

Index by Area and Country

AUSTRALASIA
ASIA & the FAR EAST

AUSTRALIA

ACI Computer Services
Australian Dept of Primary Industries & Energy
Australian Institute of Engineering Associates
Australian Institute of Petroleum Ltd
Australian Mineral Foundation
Australian Petroleum & Exploration Association
CSIRO Australia
Environmental Conservation Executive
Info-One International
New South Wales Dept of Minerals & Energy
Noble, Denton, Woodcock & Associates Pty
Offshore Oil & Gas Directory Pty Ltd
Petroleum Information
Petroleum Institute
Professional Divers' Association of Australasia
South Australian Dept of Mines & Energy
South Pacific Underwater Medicine Society
Standards Association of Australia
Underwater Training Centre
Woodside Offshore Petroleum Pty Ltd

CHINA

Bohai Oil Corporation of CNOOC
China Bohai Racal Positioning & Survey Co Ltd
China France Bohai Geoservices Co Ltd
China Nanhai Racal Positioning & Survey Co Ld
China National Offshore Oil Corporation
China Offshore Oil Development & Eng'g Corp
China Petroleum Development Corporation
China Petroleum Eng'g Construction Corp
Nanhai East Oil Corporation of China
Nanhai West Oil Corporation of China
South Huanghai Oil Corporation

HONG KONG

Corad Technology Ltd
Petroleum News Southeast Asia Ltd

INDIA

Indian Institute of Technology
Oil & Gas Commission - Inst of Petroleum Expln
Systech Consultancy Services

INDONESIA

ASEAN Council for Petroleum
Department of Mines & Energy
Indonesian Petroleum Association

JAPAN

Institute of Energy Economics
Japan Hydrographic Department
Japan Information Centre of Science & Tech'y
Japan Marine Science & Technology Centre
Japan Petroleum Development Association
Japan Petroleum Institute
Kaihatsu Computing Centre
Kyukuto Boeki Kaisha Ltd
Nihon University
Petroleum Association of Japan
Uichida Rokakuho

KOREA

Korea Ocean Research & Development Institute

NEW ZEALAND

New Zealand Oceanographic Institute
New Zealand Oil & Gas Ltd
Underwater Training Centre

PAPUA NEW GUINEA

PNG Chamber of Mines & Petroleum

PAKISTAN

Petroleum Institute of Pakistan

SINGAPORE

DTS Singapore Pte Ltd
Metes (J S) & Company (PTE) Ltd

THAILAND

Asian Institute of Technology

CENTRAL & SOUTH AMERICA

ARGENTINA

Instituto Argentino del Petroleo

BRAZIL

COPPE-UFRJ, Federal University of Rio de
    Janeiro
Instituto Brasiliero de Petroles
GCPES

CHILE

Empresa National del Petroleo

ECUADOR

Corporation Estatal Petrolera Ecuatoriana

MEXICO

Instituto Mexicano del Petroleo

PANAMA

Institute of Diving

VENEZUELA

Ministerio de Enercia y Minas

EUROPE

AUSTRIA

Geologische Bundesanstalt
International Atomic Energy Agency
Organisation of Petroleum Exporting Countries
Osterreichische Gesellschaft fur
    Erdolwissenschaften

BELGIUM

Administration des Mines Service Geologique
    de Belgique
CEC: DG XII-E: MAST
Comite National de l'Energie
CONCAWE
European Diving Technology Committee
European Petrochemical Association
Federation Petroliere Belge
Marine Clearance Divers Group

DENMARK

Association of West European Shipbuilders
Danish Energy Agency
Danish Geotechnical Institute
Danish Hydraulic Institute
Danish Maritime Institute
Danish Ministry for Energy
Danish Offshore School
Danish Oil & Natural Gas Corp
Danish Petroleum Industry Association
Danish Ship Research Laboratory
Danish Welding Institute
Dansk Boreselskab A/S
Dansk Teknisk Oplysningstjeneste
Datacentralen
Esbjerg Industrial Development Service
Federation of Danish Industries
Geological Survey of Denmark
International Council for the Exploration of the
    Sea
Lloyd's Register of Shipping
Royal Danish Navy School
Secretariat for the Nordic Energy Libraries
STOP
Technical University of Denmark
Thomson Communications (Scandinavia) AS

EIRE

Geological Survey of Ireland
Institute of Industrial Research & Standards
Irish Gas Association
Irish Offshore Services Association
Marine Survey Office

FINLAND

Finnish Institute of Marine Research
Finnish Petroleum Federation
Institute for Geological Research
Rauma-Repola Offshore Engineering
Rauma-Repola Oy
Technical Research Centre of Finland (VTT)

FRANCE

Agence Regionale de l'Energie
Arab Petroleum Research Centre
Association Aeronautique et Astronautique de
    France
Association Amicale des Anciens Eleves des
    Ecoles de Maitres-Sondeurs (AEMS-IFP)
Association Amicale des Ingenieurs Diplomes
    de L'Ecole Nationale Superieure de
    Petrole et des Moteurs (AAID-ENSPM)
Association Francaise des Techniciens du
    Petrole
Association pour la Prevention dans les
    Transports d'Hydrocarbures
Association Technique de l'Industrie du Gaz en
    France
Bureau de Normalisation de Petrole
Bureau des Recherches Geologiques et
    Minieres
Bureau Veritas - Ocean Engineering Division
CEFI
Centre de Documentation de l'Armement
Centre de Documentation Sciences Humaines
Centre de la Mer et des Eaux
Centre de Recherches & d'Etudes
    Oceanographiques
Centre Europeen de Documentation et
    d'Information Mer
Centre Francais de la Corrosion
Centre National de la Recherche Scientifique
    (CNRS)
Centre Oceanologique de Bretagne
Chambre Syndicale du Reraffinage
CMAS World Underwater Federation
Comite Consultatif de l'Utilisation de l'Energie
Comite d'Etudes Petrolieres Marine
Comite Professionnel du Petrole
Companie Francaise des Petroles TOTAL
Direction des Hydrocarbures - Ministere de
    l'Industrie
Direction Generale de l'energies et des
    Matieres Premieres
Editions Technip
Elf Aquitaine Editions
Elf-Aquitaine: Societe Nationale Elf-Aquitaine
European Fuel Information Centre
European Liquefied Petroleum Gas Assocn

Federation Francaise des Petroliers
    Independants
G-Cam-Serveur
Gaz de France
GEP-ASTEO
Gordon & Breach Science Publishers
Independent Chemical Information Services
    ICIS
INFOCEAN
Institut Francais de l'Energie
Institut Francais du Petrole
Institut National de Plongee Professionnelle et
Intergovernmental Oceanographic Commission
International Energy Agency
International Information Centre for Natural
    Gas & Hydrocarbon Gases
International Pipe Line Contractors Assocn
Mission Interministerielle de la Mer
Nantes Chambre de Commerce
Nuclear Energy Agency
Organisation for European Economic
    Development (OECD)
Questel
Secretariat d'Etat a la Mer
Telesystemes Questel
UNESCO Press
Union des Chambres Syndicales de l'Industrie
    du Petrole
Union Francaise des Geologues
Universite des Sciences Sociales de Grenoble

GERMANY

Arbeitsgemeinschaft Erdol-Gewinnung und
    Verabeitng - Association of Oil
    Extractors & Producers
Association of Crude Oil & Gas Producers
Association of German Petroleum Industry
Bibliographisches Institut AG
Bundesanstalt fur Geowissenschaften &
    Rohstoffe - Fed Inst for Geoscience &
    Natural Resources
Bundesministerium fur Wirtschaft
Cooperative Group Marine Research & Marine
    Technology Information
Deutsche Erdolversorgungs gesellschaft mbH
Fairbase Database Ltd
Federal Institute for Geosciences & Natural
    Resources
Fiz Chemie GmbH
FIZ-Karlsruhe
FIZ-Technik
Gebruder Borntraeger
GeoFiz
German Business Information
German Shipbuilding & Ocean Industries Assn
Germanischer Lloyd

GKSS Gesellschaft fur Kernenergieverwertung
    in Schiffbau und Schiffahrt mbH
Hamburg Messe Und Congress
INKADAT Fachinformationszentrum
Mineralolwirtschaftsverband ev
NOWEA
Offshore Supply Association
Oil & Gas European Magazine
Otto Veith Verlag
Rhein-Main-Donau AG
Springer Verlag
Straatliche Prufungskommission dur das
    Tauchergewe in Schleswig-Holstein
UNITI Bundsverband mittelstandischer
    Mineralol-unternehmen ev
Urban Verlag GmbH
Verein Deutscher Ingenieure

GREECE

Greek Ministry of Industry, Energy &
    Technology
Kartelias School of Diving
Public Petroleum Corporation (DEP SA)

HUNGARY

Hungarian Hydrocarbon Institute

ICELAND

National Energy Authority

ITALY

Centro Professionale Di Immersione
Consiglio Nazionale della Richerche
ESRIN (ESA/IRS)
Food & Agriculture Organisation
Istituto Technico Industriale
National Hydrocarbons Authority - Ente
    Nazionale Idrocarburi
National Mining Office for Hydrocarbons
Scuola Professionale Subacquea e Iperbarica
SLAMARK International SpA
Unione Petrolifera

LUXEMBOURG

Commission of the European Communities:
    DG XIII-C SPRINT
CEC - DG Employment & Social Affairs:
    European Diving Technology Committee

MALTA

International Ocean Institute

MONACO

Single Buoy Moorings Inc

THE NETHERLANDS

Association of the Dutch Petroleum Industry
Balkema (A A) Publishers
Centre for Professional Advancement
Deep Offshore Technology bv1
Delft Hydraulics Laboratory
Delft University of Technology
DOT bv
Eindhoven University of Technology
Elsevier Science Publishers
European Association for Grey Literature
    Exploitation
European Association of Exploration
    Geophysicists
European Association of Petroleum Geologists
Expro Science Publications
Faxon Europe
Federation of European Petroleum & Gas
    Equipment Manufacturers
Geological Survey of the Netherlands
Institute for Mechanical Construction
International Association for Hydaulic Research
International Science Services
KIVI
Kluwer Academic Publishers Group
Maritime Research Institute Netherlands
Nederlandse Aardolie Maatschappij bv
Nederlandse Assn en Gas Exploratie en
    Produktie Associatie
Netherlands Government Publishing Office
Netherlands Industrial Council for Oceanology
Netherlands Marine Research Foundation
Netherlands Maritime Information Centre
Netherlands Organisation for Applied Scientific
    Research
Netherlands Standards Institute
North Sea Monitor
North-Holland Publishing Company
RAI Exhibition Centre
Reidel (D) Publishers
SAMSOM Data Systemen bv
Stichting Coordinate Maritiem Onderzoek
TNO Technical Scientific Services
TWD/TNO
Twente University of Technology
VEG Gasinstitut Nv
Vereniging Technische Commissie Vloeibaar
    Gas

NORWAY

A S Veritas Research
Aker Engineering A/S
BP Norway Limited U A
Chr Michelsen Institute
Conoco Norway Inc
Det Norske Creditbank
Det Norske Veritas
Elf Aquitaine Norge AS
Esso Exploration & Production Norway Inc
Fabritius A S
Fridtjof Nansen Inst at Polhogda
Fridtjof Nansen Institute Library
Geolab Nor AS
Geological Survey of Norway
IKU Sintef Gruppen
Inspectorate of Explosives & Flammables
Institute for Energy Technology
Marine Technology Centre
Mobil Exploration Norway Inc
Nordisk Institut for Sjorett
Noroil Publishing House Ltd
Norsk Hydro AS
Norsk Senter for Informatik A/S
North Sea Observer
Norwegian Centre for Information
Norwegian Geotechnical Institute
Norwegian Hydrodynamic Laboratories
    - River & Harbour Laboratory
Norwegian Information Publishers AS
Norwegian Institute for Water Research
Norwegian Institute of Technology
    -(SINTEF
    - Group Oceanographic Centre
    - Marine Technology Centre
Norwegian Marine Technology Research
    Institute
Norwegian Maritime Directorate
Norwegian Ministry of Industry & Crafts
Norwegian Ministry of Local Government &
    Labour
Norwegian Petroleum Consultants AS
Norwegian Petroleum Directorate
    - The Infoil Secretariat
Norwegian Petroleum Society
Norwegian Ship Research Insitute
Norwegian Society of Automatic Control
Norwegian Society of Chartered Engineers
Norwegian Standards Association
Norwegian State Diving School
Norwegian Underwater Technology Centre
    (NUTEC)
ONS Foundation
Oslo Port Authority
Rogalandsforskning Rogaland Research
    Institute
Royal Ministry of Oil & Energy

Royal Norwegian Council for Scientific &
    Industrial Research
Saga Petroleum AS
Selvigs Publishing AS
Shell Exploration & Production (Norway) Inc
SINTEF
    - IKU
    - MARINTEK
State Pollution Control Authority (STF)
Statoil
    - Gulffaks Produksjon
STI National Institute of Technology
Technical University of Norway
Underwater Technology Conference
Universitetsforlaget
UTC
Vinn Information Centre

PORTUGAL

Associacao Portuguesa dos Gases
    Combustiveis
Bureau for Oil Exploration & Exploitation
Ministry of Industry & Energy

RUSSIA

International Centre for Scientific & Technical
    Information (ICSTI)

SPAIN

Asociacion de Geologos y Geofisicos
    Espanoles del Petroleo
Dirrecion General de Energia y Combustibles
DOC-6
Editorial Garsi
Empressa Nacional de Investigacion y
    Exploracion de Petroleo SA
Instituto Geologico y Minero de Espana
Instituto Nacional de Hidrocarburos
Ministry of Industry & Energy

SWEDEN

AB Svenska Shell
ASEA-ATOM
Avesta Projects AB
Chalmers University of Technology
National Defense Research Institute
Royal Swedish Navy Diving Schools
Swedish Corrosion Institute
Swedish Dept for Geological Research
Swedish Geotechnical Institute
Swedish National Maritime Resources Comm'n
Swedish Petroleum Institute
Swedish Trade Fair Foundation
Swedocean

SWITZERLAND

Ausbildungszentrum fur Armeetaucher
Ausbildungszentrum fur Polozei
Avia International
Eidg Verkehrs und
    Energie-Wirtschaftsdepartement
Inter Documentation Co AG
International Assn for Bridge & Structural
    Engin'g
International Federation for Information
    Processing (IFIP)
International Labour Office
Marcel Dekker AG
Petroconsultants Group SA
UN Economic Commission for Europe
United Nations Publications
World Health Organization

TURKEY

Turkieye Petrolleri Anonim

UNITED KINGDOM

Aberdeen City Libraries
Aberdeen Conference Centre
Aberdeen District Council
Aberdeen Harbour Board
Aberdeen Petroleum Publications Ltd
Aberdeen University
Aberdeen University Press
Adam Hilger
Admiralty Research Establishment
    - Dunfermline, Haslar, Holton Heath,
    Portland, Portsdown
Advisory Committee on Pollution of the Sea
Alba International Ltd
Alexanders Laing & Cruikshank
Ambersoft Oilfield Applications
American Bureau of Shipping (ABS Europe)
Arthur Andersen & Co
Association for Information Management
Association of British Directory Publishers
Association of British Independent Oil
    Exploration Companies
Association of British Offshore Industries
Association of Consulting Engineers
Association of Graduate Careers Advisory
    Service
Association of United Kingdom Oil
    Independents
Atkins Research & Development
Atomic Energy Research Establishment
Bank of Scotland (Oil Division)
Barbour Index Ltd
Barclays de Zoete Wedd Securities Ltd
Battelle Memorial Institute

Benn Business Information Services Ltd
Benn Electronics Publications Ltd
Benn Technical Books Ltd
BHR Group Ltd
Biochemical Society
Blackie and Son Limited
Blackwell Scientific Publications Ltd
BMT Cortec Ltd
BNF Metals Technology Centre
Bowker (R R) Company
Bowker Saur (UK)
BP Research Centre
BPS Exhibitions Ltd
British Cement Association
British Coal
British Constructional Steelwork Association
British Electrical & Allied Mfrs Association
British Gas Exploration & Production
British Gas plc
British Geological Survey
British Institute of Management
British Library
    - Document Supply Centre
    - Science Reference & Information
    Service
British Library Board
British Marine Equipment Council
British Maritime Technology Ltd
British Petroleum Co plc
British Standards Institution
British Steel
    - Stainless HQ
    - Technical HQ
British Technology Group (BTG)
Brown, Son & Ferguson
Brunel University
Building Research Establishment
Bureau Veritas
Business Briefings Ltd
Butterworth & Co (Publishers) Ltd
CAES
Cambridge University Press
Capital Planning Information
Centre for Petroleum & Mineral Law Studies
CEP Consultants
Chadwyck-Healey Ltd
Chapman & Hall
Charles Knight Publishing
Chartered Institute of Building Services
CHEM-INTELL Chemical Intelligence Service
Chemical Industries Association Ltd
Chemical Information System
City Exhibitions & Conferences
City University
Civil Aviation Authority
Clarkson (H) & Co Ltd
Claygate Services Ltd
Clio Press Ltd

CML Publications
College of Petroleum Studies
Confederation of British Industry
Conoco (UK) Ltd
Construction Industry Research & Information
    Assn
Construction Press Ltd
Constructional Steel Research & Development
    Organisation
Copyright Licensing Agency Ltd
Corrosion & Protection Centre, UMIST
County Natwest Woodmac
Crane Limited
Cranfield Institute of Technology
Croner Publications Ltd
Croom Helm Ltd
Data-Star Marketing Ltd
Datastream International Ltd
David & Charles Ltd
Davy Stockbrokers
Dawson Microfiche - Wm Dawson Ltd
Decompression Sickness Central Registry
Defence Research Information Centre
Department of Energy
    - Gas & Oil Measurement Branch
    - Oil Division
Department of the Environment
Department of Trade & Industry
    - Shipbuilding & Electrical Engineering
Department of Transport, Marine Library
Dept of Agriculture & Fisheries for Scotland
Derwent Publications Ltd
Det Norske Veritas
DIALOG
Directorate of Fisheries Research
Diving Medical Advisory Committee
Don Mor Productions Ltd
Drewry (H P) (Shipping Consultants) Ltd
DRI Europe Ltd
DTI Energy Library
Dun & Bradstreet
Dunstaffnage Marine Research Laboratories
Economist Intelligence Unit
Edward Arnold (Publishers) Ltd
Electricity Association
Elsevier Applied Science Publishers
Energy Business Centre
Energy Consultancy
Energy Economics Research Ltd
Energy Industries Council
Energy Publications Ltd
Energy Systems Trade Association
Engineering Committee on Oceanic Research
Engineering Equipment & Material User Assn
Engineering Industry Training Board
Environmental Data Services Ltd
ERA Technology Ltd
ESA/IRS IRS-Dialtech

ESC Publishing Ltd
ESDU International Ltd
Esso Exploration & Production UK Ltd
Esso Petroleum Co Ltd
Euromoney Publications plc
Euromonitor Publications Ltd
Europa Publications Ltd
European Chemical News
European Study Conferences
European Underwater Biomedical Society
Export Council of Norway
Filtration Society
Financial Times
Financial Times Business Information Ltd
Finsbury Data Services
Fire Research Station
Fort Bovisand Underwater Centre
Fraser Williams
Freeman (W H) & Company Ltd
FT Electronic Publishing
FT North Sea Letter
Fulmer Research Institute Ltd
Gas Consumers Concil
GEISCO
Geo Abstracts
Geological Society of London
Geologists Association
Geosystems
Germanischer Lloyd
Gilbert Eliott & Company
Global Corrosion Consultants Ltd
Goodfellow Associates
Gordon & Breach Science Publishers
Gower Publishing Company Ltd
Graham & Trotman Ltd (Kluwer)
Greenwell Montagu
Greig, Middleton & Co
Harcourt Brace Jovanovich Ltd
Health & Safety Executiv
    - Occupation Medicine & Hygiene Labs
    - Explosions & Flame Laboratory
    - Safety Engineering Laboratory
Heat Transfer & Fluid Flow Service
Heriot Watt University
Highlands & Islands Enterprise
HMSO - Publications Division
HMSO Bookshops
Hoare Govett Investment Research
HR Wallingford
Hydrographer of the Navy
Hyman Unwin Ltd
IBC Technical Services Ltd
ICI Industrial Software Products
ICIS-LOR Group Ltd
ICL
IIR Ltd
Imperial College of Science & Technology
    - Royal College of Mines

Industrial & Marine Publications
Industrial & Trade Fairs International
Industrial Promotions
Industry Department For Scotland
Infield Systems Ltd
Infonorme - London Information
Inmarsat Maritime Satellite Communications
Institute for Marine Environmental Research
Institute for Scientific Information
Institute of Corrosion Science & Technology
Institute of Energy
Institute of Information Scientists
Institute of Marine Engineers
Institute of Materials
Institute of Naval Medicine
Institute of Oceanographic Science
    - Deacon Laboratory
    - The Bidston Observatory
Institute of Petroleum
Institute of Water & Environmental Management
Institution of Chemical Engineers
Institution of Civil Engineers
Institution of Electrical Engineers
Institution of Environmental Science
Institution of Gas Engineers
Institution of Geologists
Institution of Mechanical Engineers
Institution of Mining & Metallurgy
Institution of Mining Engineers
Institution of Structural Engineers
International Association of Underwater
    Engineering Contractors
International Chamber Of Shipping
International Labour Office
International Maritime Organization
International Petroleum Exchange of London
International Research & Development Co Ltd
International Tanker Owners Pollution
    Federation
International Thomson Publishing Services Ltd
IRL Press Ltd
James Capel & Company
John Brown Constructors & Engineers Ltd
John Wiley and Sons Ltd
Kenny JP & Partners Ltd
Keynote Publications Ltd
Knighton Enterprises Ltd
Kogan Page Limited
Langham Oil Conferences Ltd
Law Society Solicitors European Group
Leith International Conferences
Library Association
Liquiefied Petroleum Gas Industry Technical
    Assn
Lloyd's Maritime Information Services Ltd
Lloyd's of London Press Ltd
Lloyd's Register of Shipping
London Oil Reports

Longman Cartermill Ltd
Longman Group Ltd
Lorne Maclean Marine
Macdonald & Janes
Mackay Consultants
Maclean Hunter
Macmillan Distribution Ltd
Macmillan Publishers Ltd
Marathon International Petroleum (GB) Ltd
Marine Biological Association of the UK
Marine Conservation Society
Marine Environmental Research
Marine Policy
Marine Publications International
Marine Technology Support Unit
Marinetech North West
Martin Dunitz
Matthew Bender & Co Inc
Maus (J M J) Ltd
MBC Information Services Ltd
McGraw-Hill Book Company (UK) Ltd
Mead Data Central
Meadowfield Press Ltd
Medical Research Council - Decompression
    Sickness Panel
Mekaniske Verksteders Landsforening
MEP Ltd
Meteorological Office
Metocean Consultancy Ltd
MicroInfo Ltd
Middlesex University
Ministry of Agriculture Fisheries & Food
Ministry of Defence (HQ)
Monks Wood Experimental Station
Morgan Grampian Book Publishing Co Ltd
MTD Ltd
Muir-Carby Bottkjaer NV
National Economic Development Office
National Engineering Laboratory
National Hyperbaric Centre
National Library of Scotland
National Physical Laboratory
National Power
National Radiological Protection Board
Natural Environmental Research Council
    (NERC)
Noroil Publishing House Ltd
Norske Hydro Oil & Gas
North East London Polytechnic, Underwater
    Technology Group
North East Scotland Development Authority
North Sea Books
North West University Consortium
Norwegian Chamber of Commerce
Offshore Certification Bureau
Offshore Conferences & Exhibitions
Offshore Engineer
Offshore Fire Training Centre

Offshore Information Literature
Offshore Petroleum Industry Training Board
Offshore Supplies Office
Oil & Colour Chemists Association
Oil & Pipelines Agency
Oil Business Information Service
Oil Companies International Marine Forum
Oil Industry International Exploration &
       Production Forum (E+P Forum)
Oil News Service
Oilfield Publications Ltd
Oilman
Open University
Optech Ltd
ORBIT Search Service
Oslo Commission
Oxford Economic Research Associates
Oxford Institute for Energy Studies
Oxford University Press
Paint Research Association
Paris Commission
PennWell Publishing Company
Pergamon Financial Data Services (PFDS)
Pergamon Press
Peter Peregrinus Ltd
Petrocompanies
Petroconsultants (CES) Ltd
Petroleum Argus Ltd
Petroleum Economist
Petroleum Information Bureau
Petroleum Information Ltd
Petroleum Review
Petroleum Science & Technology Institute
Petroleum Training Association, North Sea
Petroleum Training Validation
Phillips & Drew
Phillips Petroleum Co
Pipeline Industries Guild
Pipes & Pipelines International
Pitman Books Ltd
Plastics & Rubber Institute
Plenum Publishing Co Ltd
Plymouth Marine Laboratory
Predicasts (UK)
Prodive Centre for Underwater Technology
Profile Information
Protection
Queen's University of Belfast
RAPRA Technology Ltd
Reed International Books
Reed Information Services ltd
Register
Research Vessel Service
Reuters Ltd
Robert Gordon University
Robert Hale Ltd
Royal Aeronautical Society
Royal Bank of Scotland

Royal Institute of Navigation
Royal Institution of Chartered Surveyors
Royal Institution of Great Britain
Royal Institution of Naval Architects
Royal Society
Royal Society for the Prevention of Accidents
Royal Society of Chemistry
Royal Society of Edinburgh
Salvage Association
Savant Research Studies
SCICON
Science & Engineering Research Council
    (SERC)
Scientific Surveys Ltd
Scientific Symposia
Scotish Offshore Training Assn Limited
Scottish Enterprise
Scottish Office
Sea Fish Industry Authority
Seatrade Conferences
Selection & Industrial Training Administration
Sell's Publications Ltd
Sharnell Ltd
Sharp (I P) Associates Ltd
Shell International Petroleum Co Ltd
Shell UK Exploration & Production
Sheppards & Chase
SIA Ltd
Simon & Schuster International Group
Single Buoy Moorings Inc
SIRA Ltd
Smith Rea Energy Analysts Ltd
Society for Underwater Technology
Society of Consulting Marine Engineers & Ship
    Surveyors
Solent Exhibitions
Spearhead Exhibitions
Spon (E & F N)
Springer Verlag London Ltd
Stanley Thornes (Publishers) Ltd
STN International
Stock Beech
Strathclyde Regional Council
Submex Limited
Subsea Engineering News
Sweet & Maxwell Ltd
Taylor & Francis Ltd
Tecnon (UK) Ltd
Telerate (Europe/Gulf) Ltd
Themedia Ltd
Thomas Reed Publications Ltd
Thomas Telford Limited
Tolley Publishing Company Ltd
TWI
UK Module Constructors Assn Ltd
UK Offshore Operators Association
Underwater Association
Underwater Systems Design

Underwater Training Centre
University College of Swansea
University College of Wales
University Microfilms International
University of Aston in Birmingham
University of Bath Press
University of Birmingham
University of Bradford
University of Bristol
University of Cambridge
University of Central England in Birmingham
University of Edinburgh
University of Glasgow
University of Liverpool
University of London
  - Queen Mary College
  - University College London
University of Manchester
  - Institute of Science & Technology
  (UMIST)
University of Newcastle upon Tyne
  - Decompression Sickness Central
  Register
University of Northumbria at Newcastle
University of Plymouth
University of Salford
University of Southampton
University of Strathclyde
  - Dept of Shipbuilding & Naval Arch
University of Sunderland
University of Wales Inst of Science &
  Technology
VCH Publishers (UK) Ltd
Vikoma Ltd
Vortex
Warburg Securities
Warren Spring Laboratory
Warsash Nautical Bookshop
Water Research Centre
Wessex Institute of Technology
Wharton Econometric Forecasting Associates
  Group (WEFA)
Whittles Publishing Services
Wilson (H W) Company
Witherby & Co Ltd
Wood Group
World Microfilm Publications Ltd
Yard Ltd

MIDDLE EAST and AFRICA

ABU DHABI

Ministry of Petroleum & Mineral Resources

ALGERIA

Ministry of Industry

GABON

Ministere des Mines et des Hydrocarbures

IRAN

Ministry of Petroleum
National Iranian Petrochemical Company

IRAQ

Iraq National Oil Company

ISRAEL

Ministry of Energy & Infrastructure

KUWAIT

Kuwait Petroleum Corporation
Ministry of Oil

LIBYA

Ministry of Oil

SAUDI ARABIA

Ministry of Petroleum & Mineral Resources

NIGERIA

Federal Ministry of Petroleum & Enbergy
Nigerian National Petroleum Corporation

NORTH AMERICA

CANADA

Alberta Oil Sands Technology & Research
    Authority
Alena Enterprises
Arctic Institute of North America
Arctic Offshore Technology Conference
Association of Consulting Engineers of Canada
Axses Information Systems
Banff Centre
Bedford Institute of Oceanography
Canada Centre for Inland Waters
Canada Gas Association
Canada Institute For Scientific & Technical
    Information (CISTI)
Canada Oil & Gas Lands Administration
Canadian Association of Oilwell Drilling
    Contractors
Canadian Centre for Marine Communications
Canadian Defense Research Establishment
    Pacific
Canadian Dept of Energy, Mines & Resources
    - Centre for Remote Sensing
    - Polar Continental Shelf Project
    - Resource Management &
    Conservation Bureau
Canadian Dept of Fisheries & Oceans
    - Marine Science & Information
    Directorate
    - Canadian Committee of Oceanography
    - Physical & Chemical Sciences
    Directorate
Canadian Dept of the Environment
    - Environmental Protection Service
    - Fisheries & Marine Service
Canadian Gas Research Institute
Canadian Geotechnical Society
Canadian Government Ocean Industry
    Development Centre
Canadian Government Publishing Centre
    Supply Service
Canadian Institute of Mining & Metallurgy
Canadian Library Association
Canadian Maritime Industries Association
Canadian Petroleum Association
Canadian Petroleum Products Institute
Canadian Society of Petroleum Geologists
Canadian Standards Association
Centre for Cold Ocean Resources Engineering
CISTI, CAN-OLE (Canadian Online Enquiry
    Service)
Dataline Inc

Environment Canada
    - Arctic Biological Station
    - Atmospheric Environment Service
    - National Environmental Emergency
    Centre
Environment Directorate
Erico Petroleum Information Ltd
Geochemical Society
Geological Association of Canada
Geological Survey of Canada
GeoVision Corp
Government of Newfoundland & Labrador
Gulf Canada Resources Ltd
Holleman (A J) Engineering Ltd
Infomart Online
Institut Maurice-Lamontagne
International Union of Geological Sciences
James Lorimer & Co
Kent Marketing Services Ltd
McGill University
McLaren Micropublishing
Micromedia Limited
Mineralogical Society of Canada
National Energy Board Canada
National Research Council of Canada
National Research Council of Canada
    - Atlantic Regional Laboratory Library
Newfoundland & Labrador Dept of Mines &
    Energy
Newfoundland & Labrador Government
Newfoundland Ocean Industries Association
News$ource
Noraton Information Corporation
Ocean Engineering Information Centre
Oceans Institute of Canada
Offshore Petroleum Board Library
Pallister Resource Management Ltd
Petroleum Communication Foundation
Petroleum Industry Training Service
Petroleum Society of CIM
Precision Microfilming Services
QL Systems Ltd
Sharp (I P) Associates
Southam Business Information &
    Communications Group
Standards Council of Canada
Universite Laval
University College of Cape Breton
University of Alberta Computing Services
University of British Columbia
University of New Brunswick
Woodside Research

UNITED STATES OF AMERICA

AAA Technology & Specialities Company Inc
Aangstrom Precision Corporation
ABC - Clio
ACS
Action Instrument Inc
Adage Inc
Adams Engineering Inc
Adaptive Systems Inc
AGS Management Systems Inc
Alaska Division of Oil & Gas
Alaska Oil & Gas Association
Alaska State Library
Allegro Technology Corporation
American Association for the Advancement of
    Science
American Association of Engineering Societies
American Association of Petroleum Geologists
American Bureau of Shipping
American Chemical Society
American Concrete Institute
American Enterprise Inst for Public Policy
    Research
American Gas Association
American Geological Institute
American Geophysical Union
American Independent Refiners Association
American Institute of Aeronautics &
    Astronautics
American Institute of Chemical Engineers
American Institute of Professional Geologists
American Journal of Science
American Management Systems Inc
American Meteorological Society
American National Standards Institute
American Petroleum Institute
American Society for Non-Destructive Testing
American Society for Photogrammetry
American Society for Testing & Materials
American Society of Civil Engineers
American Society of Lubrication Engineers
American Society of Mechanical Engineers
American Society of Petroleum Operations
    Engineers
American Technology Laboratories
American Welding Society
Ametek Inc
AMS Press Inc
Analytical & Computational Research Inc
Anchor Management Systems Inc
AnchorPress / Doubleday
Ann Arbor Science (Publishers) Inc
Applied Automation Inc
Associated Press
Association of Diving Contractors
Association of Energy Professionals
Atlantic Information Services Inc

Automated Science Group
Balch Institute
Balkema (A A) Publishers
Ballinger Publishing Company
Battelle Memorial Institute
Bell & Howell Micro Photo Division
Bio-Rad Laboratories
Bonneville Market Information
Bowker A&I Publishing/Bowker Electronic Pub'g
Brady (R J) Publisher
BRS Information Technologies
BT Tymnet Inc
Bureau of Economic Geology
Burmass Publishing Company
Busby Associates
Business International Corp
Butterworth Publishers Inc
Byram (R W) & Company
Cahners Books International Inc
California State Library
California State University
Cambridge Scientific Abstracts
Capitol Publications Inc
Carl Systems Inc
Centre for Short-Lived Phenomena
Chase Econometrics
Chemical Abstracts Service
Chemical Information System
Chemical Market Associates Inc
Clay Minerals Society
Coastal Engineering Research Council (ASCE)
Coastal Zone Management Institute
Colorado School of Mines
Commercial Diving Center
Compact Cambridge
Compass Publications Inc
Compuserve Business Information Service
Computer Petroleum Corporation
Cornell Maritime Press Inc
Cornell University Press
Crane Russak & Co Inc
CRC Press Inc
Cushman Foundation for Foraminiferal
    Research
Cutter Information Group
Data Resources Inc
Datatimes Corporation
DeGolyer & MacNaughton
Delco Electronics
Department of Community & Regional Affairs
DeWitt & Company
Dialcom Inc
DIALOG Information Services Inc
Diver Equipment Information Centre
Divers Institute of Technology Inc
Divers Training Academy Inc
Douglas W Hilchie Inc
Dow Jones & Co Inc

DRI/McGraw Hill
Duke University
Dwight's Energydata Inc
Eastern Petroleum Directory Inc
Economic Geology Publishing Company
Electronic Markets & Information Systems Inc
Elsevier Science Publishing Co Inc
Energy Communications Inc
Energy Datasearch
Energy Enterprises
Energy Information Administration
Energy Information Resource Inventory
Energy Publishing Company
Engineering Information Inc
Engineering Societies Library
Environment Information Centre Inc
Environmental Data & Information Service
Environmental Protection Agency
Environmental Research Institute
ERIC Document Reproduction Service
ESL Information Services Inc
Exxon Production Research Co
Fairchild Microfilms
Fein-Marquart Associates Inc
Florida Institute of Technology
Foster Associates Inc
Fred B Rothman & Co
Freeman (W H) & Company Ltd
Gale Research Company
Gas Research Institute
GEISCO
General Microfilm Co
GeoBased Systems Inc
Geochemical Society
Geocomp Corp
Geological Society of America
Geophysical Directory Inc
Geovision Inc
Getty Research Library
Gordon & Breach Science Publishers
Gulf Publishing Company
Harcourt Brace Jovanovich Inc
Hart Publications Inc
Hemisphere Publishing Corporation
Highline Community College
HMW Enterprises Inc
Hoover Institution Press
Hotline Energy Reports
Howell Publications
Hughes Tool Company
Humbolt State University Library
Hunter Publishing Ltd
IFI/Plenum Data Company
Image Graphics Inc
IMCO Services
Independent Petroleum Association of America
Information Access Company
Information Inc

Institute for Scientific Information
Institute of Energy Development
Institution of Electrical & Electronic Engineers
International Assn of Drilling Contractors
International Assn of Geophysical Contractors
International Human Resources Dev'mnt Corp
International Oceanographic Foundation
International Palaeontological Association
International Publications Service
International Society of Offshore & Polar
    Engineers
IRL Press Inc
Canner (J S) & Company
JAI Press Inc
John M Campbell Publisher
John Wiley and Sons Inc
Johns Hopkins University
Johnson Reprint Microeditions
Kluwer Academic Publishers Group
Knowledge Systems Inc
Kraus Reprints & Periodicals
Krieger Publishing Company
Kyodo News International Inc
Learned Information Inc
Lehigh University
Lexington Books
Libraries Unlimited Inc
Library Microfilms
Library of Congress
Lincoln Electric Company
Lomond Publications
Long Beach Public Library
Louisiana State University
Lundberg Surveys Inc
Marcel Dekker Incorporated
Marine & Estuarine Management Div (NOAA)
Marine Science Communications
Marine Technology Society
Maritime Research Information Service
Massachusetts Coastal Zone Management
    Office
Massachusetts Institute of Technology
Mathematical Geologists of the United States
Matthew Bender & Co Inc
McGraw-Hill Book Company
McGraw-Hill Inc
MCI International Inc
Mead Data Central Inc
Microfilms International Marketing Company
Mid-Continent Oil & Gas Association
Midwest Oil Register
Mineralogical Society of America
Minerals Management Service
    - Office of Offshore Information Serv
    - OCS Information Program
    - Technology Assessment & Research
    Branch
Mining and Metallurgical Society of America

MIT Press
Mobil Research & Development Corp
National Academy of Engineering Science
National Academy Press
National Assn of Corrosion Engineers
National Drilling Contractors Association
National Energy Information Centre - FEDEX
National Information Services Corporation
National Institute for Occupational Safety &
 Health
National Institute for Petroleum & Energy
 Research
National Institute of Standards & Technology
National Library of Medicine
National Ocean Industries Association
National Paint & Coatings Association
National Petroleum Council
National Research Council
National Science Foundation
National Sea Grant Depository
New England River Basins Commission
New York Times Information Bank
Newport Associates
News-A-Tron Corp
Newsbank Inc
NewsNet Inc
Nichols Publishing Company
Noyes Data Corporation Inc
Occidental College
Oceana Publications Inc
Oceandril Inc
Oceanic Society
Oceanroutes Inc
OCLC Online Computer Library Center Inc
OCS Dept of Commerce & Economic
 Development
Office of Ocean & Coastal Resource
 Management
Offshore
Offshore Data Services Inc
Offshore Technology Conference
OGM Publishing Co Inc
Ohio State University Press
Oil & Gas Directory
Oil & Gas Journal
Oil and Gas Consultants Inc
Oil Daily
Oildom Publishing Company of Texas
Oklahoma Petroleum Directory
Oregon State Library
Oregon State University
Paleontological Research Institution
Paleontological Society
Parent Information Enterprises
Pasha Publications Inc
Pennsylvania State University
PennWell Publishing Company
Pergamon Press Inc

Petroconsultants Inc
Petroflash! Inc
Petroleum Engineer
Petroleum Industry Research Foundation
Petroleum Information Corporation
Petroleum Intelligence Weekly
Petroleum Management
Petroleum Publishers Incorporated
Petroscan
Phillips Petroleum Co
Plenum Publishing Corporation
Portland State University Library
Praeger Publishers Inc
Predicasts (USA)
Prentice-Hall International
Princeton Microfilm Corporation
Princeton University
Professional Assn of Diving Instructors
Professional Data Services
Professional Publications Inc
PSI Energy Software Inc
Quick Source Inc
Ray Piper Company Inc (Alaska Update)
Readex Microprint Corp
Resource Planning Consultants Inc
Resource Publications Inc
Rice University
Robert Stanger & Company
Rocky Mountain Mineral Law Foundation
Rutgers University
Sabbagh Associates Inc
Sage Data Inc
San Diego State University Library
San Francisco State University
Scarecrow Press
Scientific Press
Scripps Institution of Oceanography
Sea Technology
Seismological Society of America
Smithsonian Science Information Exchange Inc
Society for Mining, Metallurgy & Exploration
 (SME) Inc
Society of Economic Geologists
Society of Exploration Geophysicists
Society of Naval Architects & Marine Engineers
Society of Petroleum Engineers
Society of Photo-Optical Instrument Engineers
Society of Piping Engineers and Designers
Society of Professional Well Log Analysts
Softsearch Inc
Spill Control Association of America
Spokane Public Library
Springer Verlag New York Inc
SRI International
Standard Oil Company
Stanford Maritime Press
Stanford University Libraries
Stevens Institute of Technology

Telerate Systems Inc
Tennessee Microfilms
Texaco Inc
Texas A & M University
Thompson Wright Associates
Undersea & Hyperbaric Medical Society
United Communications Group
United Engineering Center
United Nations Publications
University Microfilms International
University of Alaska
    -Environment & Natural Resources
    Institute
University of California
    - Earthquake Engineering Research
    Centre
University of Chicago Press
University of Delaware
University of Florida
    - Coastal Engineering Archives
University of Hawaii
University of Michigan
    - MITS
    - University of Michigan Press
University of New Hampshire
University of North Carolina
University of Oklahoma
    - Energy Resources Center
    - University of Oklahoma Press
University of Oregon Library
University of Rhode Island
    - International Center for Marine
    Resource Development
    - Marine Advisory Service
    - University of Rhode Island - Graduate
    School of Oceanography
University of Southern California
    - Inst for Marine & Coastal Studies
University of Texas at Austin
University of Texas at Austin
    - Petroleum Extension Service (PETEX)
University of Washington Libraries
University of Wisconsin Library
US Army Coastal Engineering Research Center
US Army Corps of Engineers
US Bureau of Mines
US Coast Guard
US Department of Commerce
    - National Technical Information Service
US Dept of Energy
    - POWER
    - Energy Information Administration
    - Office of Scientific & Technical
    Information
    - Office of Oil Imports
    - Bartlesville Project Office

US Dept of Health & Human Services
US Dept of the Interior
US Dept of the Navy
    - Naval Sea Systems Command
    - Naval Construction Battalion Center
    - Office of Naval Research
    - Naval Ocean Systems Center
    - Navy Oceanographic Office
    - Naval Underwater Systems Center
    - Naval Research Laboratory
    (Underwater Sound)
US Environmental Protection Agency
US Geological Survey
    - Earthquake Data Base
US Government Printing Office
US Marine Safety Council
US National Academy of Sciences
US National Ocean Survey Map Library
US National Oceanic & Atmospheric
    Administration
    - National Oceanographic Data Center
    - Sanctuaries & Reserves Division
    - Committee on Atmospheres & oceans
    - Environment Conservation Division
    - National Geopohysical Data Center
    - Outer Continental Shelf Environmental
    Assessment Program
US Office of the Oceanographer of the Navy
Van Nostrand Reinhold Company
Vestlash Group Inc
Voight Industries Inc
Warren Publishing Inc
Washington Area Council
Washington State Library
West Publishing Company
Western Union Telegraph Company
Western Washington University
Wharton Econometric Forecasting Associates
    Group (WEFA)
Whico Atlas Company
Whitney Communication Corporation
Whole World Publishing Inc
William S Hein & Co Inc
Wilson (H W) Company
Woods Hole Oceanographic Institution
    World Federation of Pipeline Contractors
    Assn
World Information Systems
World Oil
Worldwide Videotex

Milton Keynes UK
Ingram Content Group UK Ltd.
UKHW051856071024
449327UK00025B/1985